BASIC MATHEMATICS

STEVE SLAVIN

Professor Emeritus

Union County College

GINNY CRISONINO

Associate Professor

Union County College

πr^2

Publishing Company

Brooklyn, NY

Production: Lori Kueter, Lynne Fox

Customer Service Representative: Cindy Sweeney

Cover: Michael Warrell

Compositors: F.M. Piper Printing, LLC; Carlisle Communications, Ltd.

Printer: Quebecor World Dubuque, Inc.

1 2 3 4 5 6 7 8 9 0 – 02

ISBN 0-9716544-0-9

Preface to the Instructor

Why is this text different from all other basic mathematics texts? We all cover the same work, but our book is formatted to meet the needs of today's college students. They need a less intimidating text with a steady stream of positive reinforcement. Their needs are not met by a text that is cluttered with bells and whistles, or one so verbose that it puts our students to sleep.

Our pedagogy is simple: You learn math by doing math. Students need to realize that they can do the work. By working their way through the text, checking their work, correcting their own mistakes, and mastering the material on their own, they will keep getting positive reinforcement. And we are right there with them, anticipating and then answering their questions, and, in effect, having a mathematical conversation.

Each chapter has an identical format. A section begins with a few illustrative examples, which come with full solutions. Next comes a problem set, which the student is asked to complete, followed by the full solutions to each of these problems. The next subsections are set up the same way. At the end of each section is a set of workbook questions, which can be done in class or for homework. Each chapter ends with a chapter review and two chapter tests. These can also be done in class or for homework, and provide more positive reinforcement.

Unlike all the other basic mathematics books out there, ours covers the most important topics such as order of operations and division by zero throughout the book, not just in one section of one chapter. In the back of the book, we have the step-by-step solutions, not just the answers. Instead of putting the applications at the end of each section, we have a whole section of each chapter dedicated to applications. That way, students have to figure out which operations are needed for each problem.

Basic mathematics is arguably the hardest course to teach, since you are expected to cover up to eight years of math in just one semester. Because our formatting encourages students to work their way through our text, you'll find that our text basically teaches itself. If this makes your job any easier, and if more students are prepared to tackle algebra, then this text really is different from all other basic mathematics texts.

Preface to the Student

You can't learn how to drive a car by reading a driver's manual or watching a movie in driver's ed. You need to actually get behind the wheel and start driving. Learning math is a lot like learning to drive. You learn math by doing math. We'll show you how to solve a problem. Then we'll ask you to solve similar problems. After you do math for a while, you'll get good at it. Like learning to drive, at first it looks very intimidating. But once you start doing it, you'll be amazed at how easy it is.

Learning math is like putting up a building. You need to build a strong foundation. Once you've built that foundation, there's no telling how high you can build.

Acknowledgements

As soon as we finished writing the last chapter of this book, we congratulated each other on finally completing about three years of work. But our celebration was premature. A whole new team needed to shape our manuscript into a polished textbook. There would be reviewers, professors who would test our book in their classes, students providing feedback, compositors, graphic artists, and printers. It would be impossible to acknowledge each individual who helped, but we're going to try anyway.

Jim Magliano, the math chairman at Union County College, made dozens of suggestions, ranging from finding arithmetic errors to very major omissions. Having followed virtually all of Jim's suggestions, we moved our book a lot closer to perfection. Professors Cindy Roemer, Abe Feria, Vartan Abdo, and Sam Zhang also class-tested our book, and made many valuable suggestions which we followed. We are especially appreciative of the feedback we received from our students, and especially to Anita Abat, who found still more errors, which we were able to correct before we sent the book to the printer.

Two others went through the manuscript with fine tooth combs. Tim Ombrello and Melanie Shapiro checked the solutions to every problem, and certified the book virtually error-free. So blame them for any remaining errors. Seriously, folks, we take full responsibility for any mistakes, and would appreciate your letting us know if you find any.

Mathematical typing can be a nightmare, even with an equation editor, and we are extremely grateful to Hazel Staloff (known as Hazel from Brooklyn on talk radio) for typing Chapters 3 and 4, which deal with fractions. Don't try to do this at home, folks. Hire Hazel.

Dennis Ames and Lynne Fox of Piper Printing and Lori Kueter and Susan Konzen of Carlisle Communications were our compositors. They magically converted our disks into files that could be used by our printer. Cindy Sweeney, Lisa Ludwig, and Julie Carlisle ensured that we made all our deadlines, including the impossible ones. Marcus Lancaster and Vicky Beesecker of Quebecor made sure our book got printed and shipped out.

Have you noticed the pizza pie on the cover? Our cover designer, Michael Warrell baked that pie, so to speak

Finally we want to thank our reviewers, who not only provided valuable criticism, but very welcome encouragement:

Katherine Golesorkhi, Suffolk County Community College, Brentwood, NY
Howard Stoner, Hudson Valley Community College, Troy, NY
Lisa Topolovec, Sullivan County Community College, Loch Sheldrake, NY
Herman Rosenberg, New Jersey City University, Jersey City, NJ
Carl Mancuso, William Patterson University, Wayne, NJ
Donald Burd, Monroe College, New Rochelle, NY
Sharon Testone, Onondaga Community College, Syracuse, NY 13215
Nancy Desilet, Carroll Community College, Westminster, MD
Margaret Finster, Erie Community College, Orchard Park, NY
Beth Hempleman, Mira Costa College, Oceanside, CA 92056
Larry Smyrski, Henry Ford Community College, Dearborn, MI
Bettie Truitt, Black Hawk Community College, Moline, IL
Jean Schrader, Jamestown Community College, Jamestown, NY

Mary Woestman, Broome Community College, Binghamton, NY
Sidney Samuel, John Jay College, NY, NY
Lily Christ, John Jay College, NY, NY
Helen Strassberg, York College, Queens, NY
Sven Trenholm, North Country Community College, Saranac Lake, NY
Trish Lanz, Erie Community College, Orchard Park, NY
Angela DeGroat, State University of New York, Cortland, NY
Karen S. Hale, State University of New York, Morrisville, NY
Mike Iannone, The College of New Jersey, Ewing, NJ
Susan A. Sherman, Middlesex Community College, Fords, NJ

About the Authors

Steve Slavin, Ph.D. is Professor Emeritus of Economics at Union County College in Cranford, New Jersey. He is the author of Quick Business Math, Math for Your First- and Second-Grader, and All the Math You'll Ever Need, published by John Wiley and Sons, and Economics, published by McGraw-Hill.

Ginny Crisonino has taught mathematics at Union County College since 1983, and has completed most of her work toward a Ph.D. at New York Polytechnic University. She and Steve Slavin have coauthored Precalculus, A Self-Teaching Guide and Geometry, A Self-Teaching Guide, both published by John Wiley and Sons.

Instructor's Manual

At the back of the text we provide full solutions to all the odd-numbered workbook problems and to all the chapter review problems. In the Instructor's Manual, you'll find the answers to the even-numbered workbook problems and to the chapter tests. In addition, there are four chapter tests for each chapter.

Chapter 1. Addition and Subtraction of Whole Numbers

In math we use numbers the way a language uses an alphabet. In this chapter and the next, we'll work with whole numbers such as 4, 10, 29, and 386. We'll solve hundreds of problems, and the answer to each of them will be a whole number.

To be successful at mathematics, you need to be able to add, subtract, multiply and divide without using a calculator. So put away your calculator because you won't be using it even once during this course. Back in the good old days, which both authors are old enough to remember, there *were* no calculators. Believe it or not, almost everyone got pretty good at math. So why not you? Well, what are we waiting for? Put away that calculator and let's get started!

1.1 Place Value

Before we can perform any arithmetic operations or applications with whole numbers, we need to study the value of the numbers. To do this, we must learn the following place value table.

TABLE 1.1   PLACE VALUE: ONES TO MILLIONS

Millions Column	Hundred Thousands Column	Ten Thousands Column	Thousands Column	Hundreds Column	Tens Column	Ones Column
1,000,000	100,000	10,000	1,000	100	10	1

Example 1: Let's use the place value table to determine the place value and the value of the digits in the number 2,791.

SOLUTION:

There is a 1 in the ones' place, so it has a value of .... $1 \times 1 = 1.$

There is a 9 in the tens' place, so it has a value of .... $9 \times 10 = 90.$

There is a 7 in the hundreds' place, so it has a value of .... $7 \times 100 = 700.$

There is a 2 in the thousands' place, so it has a value of .... $2 \times 1,000 = 2,000.$

Example 2: Now we'll use the place value table to determine the place value and the value of the digits in the number 2,849, and to write it in expanded form.

SOLUTION:

9 is in the ones' place, so it has a value of .... $9 \times 1 = 9.$

4 is in the tens' place, so it has a value of .... $4 \times 10 = 40.$

8 is in the hundreds' place, so it has a value of .... $8 \times 100 = 800.$

2 is in the thousands' place, so it has a value of .... $2 \times 1,000 = 2,000.$

We can say that $2,849 = 2,000 + 800 + 40 + 9.$

When we write a number this way, it's called *expanded form*. When a number is in expanded form, it's easy to see the place value of each of its digits. Now it's your turn to try one.

Example 3: Determine the place value and the value of the digits in the number 345,952, and then write it in expanded form.

SOLUTION:

2 is in the ones' place, so it has a value of	$2 \times 1 = 2.$
5 is in the tens' place, so it has a value of	$5 \times 10 = 50.$
9 is in the hundreds' place, so it has a value of	$9 \times 100 = 900.$
5 is in the thousands' place, so it has a value of	$5 \times 1,000 = 5,000.$
4 is in the ten thousands' place, so it has a value of	$4 \times 10,000 = 40,000.$
3 is in the hundred thousands' place, so it has a value of	$3 \times 100,00 = 300,000.$

Expanded form: $345,952 = 300,000 + 40,000 + 5,000 + 900 + 50 + 2.$

Example 4: Determine the place value and the value of the digits of 1,234,567, and write it in expanded form.

SOLUTION:

7 is in the ones' place, so it has a value of	$7 \times 1 = 7.$
6 is in the tens' place, so it has a value of	$6 \times 10 = 60.$
5 is in the hundreds' place, so it has a value of	$5 \times 100 = 500.$
4 is in the thousands' place, so it has a value of	$4 \times 1,000 = 4,000.$
3 is in the ten thousands' place, so it has a value of	$3 \times 10,000 = 30,000$
2 is in the hundred thousands' place, so it has a value of	$2 \times 100,000 = 200,000.$
1 is in the millions' place, so it has a value of	$1 \times 1,000,000 = 1,000,000.$

Expanded form: $1,234,567 = 1,000,000 + 200,000 + 30,000 + 4,000 + 500 + 60 + 7.$

Example 5: Determine the place value and the value of the digits of 3,580, and write it in expanded form.

SOLUTION:

0 is in the ones' place, so it has a value of	$0 \times 1 = 0.$
8 is in the tens' place, so it has a value of	$8 \times 10 = 80.$
5 is in the hundreds' place, so it has a value of	$5 \times 100 = 500.$
3 is in the thousands' place, so it has a value of	$3 \times 1,000 = 3,000.$

Expanded form: $3,580 = 3,000 + 500 + 80.$

Example 6: Determine the place value and the value of the digits, of 513,296 and write it in expanded form.

SOLUTION:

6 is in the ones' place, so it has a value of	$6 \times 1 = 6.$
9 is in the tens' place, so it has a value of	$9 \times 10 = 90.$
2 is in the hundreds' place, so it has a value of	$2 \times 100 = 200.$

3 is in the thousands' place, so it has a value of $\qquad$ 3 x 1,000 = 3,000.

1 is in the ten thousands' place, so it has a value of $\qquad$ 1 x 10,000 = 10,000.

5 is in the hundred thousands' place, so it has a value of $\qquad$ 5 x 100,000 = 500,000.

Expanded form: 513,296 = 500,000 + 10,000 + 3,000 + 200 + 90 + 6.

We would like to point out to you that the commas in the numbers separate out the "major groups": thousands, millions, billions, and so on. Within each "major group" you find a number between 0 and 1,000; therefore a comma is needed.

Problem Set:

State the place value of the 5 and the value of the 5 in each of the following.

1) 235   2) 356   3) 50,998   4) 185,934   5) 577

State the place value of the 7 in each of the following and write each number in expanded form.

6) 1,750   7) 96,047   8) 7,294,300   9) 275,300   10) 1,205,072

SOLUTIONS TO PROBLEM SET:

1)   ones' place   5

2)   tens' place   50

3)   ten thousands' place   50,000

4)   thousands' place   5,000

5)   hundreds' place   500

6)   hundreds' place   1,750 = 1,000 + 700 + 50

7)   ones' place   96,047 = 90,000 + 6,000 + 40 + 7

8)   millions' place   7,294,300 = 7,000,000 + 200,000 + 90,000 + 4,000 + 300

9)   ten thousands' place   275,300 = 200,000 + 70,000 + 5,000 + 300

10)   tens' place   1,205,072 = 1,000,000 + 200,000 + 5,000 + 70 + 2

Now that we know all about place value, let's practice reading the names for numbers in the following examples.

Example 7: **1,234,567**

Solution: This number is read: one million, two hundred thirty-four thousand, five hundred sixty-seven. Notice the word "and" is not used when reading this number. The word "and" is used only when there's a decimal point in the number. If you misread this number as, one million, two hundred thirty-four thousand, five hundred and sixty-seven it, would be written 1,234,500.67. Don't use the word "and" unless the number has a decimal point.

Example 8: 3,580

        Solution: This number is read: three thousand, five hundred eighty.

Example 9: 2,791

        Solution: This number is read: two thousand, seven hundred ninety-one.

Example 10: 54,670

        Solution: This number is read: fifty-four thousand, six hundred seventy.

Problem Set:

Write the following numbers in words.

1) 235   2) 356   3) 50,998   4) 185,934   5) 577

SOLUTIONS TO PROBLEM SET:

1) two hundred thirty-five   2) three hundred fifty-six

3) fifty thousand, nine hundred ninety-eight

4) one hundred eighty-five thousand, nine hundred thirty-four

5) five hundred seventy-seven

Now we'll do the opposite; we'll give you a number in words and ask you to write it with digits.

Example 11: Write six hundred thirty using digits.

        SOLUTION: 630

Example 12: Write six hundred three using digits.

        SOLUTION: 603

Example 13: Write sixty thousand, three hundred thirty-three using digits.

        SOLUTION: 60,333

Example 14: Write two million, three hundred forty thousand, six hundred five using digits.

        SOLUTION: 2,340,605

Problem Set:

Write the following numbers using digits.

1) Three hundred twenty-three   2) Two thousand three

3) Two thousand, three hundred four   4) Six hundred fifty-six thousand, twenty-six

SOLUTIONS TO PROBLEM SET:

1) 323   2) 2,003   3) 2,304   4) 656,026

Workbook

State the place value of the digit 6 in each of the following numbers, and then write each number in expanded form.

1) 63

2) 776

3) 1,689

4) 61,435

5) 600,980

6) 186

7) 26,897

8) 163,007

9) 506

10) 6,009,540

State the place value of the digit 3 in each of the following numbers, and then write each number in expanded form.

11) 300

12) 12,039

13) 3,721

14) 973

15) 30,721

16) 36,000,000

17) 3,946,125

18) 437,091

19) 930

20) 6,735,000

Write the following numbers in words.

21) 63   22) 76

23) 603   24) 706

25) 6,003   26) 7,006

27) 6,663,330   28) 7,007,770

29) 506   30) 245

Write the following numbers using digits.

31) eight hundred twenty-three

32) eight thousand, twenty-three

33) seventy-one

34) twenty-three thousand

35) ten million, three hundred thousand, four hundred

36) one hundred twenty-four thousand, six hundred two

1.2 Rounding

Sometimes it's not necessary to use the exact number to answer a question, so we round off the number. For example, when people ask us how many students attend our college, we say about 10,000, instead of the exact answer of 10,245. Notice we don't say 11,000, because 10,245 is closer to 10,000 than 11,000.

Example 1: Round 42 to the nearest ten.

SOLUTION: We can round off a number using one of two methods. The first method logically reasons out the answer. For example, let's look at the number 42. We're asked to round 42 to the nearest ten. Is 42 closer to 40 or to 50? Since it's closer to 40, 42 rounds to 40. The other method uses a procedure of examining the number one place to the right of the place value you're asked to round off to. If the number in that place is 5 or more, you round up by adding a 1 to the number in the place value you're rounding to. All the digits to its right are replaced by zeros. To apply this method to 42, we look one place to the right of the tens' place and find a 2. Two is not 5 or more, so we don't add a 1 to the 4, and everything to its right is replaced by zeros.

Example 2: Round 356 to the nearest ten.

SOLUTION: 356 is between 350 and 360, but it's slightly closer to 360, so 356 rounds up to 360. By the other method, if we look one place to the right of the tens place, we see a 6, which causes us to round up by adding a 1 to the 5, and the digit to its right is replaced by zeros.

$$
\begin{array}{r}
356 \\
+\ 1 \\
\hline
360
\end{array}
$$

Example 3: Round 1,247 to the nearest hundred.

SOLUTION: Is 1,247 closer to 1,200 or 1,300? It's closer to 1,200, of course. So 1,247 rounds to 1,200. If we look one place to the right of the hundreds' place, we have a 4. Four is not 5 or greater, so we don't add a 1 to the 2. It remains a 2, and the digits to its right are replaced by zeros.

Example 4: Round 95,678 to the nearest thousand.

SOLUTION: 95,678 is closer to 96,000 than 95,000, so it rounds up to 96,000. If we look one place to the right of the thousands' place, we see a 6, so we add a 1 to the 5 and replace all the digits to its right with zeros.

$$
\begin{array}{r}
95,678 \\
+\ 1 \\
\hline
96,000
\end{array}
$$

Example 5: Round 45 to the nearest ten.

SOLUTION: Wait a minute! 45 is not closer to 40 or 50, so we'll use the other technique. If we look one place to the right of the 4, we see a 5, so we round up to 50. Even though 45 is in the middle of 40 and 50, it's customary to round 45 up to 50.

Example 6: Round 7,529,006 to the nearest million.

SOLUTION: 7,529,006 is closer to 8,000,000 than to 7,000,000, so we round up to 8,000,000. If we look one place to the right of the millions' place, we see a 5, so we add a 1 to the 7 and replace the digits to the right by zeros.

$$
\begin{array}{r}
7,529,006 \\
+ 1 \\
\hline
8,000,000
\end{array}
$$

Rounding numbers with nines.

Sometimes rounding a number that contains nines can be confusing. Just remember to follow the procedure we used in the last five examples. When you add a 1 to a 9, it carries or regroups. How much is 299 rounded to the nearest hundred? The answer is 300. How much is 29,999 rounded to the nearest thousand? The answer is 30,000. If we look one place to the right of the thousands' place we see a 9, so we add 1 to the 9 in the thousands' place, and carry a one to the ten thousands' place, and everything to its right is replaced by zeros.

$$
\begin{array}{r}
29,999 \\
+ 1 \\
\hline
30,000
\end{array}
$$

Example 7: Round 55,996 to the nearest ten.

SOLUTION: One place to the right of the tens' place is a 6, which means we round up. Add a 1 to the 9 in the tens' place and we get 56,000.

$$
\begin{array}{r}
55,996 \\
+ 1 \\
\hline
56,000
\end{array}
$$

Problem Set:

Round each of the following numbers to the given place.

1) 789   tens' place
2) 1,876   thousands' place
3) 286   hundreds' place
4) 30,993   hundreds' place
5) 40,872   ten thousands' place
6) 8,543,501   millions' place
7) 89,099   hundreds' place
8) 2,099,982   ten thousands' place

SOLUTIONS TO PROBLEM SET:

1) 790
2) 2,000
3) 300
4) 31,000
5) 40,000
6) 9,000,000
7) 89,100
8) 2,100,000

Round each of the following numbers to the nearest ten, hundred, and thousand.

		TEN	HUNDRED	THOUSAND
1)	2,657	_____	_____	_____
2)	16,738	_____	_____	_____
3)	349,999	_____	_____	_____
4)	6,345	_____	_____	_____
5)	38,516	_____	_____	_____
6)	650	_____	_____	_____
7)	124,801	_____	_____	_____
8)	2,949,558	_____	_____	_____

Round each of the following numbers to the nearest ten thousand, hundred thousand, and million.

		TEN THOUSAND	HUNDRED THOUSAND	MILLION
9)	2,456,900	_____	_____	_____
10)	5,098,643	_____	_____	_____
11)	3,899,231	_____	_____	_____
12)	6,004,605	_____	_____	_____
13)	11,456,004	_____	_____	_____
14)	25,999,000	_____	_____	_____
15)	9,754,310	_____	_____	_____
16)	1,086,531	_____	_____	_____

1.3   Addition

The Addition Table

Chances are you already know the addition table, which is shown below. If not, then you *must* learn it. You need to have the entire table memorized to do nearly all of the math that follows, not just in this course, but in more advanced courses.

ADDITION TABLE

+	0	1	2	3	4	5	6	7	8	9	10
0	0	1	2	3	4	5	6	7	8	9	10
1	1	2	3	4	5	6	7	8	9	10	11
2	2	3	4	5	6	7	8	9	10	11	12
3	3	4	5	6	7	8	9	10	11	12	13
4	4	5	6	7	8	9	10	11	12	13	14
5	5	6	7	8	9	10	11	12	13	14	15
6	6	7	8	9	10	11	12	13	14	15	16
7	7	8	9	10	11	12	13	14	15	16	17
8	8	9	10	11	12	13	14	15	16	17	18
9	9	10	11	12	13	14	15	16	17	18	19
10	10	11	12	13	14	15	16	17	18	19	20

Adding Whole Numbers

Example 1: $\begin{array}{r} 24 \\ + 43 \\ \hline \end{array}$

SOLUTION: We'll do this problem in expanded form so you understand the value of the digits we're adding.

$$\begin{array}{r} 24 = 2 \text{ tens} + 4 \text{ ones} \\ + 43 = 4 \text{ tens} + 3 \text{ ones} \\ \hline 67 = 6 \text{ tens} + 7 \text{ ones} \end{array}$$

The same problem can be done much faster:

Step 1: $\begin{array}{r} 24 \\ + 43 \\ \hline 7 \end{array}$ First add the ones' column. $4 + 3 = 7$

Step 2: $\begin{array}{r} 24 \\ + 43 \\ \hline 67 \end{array}$ First add the tens' column. $2 + 4 = 6$

Example 2:   375
              + 214

SOLUTION: Again, we'll do the expanded form first:

$$375 = 3 \text{ hundreds} + 7 \text{ tens} + 5 \text{ ones}$$
$$+\ 214 = 2 \text{ hundreds} + 1 \text{ ten} + 4 \text{ ones}$$
$$589 = 5 \text{ hundreds} + 8 \text{ tens} + 9 \text{ ones}$$

And here's the short form:

Step 1:   375   add the ones' column $5 + 4 = 9$
           + 214
             9

Step 2:   375   add the tens' column $7 + 1 = 8$
           + 214
            89

Step 3:   375   add the hundreds' column $3 + 2 = 5$
           + 214
           589

Let's use the short form on the next example.

Example 3:   121 + 245 + 432

SOLUTION:

   121   Step 1: Add the ones' column ($1 + 5 = 6; 6 + 2 = 8$)
   245
 + 432   Step 2: Add the tens' column ($2 + 4 = 6; 6 + 3 = 9$)
   798   Step 3: Add the hundreds' column ($1 + 2 = 3; 3 + 4 = 7$)

Addition with Carrying (Regrouping)

So far we've been adding digits in a place value whose sums came to less than 10. What do we do if these sums add up to more than 10? We carry or regroup.

Example 4:   57
              + 25

SOLUTION: First we'll solve this problem by adding in expanded form:

$$57 = 5 \text{ tens} + 7 \text{ ones}$$
$$+\ 25 = 2 \text{ tens} + 5 \text{ ones} \quad 12 \text{ ones} = 1 \text{ ten} + 2 \text{ ones}$$
$$82 = 7 \text{ tens} + 12 \text{ ones} = 7 \text{ tens} + 1 \text{ ten} + 2 \text{ ones} = 8 \text{ tens} + 2 \text{ ones} = 80 + 2 = 82$$

Now let's do the same problem using a shortcut method.

$$\begin{array}{r} \overset{1}{}57 \\ +\ 25 \\ \hline 2 \end{array}$$ Step 1: $7 + 5 = 12$ Write down the 2 and carry the 1 over to the tens' column.

$$\begin{array}{r} \overset{1}{}57 \\ +\ 25 \\ \hline 82 \end{array}$$ Step 2: $1 + 5 + 2 = 8$ Add the tens' column.

Example 5:
$$\begin{array}{r} 198 \\ 276 \\ +\ 99 \\ \hline \end{array}$$

SOLUTION: First we'll add these numbers by using expanded form:

$$\begin{array}{l} 198 = 1 \text{ hundred} + 9 \text{ tens} + 8 \text{ ones} \\ 276 = 2 \text{ hundreds} + 7 \text{ tens} + 6 \text{ ones} \\ +\ 99 = 0 \text{ hundreds} + 9 \text{ tens} + 9 \text{ ones} \end{array}$$

23 ones = 2 tens + 3 ones; the 2 carries to the tens' column.
27 tens = 2 hundreds + 7 tens; the 2 carries to the hundreds' column.

$$\begin{array}{r} \overset{2\,2}{198} \\ 276 \\ +\ 99 \\ \hline 573 \end{array}$$

Using the expanded method on this problem may have been a little confusing. Let's simplify the same problem using the shortcut method.

$$\begin{array}{r} \overset{2}{198} \\ 276 \\ +\ 99 \\ \hline 3 \end{array}$$ Step 1: $8 + 6 + 9 = 23$ Write down the 3 and carry the 2 to the tens' column.

$$\begin{array}{r} \overset{2}{198} \\ 276 \\ +\ 99 \\ \hline 73 \end{array}$$ Step 2: $2 + 9 + 7 + 9 = 27$ Write down the 7 and carry the 2 to the hundreds' column.

$$\begin{array}{r} \overset{2}{198} \\ 276 \\ +\ 99 \\ \hline 573 \end{array}$$ Step 3: $2 + 1 + 2 = 5$ Write down the 5 in the hundreds' column.

Problem Set: Perform the following additions.

1)   35
   + 54

2)   46
    + 21

3)   62
   + 26

4)   136
    + 463

5)   59
   + 86

6)   348
   + 554

7)   657
    363
   + 437

8)   4289
   2579
  + 7895

SOLUTIONS TO PROBLEM SET:

1) 89

2) 67

3) 88

4) 599

5) 145

6) 902

7) 1,457

8) 14,763

Workbook

Perform the following additions.

1)
$$\begin{array}{r} 23 \\ + 75 \\ \hline \end{array}$$

2)
$$\begin{array}{r} 32 \\ + 54 \\ \hline \end{array}$$

3)
$$\begin{array}{r} 15 \\ + 63 \\ \hline \end{array}$$

4)
$$\begin{array}{r} 124 \\ + 45 \\ \hline \end{array}$$

5)
$$\begin{array}{r} 89 \\ + 19 \\ \hline \end{array}$$

6)
$$\begin{array}{r} 298 \\ + 109 \\ \hline \end{array}$$

7)
$$\begin{array}{r} 1928 \\ 345 \\ + 2399 \\ \hline \end{array}$$

8)
$$\begin{array}{r} 23 \\ 222 \\ + 13 \\ \hline \end{array}$$

9)
$$\begin{array}{r} 10092 \\ 77 \\ 5678 \\ + 34511 \\ \hline \end{array}$$

10)
$$\begin{array}{r} 2971 \\ 343 \\ + 8728 \\ \hline \end{array}$$

11)
$$\begin{array}{r} 23645 \\ 1111 \\ + 56780 \\ \hline \end{array}$$

12)
$$\begin{array}{r} 2009 \\ 4 \\ + 304 \\ \hline \end{array}$$

13)
$$\begin{array}{r} 34567 \\ 3788 \\ + 99999 \\ \hline \end{array}$$

14)
$$\begin{array}{r} 34 \\ 1347 \\ + 801 \\ \hline \end{array}$$

15)
$$\begin{array}{r} 346 \\ 4 \\ 567 \\ + 892 \\ \hline \end{array}$$

16)
$$\begin{array}{r} 12390 \\ 34567 \\ + 55678 \\ \hline \end{array}$$

17)
$$\begin{array}{r} 294057 \\ 300000 \\ 237 \\ + 34567 \\ \hline \end{array}$$

18)
$$\begin{array}{r} 45000 \\ 467 \\ + 67890 \\ \hline \end{array}$$

1.4   Subtraction

Subtraction means "take away." In this section, we'll subtract smaller numbers from larger numbers. For instance, how much is $8 - 5$? We could also ask, what's the difference of 8 and 5? Or we could say how much is 8 minus 5 or 8 take away 5 or five less than eight? No matter how we say it, the answer is 3. So we have a few different ways of stating a subtraction problem.

Example 1: Let's try a problem using a version of expanded form.

$$\begin{array}{r} 69 \\ - \ 54 \\ \hline \end{array}$$

SOLUTION: Here's how we would do this problem in expanded form.

$$\begin{array}{rl} 69 = & 6 \text{ tens} + 9 \text{ ones} \\ - \ 54 = & 5 \text{ tens} + 4 \text{ ones} \\ \hline 15 = & 1 \text{ ten} \ + 5 \text{ ones} \end{array}$$

We won't be doing all our subtraction problems in expanded form because it would take too long. But doing a few problems in expanded form helps us understand just what we're doing and why we can do it. Subtraction is the inverse of addition. If you deposit $50 in your bank account it's the opposite of withdrawing $50 from your account. Suppose we wanted to make sure we worked out a subtraction problem correctly. All we would have to do to check our answer is add. For example, let's check to see if $69 - 54$ really is 15.

$$\begin{array}{r} 15 \\ + \ 54 \\ \hline 69 \end{array}$$ This problem checks.

See if you can work out the next problems in both expanded form and short form and check your answer.

$$\begin{array}{r} 86 \\ - \ 32 \\ \hline \end{array} \qquad \begin{array}{r} 357 \\ - \ 242 \\ \hline \end{array}$$

$$\begin{array}{rl} 86 = & 8 \text{ tens} + 6 \text{ ones} \\ - \ 32 = & 3 \text{ tens} + 2 \text{ ones} \\ \hline 54 = & 5 \text{ tens} + 4 \text{ ones} \end{array} \qquad \begin{array}{rl} 357 = & 3 \text{ hundreds} + 5 \text{ tens} + 7 \text{ ones} \\ - \ 242 = & 2 \text{ hundreds} + 4 \text{ tens} + 2 \text{ ones} \\ \hline 115 = & 1 \text{ hundred} \ + 1 \text{ ten} \ + 5 \text{ ones} \end{array}$$

Check:
$$\begin{array}{r} 54 \\ + \ 32 \\ \hline 86 \end{array} \qquad \begin{array}{r} 115 \\ + \ 242 \\ \hline 357 \end{array}$$

Subtraction with Borrowing

So far we've been doing very straightforward subtraction problems. Now we're going to turn things up a notch.

Example 2:
$$\begin{array}{r} 64 \\ -\,37 \\ \hline \end{array}$$

SOLUTION:
$$\begin{array}{l} 64 = 6 \text{ tens} + 4 \text{ ones} \\ -\,37 = 3 \text{ tens} + 7 \text{ ones} \\ \hline \end{array}$$

Our problem here is with the ones. You can't subtract 7 ones from 4 ones. So what we'll do is borrow ten ones from the tens' column.

We know that 1 ten = 10 ones. If 64 = 6 tens + 4 ones, we can borrow one of the tens and give it to the ones. This is called regrouping.

64 = 6 tens + 4 ones = 5 tens + 1 ten + 4 ones = 5 tens + 14 ones.

Now we're ready to subtract.
$$\begin{array}{l} 64 = 5 \text{ tens} + 14 \text{ ones} \\ -\,37 = 3 \text{ tens} + 7 \text{ ones} \\ \hline 27 = 2 \text{ tens} + 7 \text{ ones} \end{array}$$

So the answer is 27.

Check:
$$\begin{array}{r} 37 \\ +\,27 \\ \hline 64 \end{array}$$

Example 3:
$$\begin{array}{r} 436 \\ -\,253 \\ \hline \end{array}$$

SOLUTION: We can't subtract 5 tens from 3 tens. So what we'll do is borrow 10 tens from the hundreds' column.

We know that 1 hundred = 10 tens. Let's borrow one of those hundreds and give it to the tens as 10 tens.

436 = 4 hundreds + 3 tens + 6 ones = 3 hundreds + 13 tens + 6 ones

$$\begin{array}{l} 436 = 3 \text{ hundreds} + 13 \text{ tens} + 6 \text{ ones} \\ -\,253 = 2 \text{ hundreds} + 5 \text{ tens} + 3 \text{ ones} \\ \hline 183 = 1 \text{ hundred} + 8 \text{ tens} + 3 \text{ ones} \end{array}$$

So the answer is 183.

Check:
$$\begin{array}{r} 253 \\ +\,183 \\ \hline 436 \end{array}$$

Example 4:   5001
                  − 3994

SOLUTION:

$$\overset{\scriptstyle 4\,9\,9}{\cancel{5}\cancel{0}\cancel{0}\overset{1}{1}}$$
−3994
―――――
1007

This problem is a little tricky because you can't subtract 4 from 1, and you can't borrow from the 0 in the tens' column or from the 0 in the hundreds' column. So you have to borrow from the 5 in the thousands' column. That's like borrowing 1 from the 500, which would leave you with 499. Don't forget to add the 1 you borrowed to the 1 in the 5,001. The 1 you borrowed is actually 1 group of 10. You add the ten you borrowed to 1 and get 11.

Check: 3994
            + 1007
            ―――――
              5001

Now see what you can do with the following problem.

Example 5:   43002
               − 7688

SOLUTION:

$$\overset{\scriptstyle 3\,12\,9\,9}{4\cancel{3}\cancel{0}\cancel{0}\overset{1}{2}}$$
−7688
―――――
35314

This problem is similar to example 4 in that you can't borrow from zeros. So you borrow 1 from 300, which leaves you with 299. Don't forget to add the 1 (which is actually 1 ten) you borrowed to the 2 which becomes 12. In this problem, you actually have to borrow twice, because you can't subtract the 7 in the thousands' column from the 2 above it. You have to borrow 1 from the 4, making the 4 a 3, and add the 1 (which is actually 1 ten thousand), to the 3 (which is actually 3,000) to make it 13,000.

Check: 7688
            + 35314
            ―――――
              43002

Problem Set:

Try the following subtraction problems. Remember, you can check your answers using addition.

1)   37   2)   56   3)   873
     − 25   − 14   − 461

4) 57234   5) 99000   6)   3006
     − 5469   − 79999   − 108

SOLUTIONS TO PROBLEM SET:

1)   12   2)   42   3)   412

4) $\overset{\scriptstyle 6\,11\,12}{57\cancel{2}\cancel{3}\overset{1}{4}}$ 5) $\overset{\scriptstyle 8\,9\,9}{99\cancel{0}\cancel{0}\overset{1}{0}}$ 6) $\overset{\scriptstyle 2\,9\,9}{3\cancel{0}\cancel{0}\overset{1}{6}}$
   −5469   −7999   −108
   ―――――   ―――――   ―――――
   51765   19001   2898

Perform the following subtraction problems.

1)   24
    -18

2)   40
    -29

3)   47
    -39

4)   28
    -19

5)   101
    -88

6)   456
   -149

7)   287
   -179

8)   500
   -307

9)   286
   -187

10)   900
    -878

11)   1234
   -1009

12)   23456
   -3499

13) 11083
  -10989

14) 129875
   -11999

15) 340001
   -9999

16)   52003
    -85

17) 909090
  -100918

18)   83857
   -20200

19)   80029
    -779

20) 918234
   -33999

1.5 Applications

Let's apply some of the things you've learned so far. Out there in the "real world," we rarely find addition and subtraction problems already set up for us. We have to set them up ourselves.

Example 1: A housing development has 24 finished homes. The builder has already started construction on 17 new homes and plans to start another 15. How many homes will there be in this development when construction is complete?

SOLUTION:

Step 1: The given information is the number of finished homes, 24; the number of homes already started, 17; and the number of homes that are planned, 15. We need to find the total number of homes.

Step 2: Since we want a *sum*, for the total we have to add.

Step 3:
$$\begin{array}{r} 1 \\ 24 \text{ homes} \\ 17 \text{ homes} \\ +\ 15 \text{ homes} \\ \hline 56 \text{ homes} \end{array}$$

Example 2: You left the house this morning with $123 and spent $58. How much money do you have left?

SOLUTION:

Step 1: The given information is that you started with $123 and spent $58. You need to find how much money you have left.

Step 2: Since we want a *difference*, we have to subtract.

Step 3:
$$\begin{array}{r} 0\ 11 \\ \$123 \\ -\ \$\ 58 \\ \hline \$\ 65 \end{array}$$

Example 3: Sammy Sosa hit a home run just over the 390 foot sign in center field. How far did the ball travel, rounded to the nearest hundred feet?

SOLUTION: We are asked to round 390 feet to the nearest hundred feet. Since 390 is closer to 400 than to 300, the answer is 400 feet.

Example 4: Farmer Jones owns 134 acres of land. His brother Lem owns 127 acres and his brother Clem owns 185 acres. How many acres do the Jones brothers own all together?

SOLUTION:
$$\begin{array}{r} 1\ 1 \\ 134 \text{ acres} \\ 127 \text{ acres} \\ +\ 185 \text{ acres} \\ \hline 446 \text{ acres} \end{array}$$

Example 5: On July 14, when the Greenville Appliance Store opened its doors at 9 A.M., it had 232 air conditioners in stock. At closing time, it had 117 left. If no new air conditioners were delivered to the store that day, how many air conditioners were sold that day?

SOLUTION:
$$
\begin{array}{r}
23\overset{2}{\cancel{3}}2 \text{ air conditioners} \\
-117 \text{ air conditioners} \\
\hline
115 \text{ air conditioners}
\end{array}
$$

Example 6: Max earned one hundred forty-eight dollars and his sister Alice earned one hundred fifty-six dollars. How much did they earn together?

SOLUTION:
$$
\begin{array}{r}
\overset{1\,1}{\$148} \\
+\,\$156 \\
\hline
\$304
\end{array}
$$

Problem Set:

1) The New England Patriots scored seven points in the first quarter, thirteen points in the second quarter, nine points in the third quarter, and three points in the fourth quarter. How many points did the Patriots score all together?

2) Irma Mendoza earned $53,166. How much are her earnings rounded to the nearest thousand dollars?

3) If you began the day with $206 and spent $177 during the day, how much money did you have left at the end of the day?

4) The area of the United States is 3,717,796 square miles. How much is the area when rounded to the nearest hundred thousand square miles?

5) The Smith brothers inherited $1,103,375. If Jason Smith inherited $592,840, how much did his brother Joshua inherit?

6) In one term Javier spent $370 on books, $1,700 for his dorm room, $1,415 for food and $4,250 for tuition. How much did it cost him to attend college that term?

SOLUTIONS TO PROBLEM SET:

1)
$$
\begin{array}{r}
7 \text{ points} \\
13 \text{ points} \\
9 \text{ points} \\
+\,3 \text{ points} \\
\hline
32 \text{ points}
\end{array}
$$

2) $53,000

3)
$$
\begin{array}{r}
\overset{1\,9}{\$20\cancel{6}} \\
-177 \\
\hline
\$\,29
\end{array}
$$

4) 3,700,000 square miles

5)
$$
\begin{array}{r}
1,1\overset{0\,\,2}{\cancel{0}}3,375 \\
-592,840 \\
\hline
510,535
\end{array}
$$

6)
$$
\begin{array}{r}
\$\quad 370 \\
1,700 \\
1,415 \\
+\,4,250 \\
\hline
\$\,7,735
\end{array}
$$

1) George Goldberg drove 106 miles to the airport and took a 538 mile plane trip. How far did he travel all together?

2) Marsha Chen took home $273 after taxes. If she paid $69 in taxes, how much was her salary?

3) The Chicago Bulls scored 22 points in the first quarter, 27 points in the second quarter, 19 points in the third quarter, and 20 points in the fourth quarter. How many points did the Bulls score in the entire game?

4) Patrick McDonnell had 675 calories for breakfast, 490 calories for lunch, 835 calories for dinner, and had a snack of 360 calories. How many calories did he have all day?

5) A ranch is 44,591 acres. What is its area rounded to the nearest ten thousand acres?

6) A woman buys jewelry costing three thousand, seven hundred fifty dollars, and clothing costing two thousand, eight hundred ninety-three dollars. How much money did she spend?

7) In the month of June there were three forest fires in California. One burned up 12,902 acres, another burned up 39,724 acres, and the third burned up 25,090 acres. How many acres were burned up all together?

8) A bowler scored 199, 168 and 189 in three games. What was his total score for these games?

9) Max bought a computer for $2,899, a printer for $329, and a scanner for $225. How much did he spend all together?

10) Susan's rent was $850, her electricity was $75, her gas bill was $20, and her phone bill was $85. How much were her monthly expenses?

11) On a business trip John drove 435 miles on Tuesday, 294 on Wednesday, 197 on Thursday, and 239 on Friday. How far did he drive over these four days?

12) One thousand, four hundred fifty-eight people saw the first act of a play, but one hundred twenty-three left at intermission. How many people saw the beginning of the second act of the play?

13) If there are one hundred seven million, three hundred forty-eight thousand, seven hundred fifty-five households in the U.S., how many households are there rounded to the nearest ten thousand?

14) Phillip purchased a painting costing $105,500, another costing $88,350, and a third costing $179,000. How much did he spend on the three paintings?

15) Here are the points scored in a game by the starting players on the New York Knicks: Camby, 14; Ward, 12; Sprewell, 19; Houston, 23; and Thomas, 15. How many points did the five starting players score?

16) Jessica scored 1215 on the SATs. After taking a review course, her score rose by 78 points. What was her new score?

17) If you earned $87,499, how much would that be rounded to the nearest thousand dollars?

18) Kyra buys a new car for $17,000 and gets $4,950 for trading in her old car. How much does she have to pay for the new car?

19) On four consecutive Sabbaths, attendance at one house of worship was 133, 171, 114, and 136. How many people attended all together?

20) Tran Minh left his house with $405 to go shopping. He spent $198. How much did he have left?

Chapter 1 Review

1) State the place value of the number 3 in each of the following numbers, and write the number in expanded form.
     a) 163   b) 300,175   c) 1,394   d) 3,190,040

2) Write the following numbers in words.
     a) 469   b) 132,098   c) 40,922   d) 1,604

3) Round each of the following numbers to the nearest ten, hundred and thousand.
     a) 4,163   b) 82,503   c) 1,619,236

4) Round each of the following numbers to the nearest ten thousand, hundred thousand, and million.
     a) 14,375,100   b) 6,504,355   c) 9,499,562

Perform the indicated operations.

5)
$$46 + 33$$

6)
$$231 + 348$$

7)
$$21 + 3 + 42 + 11$$

8)
$$29 + 46$$

9)
$$263 + 429 + 356$$

10)
$$1637 + 10524 + 19783$$

11)
$$86 - 32$$

12)
$$594 - 341$$

13)
$$14396 - 11264$$

14)
$$570 - 389$$

15)
$$6414 - 4637$$

16)
$$100136 - 98729$$

17) State the place value of the digit 7 in each of the following numbers, and write the number in expanded form.
     a) 272   b) 4,719   c) 97,460   d) 7,285,116

18) Write the number in words.
    a) 7,083   b) 893   c) 4,510,384   d) 123,086

19) Round each of the following numbers to the nearest ten, hundred, and thousand.
    a) 4,378   b) 55,498   c) 5,699

20) Round each of the following numbers to the nearest ten thousand, hundred thousand and million.
    a) 24,503,546   b) 15,675,019   c) 1,604,990

21) Do each of these problems in expanded form.

    a)   56   b)   437   c)   5689
        − 23   − 214   − 3253

Perform the indicated operations (not in expanded form).

22)   45   23)   124   24)   2321
    + 36   + 632   15796
                                             + 8529

25)   754   26)   1692   27)   26342
    − 313   − 1451   − 15111

28)   425   29)   5924   30)   113471
    − 326   − 3976   − 99593

31) If your salary is $355 and you earned $178 in overtime, how much did you earn all together?

32) A ranch has one hundred forty-six thousand, three hundred twenty-five acres. Round that to the nearest ten thousand acres.

33) If you earned $44,150 and paid $8,279 in taxes, how much money would you have left?

34) Juan Sanchez traveled 175 miles on Monday, 209 miles on Tuesday, 133 miles on Wednesday, and 214 miles on Thursday. How far did he travel?

Chapter 1 • Test 1

1) State the place value of the digit 3 in each of the following numbers and write the number in expanded form.

    a)   163   _____

    b)   300,175   _____

    c)   1,394   _____

    d) 3,190,040   _____

2) Write the following numbers in words.

    a)   469   _____

    b)   132,098   _____

    c)   40,922   _____

    d)   1,604   _____

3) Round each of the following numbers to the nearest ten, hundred, and thousand.

	TEN	HUNDRED	THOUSAND
a) 4,163	_____	_____	_____
b) 82,503	_____	_____	_____
c) 1,619,236	_____	_____	_____

4) Round each of the following numbers to the nearest ten thousand, hundred thousand and million.

	TEN THOUSAND	HUNDRED THOUSAND	MILLION
a) 14,375,100	_____	_____	_____
b) 6,504,355	_____	_____	_____
c) 9,499,562	_____	_____	_____

Perform the indicated operations.

5) $\begin{array}{r} 46 \\ + 33 \\ \hline \end{array}$

6) $\begin{array}{r} 231 \\ + 348 \\ \hline \end{array}$

7) $\begin{array}{r} 21 \\ 3 \\ 42 \\ + 11 \\ \hline \end{array}$

8) $\begin{array}{r} 29 \\ + 46 \\ \hline \end{array}$

9) $\begin{array}{r} 263 \\ 429 \\ + 356 \\ \hline \end{array}$

10) $\begin{array}{r} 1637 \\ 10524 \\ + 19783 \\ \hline \end{array}$

11) $\begin{array}{r} 86 \\ - 32 \\ \hline \end{array}$

12) $\begin{array}{r} 594 \\ - 341 \\ \hline \end{array}$

13) $\begin{array}{r} 14396 \\ - 11264 \\ \hline \end{array}$

14) $\begin{array}{r} 570 \\ - 389 \\ \hline \end{array}$

15) $\begin{array}{r} 6414 \\ - 4637 \\ \hline \end{array}$

16) $\begin{array}{r} 100136 \\ - 98729 \\ \hline \end{array}$

17) State the place value of the digit 7 in each of the following numbers, and write the number in expanded form.

    a)   272   _____

    b)   4,719   _____

    c)   97,460   _____

    d) 7,285,116   _____

18) Write each number in words.

    a)   7,083   _____

    b)   893   _____

    c) 4,510,384   _____

    d)   123,086   _____

19) Round each of the following numbers to the nearest ten, hundred and thousand.

		TEN	HUNDRED	THOUSAND
a)	4,378	_____	_____	_____
b)	55,498	_____	_____	_____
c)	5,699	_____	_____	_____

20) Round each of the following numbers to the nearest ten thousand, hundred thousand and million.

	TEN THOUSAND	HUNDRED THOUSAND	MILLION
a) 24,503,546	_____	_____	_____
b) 15,675,019	_____	_____	_____
c) 1,604,990	_____	_____	_____

21) Perform the indicated operation in expanded form.

a)   56
    − 23

b)   437
    − 214

c)   5689
    − 3253

Perform the indicated operation.

22)   45
    + 36

23)   124
    + 632

24)   2321
      15796
    + 8529

25)   754
    − 313

26)   1692
    − 1451

27)   26342
    − 15111

28)   425
    − 326

29)   5924
    − 3976

30)   113471
    − 99593

Chapter 1 • Test 2

1) State the place value of the digit 5 in each of the following numbers and write the number in expanded form.

    a)   5,678   _____

    b)   56   _____

    c)   567,090   _____

2) Write the following numbers in words.

    a)   203   _____

    b)   12,009   _____

    c) 1,009,023   _____

    d)   2,002   _____

3) Round each of the following numbers to the nearest ten, hundred and thousand.

		TEN	HUNDRED	THOUSAND
a)	6,983			
b)	6,998			
c)	79,983			

4) Round each of the following numbers to the nearest ten thousand, hundred thousand and million.

	TEN THOUSAND	HUNDRED THOUSAND	MILLION
a) 14,679,002	_____	_____	_____
b) 7,992,309	_____	_____	_____
c) 111,111,111	_____	_____	_____

5)　　27
　　+ 29

6)　　498
　　+ 666

7)　　99
　　　8
　　 27
　　+ 30

8)　　89
　　+ 99

9)　　478
　　　34
　　+ 555

10)　2980
　　　99
　　+ 889

11)　97
　　− 34

12)　398
　　− 299

13)　99976
　　− 8999

14)　789
　　− 456

15)　909
　　− 888

16)　10099
　　−　999

17) State the place value of the digit 0 in each of the following numbers, and write the number in expanded form.

a)   909   _____

b)   90,765   _____

c)   68,021   _____

d)   20,987   _____

18) Write each number in words.

a)   6,809   _____

b)   983   _____

c)   55,008   _____

d) 21,789,230   _____

19) Round each of the following numbers to the nearest ten, hundred and thousand.

		TEN	HUNDRED	THOUSAND
a)	5,679	_____	_____	_____
b)	987	_____	_____	_____
c)	62,999	_____	_____	_____

20) Round each of the following numbers to the nearest ten thousand, hundred thousand and million.

	TEN THOUSAND	HUNDRED THOUSAND	MILLION
a) 27,876,902	_____	_____	_____
b) 99,999,999	_____	_____	_____
c) 203,781,999	_____	_____	_____

21) Perform the indicated operation in expanded form.

a)
```
   45
 − 32
```

b)
```
   456
 − 198
```

c)
```
   2234
 −  657
```

Perform the indicated operations (not in expanded form).

22)
```
   89
 + 45
```

23)
```
   1987
 +  237
```

24)
```
    785
   2234
 +   99
```

25)
```
   334
 − 277
```

26)
```
   8901
 − 7777
```

27)
```
   24919
 −  8890
```

28)
```
   367
 −  59
```

29)
```
   6802
 − 5678
```

30)
```
   66928
 −  8909
```

Chapter 2. Multiplication and Division of Whole Numbers

In the last chapter you saw that addition and subtraction were inverse mathematical operations. For example, a pay raise of 50 cents an hour is the opposite of a 50 cents an hour pay cut. When you have completed this chapter, you'll understand that multiplication and division are also inverse mathematical processes.

2.1 Multiplication with Whole Numbers

The Multiplication Table

Learning the multiplication table shown below is a basic skill that must be mastered. Do you have to memorize this table? Yes! Can't you just use a calculator? No! You must know this table by heart to be able to multiply numbers, to do division, and to do algebra. To be blunt, until you memorize this entire table, you won't be able to progress further than this page.

MULTIPLICATION TABLE

×	0	1	2	3	4	5	6	7	8	9	10	11	12
0	0	0	0	0	0	0	0	0	0	0	0	0	0
1	0	1	2	3	4	5	6	7	8	9	10	11	12
2	0	2	4	6	8	10	12	14	16	18	20	22	24
3	0	3	6	9	12	15	18	21	24	27	30	33	36
4	0	4	8	12	16	20	24	28	32	36	40	44	48
5	0	5	10	15	20	25	30	35	40	45	50	55	60
6	0	6	12	18	24	30	36	42	48	54	60	66	72
7	0	7	14	21	28	35	42	49	56	63	70	77	84
8	0	8	16	24	32	40	48	56	64	72	80	88	96
9	0	9	18	27	36	45	54	63	72	81	90	99	108
10	0	10	20	30	40	50	60	70	80	90	100	110	120
11	0	11	22	33	44	55	66	77	88	99	110	121	132
12	0	12	24	36	48	60	72	84	96	108	120	132	144

Let's get a couple of things out of the way. First, any number times 0 is 0. When we multiply two numbers, we call our answer the *product* of those two numbers. The numbers that are multiplied together are called the *factors* of the product. How much is the product of 0 and 1? $0 \times 1 = 0$.

If you don't already know the product of every combination of the numbers in the table, what you should definitely do is memorize those you don't know.

There are a few ways a multiplication problem can be expressed. Let's consider how to express the multiplication of 23 by 8. We can do it each of these ways:

The product of 23 and 8   23 times 8   23 × 8   23
                                                                      × 8

23 • 8   23(8)   (23)(8)

When multiplying two-digit, three-digit and larger numbers, we almost always express multiplication vertically because that's the easiest way to find the product. The numbers we multiply together are called the *factors* and the answer is called the *product.*

Example 1:
$$\begin{array}{r} 23 \\ \times\, 8 \\ \hline \end{array}$$

SOLUTION: There's a long method and a short method to working out multiplication problems. First we'll do the long method:

The long method helps you to see how the place value of the digit affects the multiplication process.

$$23 = 20 + 3$$
$$\begin{array}{r} \times\, 8 \\ \hline 24 \quad (3\times8) \\ +160 \quad (20\times8) \\ \hline 184 \quad (24+160) \end{array}$$

The short method is the method we'll use from now on.

$$\begin{array}{r} {}^{2}\;\; \\ 23 \\ \times 8 \\ \hline 184 \end{array}$$

Here's what we do: $3\times8 = 24$. Write down the 4 and carry the 2; $2\times8 = 16$. $16 + 2 = 18$. Write down the 18 to get the answer 184.

Example 2:
$$\begin{array}{r} 465 \\ \times\, 7 \\ \hline \end{array}$$

SOLUTION: Here's how we do this problem using the long method. Pay attention to the place values.

$$465 = 400 + 60 + 5$$
$$\begin{array}{r} \times\, 7 \\ \hline 35 \quad (7 \times 5) \\ 420 \quad (7 \times 60) \\ 2800 \quad (7 \times 400) \\ \hline 3255 \quad (35 + 420 + 2800) \end{array}$$

Here's the short method:

$$\begin{array}{r} {}^{+3}\;\;\; \\ 465 \\ \times 7 \\ \hline 3255 \end{array}$$

Here's what we do: $5\times7 = 35$; write down the 5 and carry the 3. $6\times7 = 42$ plus the 3 we carried is 45. Write down the 5 and carry the 4. $4\times7 = 28$ plus the 4 we carried is 32.

Problem Set:

1) Find the product of 36 and 5.

2)   29
    ×9

3)   162
    ×4

4) Find the product of 396 and 3.

SOLUTIONS TO PROBLEM SET:

1) $\overset{3}{3}6$
    ×5
    180

2) $2\overset{8}{9}$
    ×9
    261

3) $\overset{2}{1}62$
    ×4
    648

4) $\overset{2\,1}{3}96$
    ×3
    1188

Let's try some more complicated problems.

Example 3:   38
             × 46

SOLUTION:

       38
     × 46
      228
      152
     1748

We begin this problem just the way we did the previous problems.

We multiple 38 by 6. Now comes the tricky part. When we start multiplying by 4, we need to move one space to the left, because 4 really represents 40, since 46 = 40 + 6. When multiplying by 4 start writing your answer directly under 4.

The main thing to remember is to move one space to the left when you put your answer on the second row of numbers.

Example 4:   275
             × 394

SOLUTION:

       275
     × 394
      1100
      2475
       825
    108350

Once again the important thing to remember is to start one place to the left when you begin your second row of numbers, and another space to the left when you begin your third row of numbers. First you multiply 275 by 4 and get 1100. When you multiply by 9, you must place the 5 one space to the left, under the second 0 rather than under the first 0. And when you multiply by 3, place the 5 another one place to the left, directly under the 7.

Problem Set:

1)   71
    × 28

2)   85
    × 49

3)   240
    × 65

4)   8623
    × 539

1)	71	2)	85	3)	240	4)	8623
	×28		×49		×65		×539
	568		765		1200		77607
	142		340		1440		25869
	1988		4165		15600		43115
							4647797
	1,988		4,165		15,600		4,647,797

Multiplying Whole Numbers by Powers of Ten

Quick, what is the product of 145 and 10? The answer is 1,450. How did we do it that fast? We just tacked a 0 on to 145. Next problem: How much is 4,280×10? The answer is 42,800. Again, we just added a zero to the end of 4,280. Are you ready for a sweeping generality? To multiply any whole number by ten, just add a zero to the right of the number. What is the product of 145 and 100? The answer is 14,500. What did we do? We added two zeros to 145. To multiply any whole number by one hundred, just add two zeros to the right of the number. Can you guess what we do if we multiply a whole number by 1,000? We just add three zeros to the right of the number. For example, 145×1,000 = 145,000. So we can make a general statement: To multiply any whole number by a power of ten we just add the same number of zeros to the right of the number, as are in the power of ten. If we multiply by 10 we add one zero. If we multiply by 100 we add two zeros. If we multiply by 1,000 we add three zeros, and so on.

Example 5: Multiply 456 by   a) 10   b) 100   c) 1,000   d) 10,000

SOLUTION: a) 4,560   b) 45,600   c) 456,000   d) 4,560,000

Problem Set:

1) 23×1,000   2) 230×10   3) 500(100)   4) 1,237(1,000)

1) 23,000   2) 2,300   3) 50,000   4) 1,237,000

Multiplying by Numbers that End in Zero

The following problem is an example of a shortcut we can use to multiply numbers that end in zeros.

Example 6:   586
               × 500   Multiply 586×5, then add the two zeros to the right of 2930 to get the
             293,000   product 293,000.

SOLUTION: Here's another way to look at this problem.

                586   To multiply this as fast and easily as possible we'll rewrite this problem.
              × 500   We'll just drop down the two zeros, and multiply 586 by 5.
             293,000

Example 7:   739
            $\times$ **3000**

SOLUTION: Multiply 739 by 3; then add the three zeros to the right of 2,217 to get the product 2,217,000.

    739
$\times$3000
2,217,000

Problem Set:

1)	427	2)	872	3)	948	4)	5387
	$\times$400		$\times$600		$\times$800		$\times$9000

SOLUTIONS TO PROBLEM SET:

1)	427	2)	872	3)	948	4)	5387
	$\times$400		$\times$600		$\times$800		$\times$9000
	170,800		523,200		758,400		48,483,000

Perform the following multiplication problems.

1) 4×25 2) 8×19 3) 19×8

4) 25×12 5) 13×13 6) 91×88

7) $\begin{array}{r} 2,345 \\ \times\ 123 \\ \hline \end{array}$ 8) $\begin{array}{r} 34,569 \\ \times\ 1,456 \\ \hline \end{array}$ 9) $\begin{array}{r} 28,009 \\ \times\ 20,111 \\ \hline \end{array}$ 10) $\begin{array}{r} 199,909 \\ \times\ 68,735 \\ \hline \end{array}$

Multiply each of these numbers by 10:

11) 4   12) 600   13) 70   14) 9,010

15) 145   16) 100   17) 7   18) 222

19) 13   20) 2,000   21) 80   22) 1,000

Multiply each of these numbers by 100:

23) 72   24) 587   25) 367   26) 10

27) 109   28) 0   29) 12,999   30) 47

31) 100   32) 1,000   33) 10,000   34) 100,000

Multiply each of these sets of numbers:

35)   39   36)   199   37)   3490   38)   367
    $\times 10$ $\times 100$ $\times 1000$ $\times 10$

39)   45   40)   459   41) 69999   42)   22
    $\times 40$ $\times 300$ $\times 400$ $\times 90$

43) 230   44) 4876   45)   500   46) 1000
    $\times 60$ $\times 600$ $\times 400$ $\times 1000$

2.2   Exponents

$2^2 = 2(2) = 4$ $\qquad$ $2^3 = 2(2)(2)$ $\qquad$ $2^4 = 2(2)(2)(2)$ $\qquad$ $2^5 = 2(2)(2)(2)(2)$

$\qquad\qquad\qquad\qquad$ $4(2)$ $\qquad\qquad\qquad$ $4(2)(2)$ $\qquad\qquad\qquad$ $4(2)(2)(2)$

$\qquad\qquad\qquad\qquad$ 8 $\qquad\qquad\qquad\quad$ $8(2)$ $\qquad\qquad\qquad\quad$ $8(2)(2)$

$\qquad\qquad\qquad\qquad\qquad\qquad\qquad\qquad$ 16 $\qquad\qquad\qquad\quad$ $16(2)$

$\qquad\qquad\qquad\qquad\qquad\qquad\qquad\qquad\qquad\qquad\qquad\qquad$ 32

$3^2 = 3(3) = 9$ $\qquad$ $3^3 = 3(3)(3)$ $\qquad$ $6^3 = 6(6)(6)$ $\qquad$ $1^6 = 1(1)(1)(1)(1)(1) = 1$

$\qquad\qquad\qquad\qquad$ $9(3)$ $\qquad\qquad\qquad$ $36(6)$

$\qquad\qquad\qquad\qquad$ 27 $\qquad\qquad\qquad\quad$ 216

All these problems have one thing in common; they're all multiplying the same number times itself. There's a way to show repeated multiplication of a number times itself. Look at this problem: $4(4)(4)$. This can be written as 4^3. This is read four to the third power, or we could say four cubed. The 4 is called the *base* and the 3 is called the *power* or the *exponent*. The base is the number you multiply times itself. The exponent, or power, tells you how many of the base you multiply together. Three to the fourth power is written 3^4, and means $3(3)(3)(3)$, which is equal to 81.

Let's look at a few more examples.

Example 1: 1^4

$\qquad\qquad$ SOLUTION: $1^4 = 1(1)(1)(1) = 1$ $\qquad$ This is read one to the fourth power. 1 is the base; 4 is the exponent. One to any power is 1.

Example 2: 5^2

$\qquad\qquad$ SOLUTION: $5^2 = 5(5) = 25$ $\qquad$ This is read five squared. The 5 is the base and the 2 is the exponent.

Example 3: 0^3

$\qquad\qquad$ SOLUTION: $0^3 = 0(0)(0) = 0$ $\qquad$ This is read zero cubed. The 0 is the base, the 3 is the exponent. Zero to any power (except 0) is 0.

Example 4: 2^6

$\qquad\qquad$ SOLUTION: $2^6 = 2(2)(2)(2)(2)(2) = 64$ $\quad$ This is read two to the sixth power. The 2 is the base and the 6 is the power.

Let's look at a special case:

$\quad$ $2^0 = 1$

$\quad$ $3^0 = 1$ $\quad$ A basic rule of exponents is: *Anything except 0 to the zero power is 1* (0^0 is not defined).

$\quad$ $10^0 = 1$

We can't explain this exception to you until you've taken more math courses.

Problem Set:

Simplify the following problems and state which number is the base and which is the exponent or power.

1) 8^2

2) 7^1

3) 4^0

4) 2^4

5) 1^6

6) 3^4

SOLUTIONS TO PROBLEM SET:

1) $8^2 = 8(8) = 64$
8 is the base.
2 is the exponent.

2) $7^1 = 7$
7 is the base.
1 is the exponent.

3 $4^0 = 1$
4 is the base.
0 is the exponent.

4) $2^4 = 2(2)(2)(2) = 16$
2 is the base.
4 is the exponent.

5) $1^6 = 1(1)(1)(1)(1)(1) = 1$
1 is the base.
6 is the exponent.

6) $3^4 = 3(3)(3)(3) = 81$
3 is the base.
6 is the exponent.

Simplify the following problems.

1) 5^2

2) 8^2

3) 2^5

4) 2^8

5) 2^3

6) 3^3

7) 5^4

8) 1^{10}

9) 1^{100}

10) 0^5

11) 8^0

12) 6^0

13) 6^2

14) 11^2

15) 11^3

16) 10^1

17) 10^2

18) 10^3

19) 10^4

20) 10^5

2.3 Squares and Roots of Whole Numbers

In the last section we worked with exponents. Some of the problems had exponents of 0, 1, 2, 3, 4, etc. In this section we'll work on problems whose exponent is 2 and how to reverse the process of raising a number to a power by using radicals. The product of a number times itself is called a *perfect square*. The product of 3 and 3 is 9. Nine is a perfect square. Another way to show 3×3 is 3^2. This can be read, three to the second power or three squared. The word squared is commonly used instead of saying to the second power. In words we would say three squared is nine. In mathematical symbols it would look like this: $3^2 = 9$. The 3 is called the base and the 2 is called the exponent or power. When the power is 2 the base is multiplied times itself.

Example 1: Find the value for 4 squared. Name the base and the power.

SOLUTION: $4^2 = 4(4) = 16$ The four is the base and the two is the power. 16 is called the perfect square.

Example 2: Find the value for 10 squared. Name the base and the exponent.

SOLUTION: $10^2 = 10(10) = 100$ The ten is the base and the two is the exponent. 100 is called the perfect square.

Example 3: See if you can fill in the following table.

PROBLEM	BASE	POWER	PERFECT SQUARE
0^2			
1^2			
2^2			
3^2			
4^2			
5^2			
6^2			
7^2			
8^2			
9^2			
10^2			
11^2			
12^2			

SOLUTION:

PROBLEM	BASE	POWER	PERFECT SQUARE
0^2	0	2	0
1^2	1	2	1
2^2	2	2	4
3^2	3	2	9
4^2	4	2	16
5^2	5	2	25
6^2	6	2	36
7^2	7	2	49
8^2	8	2	64
9^2	9	2	81
10^2	10	2	100
11^2	11	2	121
12^2	12	2	144

Example 4:   Find the value of 15^2

SOLUTION:
$$
\begin{array}{r}
15 \\
\times 15 \\
\hline
75 \\
15 \\
\hline
225
\end{array}
$$

$15^2 = 225$

Problem Set:

Find the perfect square for each of the following numbers.

1) 7   2) 11   3) 5   4) 23   5) 100   6) 0

SOLUTIONS TO PROBLEM SET:

1) $7^2 = 7(7) = 49$ 2) $11^2 = 11(11) = 121$ 3) $5^2 = 5(5) = 25$

4) $23^2 = 23(23) = 529$ 5) $100^2 = 100(100) = 10,000$ 6) $0^2 = 0(0) = 0$

Let's take a look at a type of problem that is the reverse of what we just did.

Suppose we asked you what number times itself is 9. We're sure you could tell us the number is 3, because 3 times itself is 9. Instead of writing this type of problem in words, we can write it in symbols in the following way: $\sqrt{9}$. The $\sqrt{}$ symbol is called a *radical*. If we use this symbol, $\sqrt[2]{}$, the little 2 means square root. Square root asks us what number times itself equals the number under the radical. The 2 is used to indicate the base is written twice and multiplied together to get the number under the radical. If no number is written where the 2 is it's assumed the $\sqrt{}$ means square root. The number under the radical is called the *radicand*. In the term $\sqrt{9} = 3$, the 9 is the radicand and the root is assumed to be 2 for square root.

Example 5: Find $\sqrt{64}$.

SOLUTION: What number times itself is 64? By now you should know $8(8) = 64$, so the answer is 8. $\sqrt{64} = 8$.

Suppose you didn't remember that $8(8) = 64$. To figure this out, you could refer to the table you filled in Example 3 or you could try using trial and error. Suppose we guessed the answer is 10: $10^2 = 100$; 100 is too big. Suppose we guessed the answer is 7: $7^2 = 49$; 49 is a little too small. Next we'll guess 8: $8^2 = 64$. It looks like we have our answer.

Example 6: Find $\sqrt{784}$.

SOLUTION: What number times itself is 784? This is one most people (including us) don't have memorized, so we'll have to use the trial and error method. This number's pretty big so we'll guess 20. $20^2 = 400$; 400 is too small. Let's try 30. $30^2 = 900$; 30 is too big. 20 is too small and 30 is too big, so let's try 25. $25^2 = 25(25) = 625$, which is a little too small. Let's try 28. $28^2 = 28(28) = 784$. Now we know $\sqrt{784} = 28$.

Later in this book we'll show you another method for finding roots. The other method uses prime factorization.

Example 7:   Fill in the following table.

PROBLEM	ROOT	RADICAND	SQUARE ROOT
$\sqrt{0}$			
$\sqrt{1}$			
$\sqrt{4}$			
$\sqrt{9}$			
$\sqrt{16}$			
$\sqrt{25}$			
$\sqrt{36}$			
$\sqrt{49}$			
$\sqrt{64}$			
$\sqrt{81}$			
$\sqrt{100}$			
$\sqrt{121}$			
$\sqrt{144}$			

SOLUTION:

PROBLEM	ROOT	RADICAND	SQUARE ROOT
$\sqrt{0}$	2	0	0
$\sqrt{1}$	2	1	1
$\sqrt{4}$	2	4	2
$\sqrt{9}$	2	9	3
$\sqrt{16}$	2	16	4
$\sqrt{25}$	2	25	5
$\sqrt{36}$	2	36	6
$\sqrt{49}$	2	49	7
$\sqrt{64}$	2	64	8
$\sqrt{81}$	2	81	9
$\sqrt{100}$	2	100	10
$\sqrt{121}$	2	121	11
$\sqrt{144}$	2	144	12

Now that we know how to find square roots, let's take a look at some other roots. Suppose we asked you what number times itself, times itself again is 8? Let "?" represent the number we're trying to find. In other words, $(?)(?)(?) = 8$. Notice that the same number is written three times and multiplied. In mathematical symbols the way we would represent what number times itself, times itself is 8 is $\sqrt[3]{8}$. The answer to the question is $\sqrt[3]{8} = 2$, because $2(2)(2) = 4(2) = 8$. This problem is read the cube root of 8 is 2.

Example 8: Find $\sqrt[3]{64}$.

SOLUTION: What number times itself, times itself is 64? This is read the cube root of 64. $4(4)(4) = 16(4) = 64$ so $\sqrt[3]{64} = 4$.

Example 9: Find $\sqrt[4]{16}$.

SOLUTION: $(?)(?)(?)(?) = 16$ This is read the fourth root of 16. To find the correct number we can use the trial and error method. Let's start with 3. $3(3)(3)(3) = 9(3)(3) = 27(3) = 81$. 81 is larger than 64 so we have to try a number smaller than 3. Let's try 2 and see what happens. $2(2)(2)(2) = 4(2)(2) = 8(2) = 16$. Now we know $\sqrt[4]{16} = 2$.

Example 10: Find $\sqrt[5]{1,024}$.

SOLUTION: $(?)(?)(?)(?)(?) = 1,024$. This is read the fifth root of 1,024. Again we'll use the trial and error method. Let's try 4. $4(4)(4)(4)(4) = 16(4)(4)(4) = 64(4)(4) = 256(4) = 1,024$. Now we know $\sqrt[5]{1,024} = 4$.

Problem Set

Simplify the following.

1) $\sqrt{81}$ 2) $\sqrt{961}$ 3) $\sqrt[3]{1,000}$ 4) $\sqrt[4]{256}$ 5) $\sqrt[5]{243}$

SOLUTIONS TO PROBLEM SET:

1) 9   2) 31   3) 10   4) 4   5) 3

Simplify the following problems. State the base, power, and perfect square.

1) 6^2

2) 11^2

3) 20^2

4) 25^2

5) 30^2

6) 0^2

7) 27^2

8) 48^2

Simplify the following problems. State the root and radicand.

9) $\sqrt{25}$

10) $\sqrt{144}$

11) $\sqrt{100}$

12) $\sqrt{169}$

13) $\sqrt{81}$

14) $\sqrt{16}$

15) $\sqrt{324}$

16) $\sqrt{961}$

17) $\sqrt{289}$

18) $\sqrt{361}$

19) $\sqrt[3]{27}$

20) $\sqrt[3]{64}$

21) $\sqrt[3]{125}$

22) $\sqrt[3]{216}$

23) $\sqrt[4]{16}$

24) $\sqrt[4]{81}$

25) $\sqrt[4]{625}$

26) $\sqrt[4]{1,296}$

27) $\sqrt[5]{32}$

28) $\sqrt[5]{243}$

29) $\sqrt{1}$

30) $\sqrt{0}$

31) $\sqrt[3]{1}$

32) $\sqrt[3]{0}$

33) $\sqrt[4]{1}$

34) $\sqrt[4]{0}$

2.4   Division with Whole Numbers

Four friends share an eight-slice pizza. How many slices does each person get? The answer is 2. What we did is divide 8 by 4. There are several ways we can express this division:

$$8 \div 4 \qquad 4\overline{)8} \qquad \text{8 divided by 4} \qquad \text{4 into 8} \qquad \tfrac{8}{4} \qquad \frac{8}{4}$$

We would read this as eight divided by four, or we could say the *quotient* of eight and four. We call our answer 2, the *quotient* of 8 divided by 4. The 8 is called the *dividend* and the 4 is called the *divisor*. The dividend always goes inside the division box and the divisor always goes on the outside of the division box. The quotient is always on the top of the box, so it should look like this:

$$\frac{quotient}{divisor\,\overline{)dividend}} \qquad\qquad 4\overline{)\overset{2}{8}}$$

Example 1:   What is the quotient of 12 and 4?

SOLUTION: Division is the inverse of multiplication, so this question is really asking us what number times 4 is 12? The quotient of 12 and 4 is 3.

$$4\overline{)\overset{3}{12}}$$

We can always check our division by multiplication. The quotient times the divisor is equal to the dividend.

$$quotient \times divisor = dividend \qquad\qquad 3(4) = 12$$

Example 2:   What is the quotient of 15 and 3?

SOLUTION: $15 \div 3 = 5 \ or \ \dfrac{15}{3} = 5$ The quotient of 15 and 3 is 5.

Example 3:   576 ÷ 8

SOLUTION: Ask yourself, does 8 go into 5? The answer is no, so we do not begin writing our answer over the 5. Does 8 go into 57? Yes, so we start writing our answer over the 7. We know 8(7) = 56. So we write 7 over the 7 in the 57 and 56 below 57. Next, subtract 56 from 57 and get 1, then bring down the next number, 6. How many times does 8 go into 16? It goes twice, so we write a 2 over the 6 in 576. Now subtract.

$$\begin{array}{r} 7 \\ 8\overline{)576} \\ -\ 56 \end{array}$$

$$\begin{array}{r} 72 \\ 8\overline{)576} \\ -\ 56 \\ \hline 16 \\ -\ 16 \\ \hline 0 \end{array}$$

Let's check our answer; quotient × divisor = dividend.

Check:
$$\begin{array}{r} 72 \\ \times\ 8 \\ \hline 576 \end{array}$$

Three friends share an eight-slice pizza. How many slices does each person get? This time the friends can't have the same number of slices of pizza, because three doesn't divide into eight evenly. If they each have two slices, there will be two slices left over. The two leftover slices are called the *remainder*. Sometimes a number doesn't divide into another number evenly. When this happens, the number we're left with is called the *remainder*. Let's take a look at some division problems that have remainders.

Example 4: 2456 ÷ 25

SOLUTION:

$$\begin{array}{r} 9 \\ 25\overline{)2456} \\ -225 \\ \hline 20 \end{array}$$

25 does not go into 24, so we don't begin writing the quotient over the four. 25 does go into 245 nine times.

Multiply 9 times 25 and put 225 under the 245. Subtract and bring down the next number.

$$\begin{array}{r} 98 \text{ R6} \\ 25\overline{)2456} \\ -225 \\ \hline 206 \\ -200 \\ \hline 6 \end{array}$$

25 goes into 206 eight times. Multiply 25 times 8. Subtract 200 from 206. This leaves us with a remainder of 6.

To check our answer when there's a remainder, just multiply the quotient times the divisor and then add the remainder. You should get the dividend.

quotient × divisor + remainder = dividend

Check: 98 × 25 + 6 = 2,456.

We have found that our students get confused when a quotient has zeros in it. They're not sure where to put the zeros or how many there should be. Hopefully the next example will make it easier for you to understand this type of problem.

Example 5: 30045 ÷ 15

SOLUTION:

$$\begin{array}{r} 2 \\ 15\overline{)30045} \\ -30 \\ \hline 0 \end{array}$$

The key to getting the correct answer is being very careful where you write the digits of the quotient. You also have to write zeros in the quotient when a division can't be done. 15 goes into 30 twice.

When we multiply and subtract, this leaves us with 0. The next number we bring down is also a 0. 15 doesn't go into 00. We have to show this by writing a 0 in the quotient.

$$\begin{array}{r} 200 \\ 15\overline{)30045} \\ -30 \\ \hline 004 \end{array}$$

Now we bring down the next number, 4, but 15 doesn't go into 4 either, so we have to put another 0 in the quotient. Now divide 15 into 45 and we're finished.

Notice that each digit in the quotient goes directly over the last digit in its corresponding product.

$$\begin{array}{r} 2003 \\ 15\overline{)30045} \\ -30 \\ \hline 045 \\ -45 \\ \hline 0 \end{array}$$

Example 6: $4 \div 4$

> SOLUTION: *Any number (except 0) divided by itself is 1.*

$$\begin{array}{r} 1 \\ 4{\overline{\smash{\big)}\,4}} \\ \underline{-4} \\ 0 \end{array}$$

Problem Set:

Now it's time for you to try a few. Be careful where you put your zeros. Also, we suggest you check your answers.

1) $46{\overline{\smash{\big)}\,9200}}$ 2) $21{\overline{\smash{\big)}\,6405}}$ 3) $20{\overline{\smash{\big)}\,11592}}$ 4) $51{\overline{\smash{\big)}\,515151}}$

SOLUTIONS TO PROBLEM SET:

1)
$$\begin{array}{r} 200 \\ 46{\overline{\smash{\big)}\,9200}} \\ \underline{-92} \\ 000 \\ \underline{-00} \\ 0 \end{array}$$

2)
$$\begin{array}{r} 305 \\ 21{\overline{\smash{\big)}\,6405}} \\ \underline{-63} \\ 105 \\ \underline{-105} \\ 0 \end{array}$$

3)
$$\begin{array}{r} 579 \ \text{R12} \\ 20{\overline{\smash{\big)}\,11592}} \\ \underline{-00} \\ 159 \\ \underline{-140} \\ 192 \\ \underline{-180} \\ 12 \end{array}$$

4)
$$\begin{array}{r} 10101 \\ 51{\overline{\smash{\big)}\,515151}} \\ \underline{-51} \\ 051 \\ \underline{-51} \\ 051 \\ \underline{-51} \\ 0 \end{array}$$

Division with Zeros

Let's take a look at two very important types of division problems:

- $0 \div 4$, which is read 0 divided by 4, or the quotient of 0 and 4.
- $4 \div 0$, which is read 4 divided by 0, or the quotient of 4 and 0.

These two problems look very similar, but have very different answers.

Let's take a closer look at $0 \div 4$. The 0 is the dividend, so it goes inside the division box, and 4 is the divisor, so it goes outside the box.

$4{\overline{\smash{\big)}\,0}}$ We ask ourselves, what number times 4 would give us 0? 0, of course.
So 0 divided by 4 is 0.

$$\begin{array}{r} 0 \\ 4{\overline{\smash{\big)}\,0}} \end{array}$$ Check: $0 \cdot 4 = 0$

How about $0 \div 7$? What number times seven would give us 0? 0, of course.

$$\begin{array}{r} 0 \\ 7{\overline{\smash{\big)}\,0}} \end{array}$$ Check: $0 \cdot 7 = 0$

So, we can say that *zero divided by any number is zero.* The only exception to this rule is $0 \div 0$, which is undefined.

$0 \div 8 = 0$ $0 \div 10 = 0$ $0 \div 345 = 0$ $\dfrac{0}{10} = 0$

Now, let's look at the other type of problem, $4 \div 0$. Using a division box, the 0 is on the outside and the 4 is inside the box, like this: $0\overline{)4}$. Let's ask ourselves, what number times 0 is 4? You can't think of a number times 0 that will give us 4, because there isn't any. Why not? Because 0 times any number is always 0, and never that number. Whenever we divide by 0, the answer is called *undefined*. So, $4 \div 0$ is undefined. This is a very important concept in mathematics. But don't worry; we'll help you remember this by giving you many examples of this type throughout this book.

So, we can say that *any number divided by zero is undefined.*

$7 \div 0$ is undefined.　　$8 \div 0$ is undefined.　　$10 \div 0$ is undefined.　　$\frac{10}{0}$ is undefined.

Problem Set:

1) $0 \div 3$　　　　2) $3 \div 0$　　　　3) $24 \div 0$　　　　4) $0 \div 13$

SOLUTIONS TO PROBLEM SET:

1) $0 \div 3 = 0$　　　2) $3 \div 0$ is undefined　　3) $24 \div 0$ is undefined　　4) $0 \div 13 = 0$

Dividing Whole Numbers by Powers of Ten

How much is 9,000 divided by 10? The answer is 900. All we did was remove the last zero from 9000. How much is 9,000 divided by 100? The answer is 90. This time we removed the last two zeros from 9000. What's 9,000 divided by 1000? It's 9. We removed three zeros from 9,000. To divide any whole number by a power of 10 we just remove the same number of zeros from the end of the number as there are zeros in the power of ten.

If the number doesn't end in zero, we don't use this method.

Example 7:　**Divide 560,000 by**　　a) 10　　　b) 100　　　c) 1,000　　　d) 10,000

SOLUTION:　a) $560,000 \div 10 = 56,000$　　10 has one zero, so we removed one zero from 560,000.

b) $560,000 \div 100 = 5,600$　　100 has two zeros, so we removed two zeros from 560,000.

c) $560,000 \div 1,000 = 560$　　1,000 has three zeros, so we removed three zeros from 560,000.

d) $560,000 \div 10,000 = 56$　　10,000 has four zeros, so we removed four zeros from 560,000.

Problem Set:

1) Divide 500 by 10　　　　　　　　2) $1,200 \div 100$

3) $90,000 \div 10,000$　　　　　　　4) Divide 1,000,000 by 100,000

SOLUTIONS TO PROBLEM SET:

1) 50　　　　　　2) 12　　　　　　3) 9　　　　　　4) 10

Averages

Now that you know how to divide, we can work on averages. An average is supposed to be a "representative" score. To find the average of a set of numbers, add them and divide their sum by the amount of numbers you added.

Example 8: If you had exam scores of 80, 70 and 78, what was your average score?

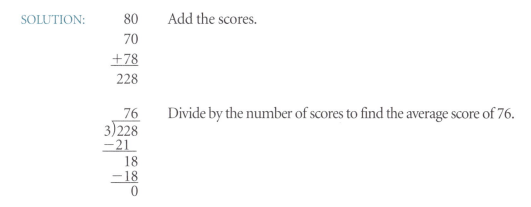

SOLUTION:

$$\begin{array}{r} 80 \\ 70 \\ +78 \\ \hline 228 \end{array}$$

Add the scores.

$$\begin{array}{r} 76 \\ 3\overline{)228} \\ -21 \\ \hline 18 \\ -18 \\ \hline 0 \end{array}$$

Divide by the number of scores to find the average score of 76.

Example 9: Over a five year period you earned $15,000, $22,500, $29,000, $15,700 and $15,700. What was your average income over this period?

SOLUTION:

$$\begin{array}{r} 15,000 \\ 22,500 \\ 29,000 \\ 15,700 \\ +15,700 \\ \hline 97,900 \end{array}$$

We'll add up the earnings and divide by five.

The sum is $97,900.

$$\begin{array}{r} 19580 \\ 5\overline{)97900} \\ -5 \\ \hline 47 \\ -45 \\ \hline 29 \\ -25 \\ \hline 40 \\ -40 \end{array}$$

Divide the sum by 5.

The average income for the five years is $19,580.

Example 10: A class's test grades on exam 1 in a statistics course are 100, 92, 85, 20, 0, 17, 83, 88, 50 and 45. What is the class's average score on the test?

SOLUTION: The sum of the exam grades is 580. There are ten grades so we'll divide 580 by 10. Since we're dividing by a power of ten, we don't need a division box. We'll take a short cut and just drop off one zero from 580 and find their average is 58. Here's a thought question for you. Look at the test scores again. Do you think 58 is a good representation of the test scores? Why or why not?

Problem Set:

Find the average of the following sets of numbers.

1) 23, 65, 77
2) 100, 400, 500, 700, 300

3) The following is a list of IQs of a sample of residents of a particular neighborhood. What's the average IQ? 90, 98, 101, 120, 100, 112, 120, 99, 128 and 132.

SOLUTIONS TO PROBLEM SET:

1) $165 \div 3 = 55$
2) $2,000 \div 5 = 400$
3) $1,100 \div 10 = 110$

Estimation

Sometimes it's not necessary to find the exact answer to a mathematical problem. When the exact answer is not necessary, estimating an answer is generally faster than actually working out the problem. For example, suppose we wanted to estimate the number of textbooks to order for your classes next semester. You might want to estimate the number of hours you'll work by the end of this year. Estimation can also be used to check the answers to arithmetic problems. For example, if we were asked to find the product of 185 and 291, we could estimate the product to be 200×300, which is 60,000. If we incorrectly multiplied 185 times 291 and got an answer of 503,835, our estimation of 60,000 would tell us to go back and try the problem again. The exact answer is 53,835. By the time you've finished this section, you'll be able to estimate the value of the answer to a problem. Now that you know how to round a number, add, subtract, multiply, and divide, it's time for you to learn how to estimate your answer.

Example 11: **Estimate the product of 324 and 546.**

SOLUTION: Before we can estimate the product we have to decide how accurate an estimation we need. We'll work this problem out in two ways. The first one will not be as accurate an estimate as the second.

First: We'll begin by rounding 324 and 546 to the nearest hundred.

$300 \times 500 = 150,000$

Second: We'll begin by rounding 324 and 546 to the nearest ten.

$320 \times 540 = 172,800$

The actual product is $324 \times 546 = 176,904$

Example 12: **Estimate the quotient of 868 and 31.**

SOLUTION: We'll round 868 and 31 to 900 and 30.

$900 \div 30 = 30$

The quotient is approximately 30.

Example 13: Estimate the sum of the following numbers: 39, 41, 62 and 11.

SOLUTION: We'll round the numbers to the nearest ten and then add.

$40 + 40 + 60 + 10 = 190$

The sum is approximately 190.

Example 14: If you make $11 an hour, estimate your paycheck for 48 hours of work.

SOLUTION: $10 \times 50 = 500$

Your paycheck should be approximately $500.

Example 15: Assume your car gets 21 miles per gallon. If you drove 77 miles one day, 82 miles the next day and 41 miles the third day, approximately how many gallons of gas did you use?

SOLUTION: First we'll round the miles to 80, 80, and 40. Now we'll add them.

$80 + 80 + 40 = 200$ miles.

Next we'll round 21 miles per gallon to 20 miles per gallon and divide the number of miles, 200 by the number of miles per gallon, 20, to find the number of gallons used.

$200 \div 20 = 10$

Approximately 10 gallons were used.

Problem Set:

Estimate your answer for each of the following.

1) 28×52
2) $621 \div 22$
3) $62 + 41 + 81 + 99$

4) A bookstore ordered 39 cases of a book. Each case contains 28 copies of the book. Approximately how many books did they order?

5) The employees of a particular company received a $2 hourly increase in pay. If the company has 212 employees, each of whom works 36 hours a week, approximately how much will the increase cost the company per week?

SOLUTIONS TO PROBLEM SET:

1) $30 \times 50 = 1{,}500$
2) $600 \div 20 = 30$
3) $60 + 40 + 80 + 100 = 280$

4) $40 \times 30 = 1{,}200$
5) $2(200)(40) = \$16{,}000$

Perform the following divisions. Check your answers.

1) $396 \div 3$

2) $5\overline{)1235}$

3) $16\overline{)368}$

4) $4{,}844 \div 44$

5) Find the quotient of 234 and 24.

6) Find the quotient of 5,567 and 123.

7) Find the quotient of 0 and 10.

8) Find the quotient of 10 and 0.

9) What number divided by 6 is 7?

10) What number divided by 7 is 6?

11) $37{,}030 \div 46$

12) $83\overline{)83{,}249}$

13) $55\overline{)110{,}220}$

14) $360{,}024 \div 12$

15) $18\overline{)10,911}$

16) $505,704 \div 101$

17) $23,457 \div 456$

18) $768\overline{)88,990}$

19) $0 \div 23$

20) $0 \div 1$

21) $23 \div 0$

22) $1 \div 0$

23) $0 \div 7$

24) $7 \div 0$

Divide each of the following numbers by 10.

25) 130

26) 7,000

27) 1,010

28) 0

29) 80

30) 14,600

Workbook

Divide each of the following numbers by 100.

31) 125,000   32) 600   33) 0

34) 9,000   35) 100

Divide each of the following numbers by 1,000.

36) 10,000   37) 0   38) 20,000

39) 85,000   40) 3,400,000

41) Find the average of the following: 7, 12, 9 and 8.

42) If Kim had exam grades of 84, 77, 65, 80 and 94, what was his average?

43) Over a six week period your grocery bills were $77, $59, $81, $70, $48 and $67. What was your average grocery bill?

44) The Fly-by-Night flying school made profits of $67,200 in the first quarter, $39,500 in the second quarter, $52,100 in the third quarter, and $45,300 in the fourth quarter. How much did their profits average quarterly?

45) Kevin worked 261 hours in January, 198 hours in February, 202 hours in March, 244 hours in April, 185 hours in May, and 224 hours in June. What were the average monthly hours during this six month period?
(Estimate the answers 46 through 48)

46) 512×488 47) $42 \div 22$ 48) $22 + 32 + 99 + 11$

49) 79 children ate 12 M&M candies each. Approximately how many candies did they eat all together?

50) Jerry drove 61 miles on Monday, 77 miles on Tuesday, 99 miles on Wednesday. About how many miles did he average in those three days?

2.5 Arithmetic Properties

Now we'll make some general statements about the arithmetic operations. These statements are called arithmetic properties. The first property is called the *Addition Property of Zero* (also called the Additive Identity). The Addition Property of Zero says that if you add a 0 to any number, its value stays the same. The way to state this formally follows.

Addition Property of Zero

If a is any number, then: $\qquad\qquad a + 0 = a \text{ and } 0 + a = a$

Some examples of the Addition Property of Zero are:

$$0 + 3 = 3 \qquad\qquad 3 + 0 = 3 \qquad\qquad 0 + 4 = 4 \qquad\qquad 0 + 16 = 16$$

Notice that, in every case, adding 0 doesn't change the number.

The next property is called the *Commutative Property of Addition*. The Commutative Property of Addition says the order in which you add two numbers doesn't change the sum. For example, $5 + 6$ and $6 + 5$ both equal 11. If you reverse the order in which you add two numbers, the answer doesn't change; it stays the same. The way to state this formally follows.

Commutative Property of Addition

If a and b are any two numbers, then: $\qquad a + b = b + a$

Some examples of the Commutative Property of Addition are:

$$2 + 3 = 3 + 2 \qquad 4 + 6 = 6 + 4 \qquad 1 + 2 = 2 + 1 \qquad 10 + 20 = 20 + 10$$
$$5 = 5 \qquad\qquad 10 = 10 \qquad\qquad 3 = 3 \qquad\qquad 30 = 30$$

Notice the order in which we add the numbers doesn't change the answer.

The next property is called the *Commutative Property of Multiplication*. The Commutative Property of Multiplication says the order in which you multiply two numbers doesn't change their product. For example $5(6)$ and $6(5)$ both equal 30. If you reverse the order in which you multiply two numbers, the answer doesn't change. It stays the same. The way to state this formally follows.

Commutative Property of Multiplication

If a and b are any two numbers, then: $\qquad a(b) = b(a)$

Some examples of the Commutative Property of Multiplication are:

$$2(3) = 3(2) \qquad 4(6) = 6(4) \qquad 1(2) = 2(1) \qquad 10(20) = 20(10)$$
$$6 = 6 \qquad\qquad 24 = 24 \qquad\qquad 2 = 2 \qquad\qquad 200 = 200$$

Notice the order in which we multiply two numbers doesn't change the answer.

Remember, in the Commutative Properties of Addition and Multiplication, the order of the numbers changes, they commute positions.

The next property is the *Associative Property of Addition*. The Associative Property says that if we change which numbers are grouped together, it doesn't change the answer. For example, $2 + (3 + 4) = (2 + 3) + 4$. The order of the numbers hasn't changed, so it's not an example of the Commutative Property of Addition. What has changed is which numbers are grouped together. On the left side of the equal sign the 3 and 4 are grouped together. On the right side of the equal sign the 2 and 3 are grouped together. Now we'll work out the problem to prove to you that we really will get the same answer.

$$2 + (3 + 4) = (2 + 3) + 4$$
$$2 + 7 = 5 + 4$$
$$9 = 9$$

Notice the order in which the numbers are grouped doesn't change the answer.

$2 + (3 + 4) = 9$ and $(2 + 3) + 4 = 9$

Associative Property of Addition

If a, b and c are real numbers, then: $a + (b + c) = (a + b) + c.$

Some examples of the Associative Property of Addition are:

$$(1 + 5) + 9 = 1 + (5 + 9)$$
$$6 + 9 = 1 + 14$$
$$15 = 15$$

$$6 + (7 + 8) = (6 + 7) + 8$$
$$6 + 15 = 13 + 8$$
$$21 = 21$$

Notice that the order in which the numbers are grouped doesn't change the answer.

The next property is the *Associative Property of Multiplication*. This is the same idea as the Associative Property of Addition, except now it's multiplication. The order in which we group numbers when we're multiplying wouldn't change the outcome. For example,

$$2(3 \cdot 4) = (2 \cdot 3)4$$
$$2(12) = 6(4)$$
$$24 = 24$$

$$5(4 \cdot 6) = (5 \cdot 4)6$$
$$5(24) = 20(6)$$
$$120 = 120$$

Notice the order in which the numbers are grouped doesn't change the answer.

Associative Property of Multiplication

If a, b and c are real numbers, then: $a(b \cdot c) = (a \cdot b)c.$

Some examples of the Associative Property of Multiplication are:

$$1(2 \cdot 3) = (1 \cdot 2)3 \qquad 6(7 \cdot 8) = (6 \cdot 7)8 \qquad (5 \cdot 4)3 = 5(4 \cdot 3)$$
$$1(6) = 2(3) \qquad 6(56) = (42)8 \qquad 20(3) = 5(12)$$
$$6 = 6 \qquad 336 = 336 \qquad 60 = 60$$

Notice that the order in which the numbers are grouped doesn't change the answer.

Multiplication Property of One

The next property is the *Multiplication Property of One* (also called the Multiplicative Identity). The Multiplication Property of One means any number times 1 is itself. In other words, any number times one is itself. The formal way to state that follows.

If a is any number, then: $a(1) = a$ and $1(a) = a.$

Some examples of the Multiplication Property of One are:

$$2(1) = 2 \qquad 1(9) = 9 \qquad 12(1) = 12 \qquad 1(12) = 12$$

Multiplication Property of Zero

If a is any number, then: $a(0) = 0$ and $0(a) = 0.$

Some examples of the Multiplication Property of Zero are:

$$2(0) = 0 \qquad 0(2) = 0 \qquad 0(3) = 0 \qquad 3(0) = 0 \qquad 5(0) = 0 \qquad 0(5) = 0$$

We've given you a lot to think about, so let's look at some examples together.

Example 1: $2 + 4 = 4 + 2$

This is an example of the Commutative Property of Addition because the order of the numbers has changed.

Example 2: $3(1) = 3$

This is an example of the Multiplication Property of One.

Example 3: $0 + 5 = 5$

This is an example of the Addition Property of Zero.

Example 4: $4(3 \cdot 2) = (4 \cdot 3)2$

This is an example of the Associative Property of Multiplication, not the Commutative Property because the order of the numbers hasn't changed; it's still 4 then 3 then 2, on both sides of the equal sign. What has changed is which numbers are grouped together inside the parentheses.

Example 5: $(7 + 8) + 5 = 7 + (5 + 8)$

This is an example of both the Commutative and the Associative Properties of Addition, because the order of the numbers has changed. 7, 8, 5 changed to 7, 5, 8, and, there are different numbers grouped together. On the left side of the equation the 7 and 8 is grouped together, but on the right side of the equation the 5 and 8 is grouped together.

Now it's time for you to try some problems.

Problem Set:

State the property or properties illustrated in the following examples.

1) $17 + 23 = 23 + 17$

2) $5 + (6 + 7) = (5 + 6) + 7$

3) $9 + 0 = 9$

4) $9(0) = 0$

5) $2 + (8 + 3) = (2 + 8) + 3$

6) $(2 \cdot 3)4 = 2(4 \cdot 3)$

SOLUTIONS TO PROBLEM SET:

1) This is an example of the Commutative Property of Addition because the order of the numbers changed.

2) This is not an example of the Commutative Property of Addition because the order of the numbers has not changed. It's an example of the Associative Property of Addition because there are different numbers inside the grouping.

3) This is an example of the Addition Property of Zero.

4) This is an example of the Multiplication Property of Zero.

5) This is an example of the Associative Property of Addition, because there are different numbers inside the grouping. This is not an example of the Commutative Property of Addition because the order of the numbers hasn't changed.

6) This is an example of both the Commutative and Associative Properties of Multiplication, because the order of the numbers has changed and there are different numbers inside the groupings.

State the arithmetic properties each of these problems demonstrates.

1) $2 + 11 = 11 + 2$

2) $0 + 5 = 5$

3) $2(11) = 11(2)$

4) $0(5) = 0$

5) $2 + (7 + 4) = (2 + 7) + 4$

6) $(2 + 7) + 4 = 4 + (2 + 7)$

7) $(2 + 7) + 4 = (4 + 2) + 7$

8) $9 + (3 + 2) = 9 + (2 + 3)$

9) $9 + 0 = 9$

10) $5(6 \cdot 9) = 5(9 \cdot 6)$

11) $5(6 \cdot 9) = (5 \cdot 6)9$

12) $88(1) = 1(88)$

13) $16 + 5 = 5 + 16$

14) $2(5 \cdot 6) = (2 \cdot 6)5$

15) $(4 + 5) + 6 = (5 + 4) + 6$

16) $(4 \cdot 11)3 = 3(4 \cdot 11)$

17) $(4 \cdot 11)3 = (3 \cdot 11)4$

18) $0(18) = 0$

19) $100 + 500 = 500 + 100$

20) $100(500) = 500(100)$

21) $0(5) = 0$

22) $0(10) = 0$

23) $1(5) = 5$

24) $1(10) = 10$

2.6   Order of Operations

The order in which you choose to work out a problem often changes the answer. Let's look at a problem where we'll get two different answers to the same problem because we approached the problem in two different orders.

$$2 + 4(3)$$
$$6(3)$$
$$18$$

$$2 + 4(3)$$
$$2 + 12$$
$$14$$

Notice we got two different answers to the same problem. In the problem on the left, we added first $2 + 4 = 6$, then multiplied $6(3) = 18$, and got the wrong answer, 18. In the problem on the right we multiplied first $4(3) = 12$, then added $2 + 12 = 14$ to get the correct answer 14. Now you're probably wondering how you're supposed to know to multiply first, and then add. We're glad you asked. Here's a step-by-step outline of the correct order to simplify problems.

Order of Operations

Step 1:   Simplify everything inside groupings first. Some grouping symbols are ( ) , [ ], { }, and | |. When there's more than one grouping symbol in a problem, work from the inside of the problem to the outside.

Step 2:   Exponents.

Step 3:   Multiplication/Division—If there's a multiplication and a division in one problem, work from left to right.

Step 4:   Addition/Subtraction—If there's an addition and a subtraction in one problem, work from left to right.

Now let's simplify that last problem using the order of operations.

$$2 + 4(3)$$ Look at the problem to see which operations have to be done. Here we
$$2 + 12$$ have multiplication and addition. Multiplication comes before addition in the
$$14$$ order of operations, so we'll multiply first, and then add.

The order of operations is a very important topic, so just to make sure you learn it, we'll give you order of operations problems in a few other chapters in this book.

Let's work out some more problems together.

Example 1: $6 + 5(4 - 2)$

SOLUTION: According to the order of operations, we should do inside the group-
$$6 + 5(4 - 2)$$ ing first: $4 - 2 = 2$. Next multiply $5(2) = 10$. Lastly, add $6 + 10 = 16$.
$$6 + 5(2)$$
$$6 + 10$$
$$16$$

Example 2: $7 + 2(5 - 2)$

SOLUTION:
$7 + 2(5 - 2)$
$7 + 2(3)$
$7 + 6$
13

First do inside the grouping: $5 - 2 = 3$. Then multiply $2(3) = 6$. Add $7 + 6 = 13$.

Example 3: $4 + 15 \div 2^3$

SOLUTION:
$4 + 16 \div 2^3$
$4 + 16 \div 8$
$4 + 2$
6

Step 1 is inside groupings, but there aren't any, so we'll go to step 2, exponents: $2^3 = 2(2)(2) = 8$. Now it's time for step 3, division: $16 \div 8 = 2$. Last is step 4, addition: $4 + 2 = 6$.

Let's try some more challenging problems. The next couple of problems may look tough, but all you have to do is follow the order of operations and you'll be fine.

Example 4: $6 + 2(5 + 40 \div 5) - 2^4$

SOLUTION:
$6 + 2(5 + 40 \div 5) - 2^4$
$6 + 2(5 + 8) - 2^4$
$6 + 2(13) - 2^4$
$6 + 2(13) - 16$
$6 + 26 - 16$
$32 - 16$
16

First on the order of operations is inside groupings. Inside the grouping we have a choice of addition or division. We'll divide first: $40 \div 5 = 8$. Next add $5 + 8 = 13$. Now that we've done everything that could be done inside the grouping, we'll look outside the grouping where we have a choice of addition, multiplication or exponents. Exponents are next on the order of operations: $2^4 = 16$. Now multiply, $2(13) = 26$. Add $6 + 26 = 32$ then subtract, $32 - 16 = 16$.

Example 5: $(18)(3) \div (2)(3)$

SOLUTION:
$(18)(3) \div (2)(3)$
$54 \div (2)(3)$
$27(3)$
81

Be very careful not to make a very common error on this problem. $(18)(3) \div (2)(3)$ is not the same as $(18 \cdot 3) \div (2 \cdot 3)$. In $(18)(3) \div (2)(3)$, parentheses are used to show multiplication—not as a grouping symbol, so we would work left to right. In $(18 \cdot 3) \div (2 \cdot 3)$, parentheses are used to show a grouping, so we would work inside the grouping first. This changes the order in which we should begin this problem. The parentheses in this problem are not used as a grouping symbol, but to show multiplication. This example has multiplication and division, so we'll work left to right, $(18)(3) = 54$, $54 \div 2 = 27$, $27(3) = 81$.

Problem Set:

Try the following order of operation problems.

1) $6 + 4(5 - 1)$

2) $8 + 2^3$

3) $40 \div 10 - 2$

4) $40 \div (10 - 2)$

5) $(16 - 4^2) \div (10 - 2 \cdot 3)$

SOLUTIONS TO PROBLEM SET:

1)
$$6 + 4(5 - 1)$$
$$6 + 4(4)$$
$$6 + 16$$
$$22$$

Inside the grouping first, $5 - 1 = 4$.
Multiply $4(4) = 16$.
Add $6 + 16 = 22$.

2)
$$8 + 2^3$$
$$8 + 8$$
$$16$$

Exponents first: $2^3 = 2(2)(2) = 8$.
Add $8 + 8 = 16$.

3)
$$40 \div 10 - 2$$
$$4 - 2$$
$$2$$

Divide first: $40 \div 10 = 4$.
Subtract $4 - 2 = 2$.

4)
$$40 \div (10 - 2)$$
$$40 \div 8$$
$$5$$

Inside the grouping first, $10 - 2 = 8$.
Divide $40 \div 8 = 5$.

5)
$$(16 - 4^2) \div (10 - 2 \cdot 3)$$
$$(16 - 16) \div (10 - 6)$$
$$0 \div 4$$
$$0$$

Inside groupings first; exponents before subtraction: $4^2 = 16$.
Multiplication before subtraction, $2 \cdot 3 = 6$.
Finish inside the groupings, then divide.
Be careful, $0 \div 4 = 0$, but $4 \div 0$ is undefined.
This is carefully explained in Section 2.3.

Apply the order of operations to the following problems.

1) $24 + 6 \div 6$

2) $5 - 5 \div 5$

3) $16 \div 4 \div 4$

4) $100 \div 5 \div 5$

5) $16 \div (4 \div 4)$

6) $100 \div (5 \div 5)$

7) $16 \div 4^2$

8) $100 \div 5^2$

9) $5(4)^2$

10) $3(10)^2$

11) $2 + 4(5 + 6 \div 6)$

12) $100 - 50 \div [10 - (2 \cdot 3) + 1]$

13) $(8 \div 8) \div (4 - 2^2)$

14) $(2 \div 2) \div (16 - 2^4)$

15) $(4 - 2^2) \div (8 \div 8)$

16) $(16 - 2^4) \div (2 \div 2)$

17) $8 \div (8 \div 4) - 2^2$

18) $(10 \div 2)^2 - 2^4$

19) $5^3 - 5^2 \div 5$

20) $3 + [4(15 \div 5) \div 6]$

21) $(7 + 5) \div (5 - 3)$

22) $40 + 5(6 + 4 \div 2)$

23) $7 + (5 \div 5) - 3$

24) $4 + 5(6 + 4) \div 2$

25) $5(3)^4 + 16 \div 4 - 2^2$

26) $3 + 2[8 + 5(6 - 4)]$

27) $8^2 \div (9 - 1)^2$

28) $7^2[6 + 4(10 + 12)]$

29) $9^2 + 4^2 \div 2^3$

30) $[(6 \div 3)(8 \div 2)] \div 8$

2.7　Applications

Now we're going to apply a lot of what you've learned in chapters 1 and 2 about adding, subtracting, multiplying, and dividing. The question students always ask about these applications is: how do you know what to do? In other words, how do you know when you're supposed to add, subtract, multiply, or divide? By the time you've completed this section, you'll no longer need to ask that question.

Applications in Addition and Subtraction

So far we've done straightforward numerical problems. Now we'll do some word problems. You'll need to figure out whether each of these problems calls for addition, subtraction, or a combination of addition and subtraction.

Example 1:　Josephina earned $48 on Monday, $39 on Tuesday, and $53 on Wednesday. How much did she earn all together?

　　　　SOLUTION:　　Whenever a question asks for "all together," we have to add. $48 + $39 + $53 = $140.

Example 2:　Max had $206 with him when he got to the mall. When he left the mall he had $97. How much did he spend at the mall?

　　　　SOLUTION:　　Max has less than he started with, so this is a subtraction problem. Whenever we have to find the difference between the original amount and the remaining amount we subtract. $206 − $97 = $109.

Example 3:　Paul had $571 in his checking account at the beginning of the month. A few days later he deposited $359 and wrote a check for $175. How much money did he have left in his account at the end of the month if he didn't write any more checks or make any more deposits?

　　　　SOLUTION:　　Paul began the month with $571 in his account. His $359 deposit raised his account balance to $930 ($571 + $359 = $930). Then his balance went down to $755 after he wrote a check for $175 ($930 − $175 = $755).

So far it has been pretty easy to tell from the wording of the problems whether to add, subtract, or to add and subtract. Let's try some problems that are a bit harder.

Example 4:　Mark's score on the exam was 92. Sarah's score was 26 points lower. What was Sarah's score on the exam?

　　　　SOLUTION:　　To get a lower score you have to subtract. Start with Mark's score of 92 and subtract 26 to get Sarah's score. 92 − 26 = 66.

Example 5: The difference between two numbers is 149. If the smaller number is 568, how much is the larger number?

SOLUTION: To find the larger number add 149 to the smaller number, 568.
568 + 149 = 717.

Example 6: When you started climbing a mountain you were 5,844 feet above sea level. When you got to the top, you were 11,909 feet above sea level. How far did you climb?

SOLUTION: To find the difference in the feet above sea level, subtract.
11,909 feet − 5,844 feet = 6,065 feet.

Problem Set:

1) The Chicago Cubs scored two runs in the third inning, one run in the fourth inning, five runs in the sixth inning, and three runs in the ninth inning. How many runs did they score in the entire nine inning game?

2) The boss brought $450 to work on Friday. She paid one of her workers $225 and she paid her other worker $190. How much money did the boss have left?

3) Jason left home with 26 marbles. He gave 9 marbles to Melissa and Sue gave him 15. How many marbles did Jason end up with?

4) Hans weighs 126 pounds and Kerry weighs 89 pounds. What is the difference in their weights?

5) When you were leaving on a trip, your car's odometer read 14,199. When you arrived at your destination, your odometer read 16,056. How far did you travel?

SOLUTIONS TO PROBLEM SET:

1) 2 + 1 + 5 + 3 = 11 runs

2) 450 − 225 − 190 = $35

3) 26 − 9 + 15 = 32 marbles

4) Difference means subtract. 126 − 89 = 37 pounds

5) 16,056 − 14,199 = 1,857 miles

Applications in Multiplication and Division

Example 7: Roberto has 48 folders. Each folder holds 39 pages. How many pages are in the folders?

SOLUTION: To find the total number of pages, we need to multiply the number of pages in each folder (39) by the number of folders (48):

$$
\begin{array}{r}
39 \\
\times\ 48 \\
\hline
312 \\
156 \\
\hline
1872 \ \text{pages}
\end{array}
$$

Example 8: An end-of-the-year bonus of $967,486 is divided equally among 26 employees. How much is each person's share?

SOLUTION: To find out how much each person's share is, we have to divide the bonus, $967,486, by the number of employees, 26.

$$
\begin{array}{r}
37211 \\
26\overline{)967486} \\
-78 \\
\hline
187 \\
-182 \\
\hline
54 \\
-\ 52 \\
\hline
28 \\
-\ 26 \\
\hline
26 \\
-\ 26 \\
\hline
\end{array}
$$

Each person's share is $37,211.

Problem Set:

1) There are 23 girls in a club. Each girl brings 72 cookies to a party. How many cookies did they bring all together?

2) If Jim Panepinto hit an average of 49 home runs each year and his career lasted 23 years, how many home runs did he hit during his career?

3) Every day Sophie brings 140 pennies to her grandmother's house. How many pennies does she bring in 7 days? How many days will it take Sophie to save up $28?

4) If there are 55 M&Ms in a bag, how many bags can be formed from a vat of 5,555 M&Ms?

SOLUTIONS TO PROBLEM SET:

1) There are 23 girls. Each brought 72 cookies. How many cookies did they all bring? We have to multiply.

$$
\begin{array}{r}
72 \\
\times\ 23 \\
\hline
216 \\
144 \\
\hline
1656 \ \text{cookies}
\end{array}
$$

2) Multiply

$$
\begin{array}{r}
49 \\
\times\ 23 \\
\hline
147 \\
98\ \ \\
\hline
1127 \quad \text{home runs}
\end{array}
$$

3)

$$
\begin{array}{r}
140 \\
\times\ 7 \\
\hline
980 \quad \text{pennies}
\end{array}
$$

There are 100 pennies in one dollar. To change the $28 to pennies, we multiply by 100: $28 \times 100 = 2800$ pennies. She saves 140 pennies a day, so we divide the 2800 pennies by 140 pennies and get our answer, 20 days.

$$
\begin{array}{r}
20 \\
140\overline{)2800} \\
-\ 280\ \ \\
\hline
00
\end{array}
$$

4)   This is a division problem because we're forming smaller groups taken from a larger group. 101 bags of M&Ms can be formed.

$$
\begin{array}{r}
101 \\
55\overline{)5555} \\
-\ 55\ \ \ \ \\
\hline
055 \\
-\ 55\ \\
\hline
0
\end{array}
$$

1) The Philadelphia Eagles scored seven points in the first quarter, ten points in the second quarter, three points in the third quarter, and fourteen points in the fourth quarter. How many points did they score in the entire game?

2) Nancy is 19 years older than Ann. If Nancy is 45, how old is Ann?

3) The ABC Corporation paid each of its 492 employees $673. How much was its payroll?

4) 4,935 cookies were divided equally among 235 wedding guests. How many cookies did each guest receive?

5) A plane flew 808 miles from New York to Chicago, and 2,166 miles from Chicago to San Francisco. What was the entire distance the plane flew? How much closer is Chicago to New York than to San Francisco?

6) A truck left the bakery with 241 loaves of bread. It dropped off 137 loaves at one location and picked up 55 loaves at another location. How many loaves of bread were now on the truck?

7) Mrs. Hashimoto had a checking account balance of $2,137. If she wrote checks for $275 and $458, how much money was left in her account?

8) Ms. Romano paid $4,712 for 248 ladies' watches. How much did each watch cost?

9) Barbara's score on an exam was 83. Mike's score was 19 points lower. What was Mike's score on the exam?

10) A bus left Billings with 42 passengers. At Helena 13 people got off the bus. At Spokane 9 people got on the bus. How many people were still on the bus as it left Spokane?

11) A total of 15,246 VCRs were shipped to 7 stores. If each store received the same number of VCRs, how many did each store receive?

12) 54 counties each sent the state government $11,250. How much money did the government receive?

13) The difference between two numbers is 382. If the larger of the two numbers is 601, how much is the smaller number?

14) Jose left the house with $317. He spent $129 on clothes and $84 on groceries. How much money did he have left?

15) At James Madison High School there are 1,816 freshmen, 2,003 sophomores, 1,919 juniors, and 1,793 seniors. How many students go to James Madison High School?

16) Bob is 23 years older than Sue. If Sue is 49, how old is Bob?

17) Phillip worked 13 hours of overtime in January, 19 hours of overtime in February, 12 hours of overtime in March, and 8 hours of overtime in April. How many hours of overtime did he work during these four months?

18) 656 gold coins are shared equally by 8 people. How many coins does each person get?

19) Kyra left for work with $283. She spent $9 on lunch, $54 on shoes, and $8 on a watch. How much money did she have left?

20) A train traveled 258 miles from Montreal to Albany, and 143 miles from Albany to New York City. What was the total distance the train traveled?

21) At the Chinese Communist Party Congress, the delegates from each of 19 provinces cast 275 votes each for the new Prime Minister. How many votes were cast for the Prime Minister?

22) If a shipment containing 36,248 cans of tuna fish arrived at a warehouse, and if the shipment was packed in 788 cartons, how many cans were in each carton?

23) At James Monroe High School there are 2,017 freshmen, 1,932 sophomores, 1,819 juniors, and 1,687 seniors. How many students go to James Monroe High School?

24) If Hans is 34 years older than Joe, how old is Joe if Hans is 61?

25) There are 214 passengers on a plane bound for Detroit. At Detroit, 98 passengers get off the plane and 77 get on. How many people are now on the plane?

26) A family has an annual income of $42,180. How much is its monthly income?

27) A shoe store orders 123 pairs of shoes for a total of $5,658. If the store paid the same amount for each pair of shoes, how much was each pair of shoes?

28) A truck with 311 cases of soda dropped off 154 cases at one location and picked up 38 cases at another. How many cases of soda were now on the truck?

29) Mr. O'Boyle had a checking account balance of $5,071. If he made a deposit of $1,250, and then wrote a check for $425, how much money would be in his account?

30) The difference between two numbers is 457. If the smaller of the two numbers is 366, how much is the larger number?

31) How many four ounce cups can be filled from a 256-ounce bottle?

32) A store was open 16 hours on Monday, 13 hours on Tuesday, 17 hours on Wednesday, 10 hours on Thursday, and 15 hours on Friday. How many hours was the store open during these five days?

33) During the 1990s the population of Los Alamos rose by 4,199. If the population of the city was 16,394 in the year 2000; how much was the population of Los Alamos in 1990?

34) Ms. Hirsch earned $61,086 in the year 2001. In 2002 she earned $98,175. How much less did she earn in 2001 than in 2002?

35) Karen left home with 36 marbles. If she gave 14 to John and received 19 from Andy, how many marbles did Karen have now?

Chapter 2 Review

1)   85
    ×6

2)   162
    × 8

3)   942
    × 5

4)   235
    × 64

5)   837
    × 425

6)   7496
    × 563

7) Multiply each of these numbers by 10, 100, and 1,000.
    a) 2   b) 10   c) 360   d) 3,100

8)   317
    × 200

9)   928
    × 4000

10) Simplify each of the following.
    a) 6^2 b) 2^6 c) 4^0 d) 3^3

11) $5\overline{)640}$ 12) $7\overline{)1533}$

13) $25\overline{)1875}$ 14) $49\overline{)9641}$

15) $40\overline{)7306}$ 16) $85\overline{)17340}$

17) Divide each of these numbers by 10, 100, and 1,000.
    a) 416,000   b) 1,000,000   c) 82,000   d) 9,000

18) Find the average of 79, 24, 90, 61, and 56.

19) State the arithmetic properties each of these problems demonstrates.
    a) $2(8) = 8(2)$ b) $3 + (1 + 6) = (3 + 1) + 6$
    c) $(5 \cdot 9) \cdot 2 = (2 \cdot 9) \cdot 5$ d) $4(0) = 0$

20) How much is $8 + 3(7 - 2)$?

21) How much is $2 + 3^3$?

22) How much is $20(6 - 2)$?

23) How much is $(16 \div 2^2) - 4$?

24) A bus with 29 passengers made two stops. At the first stop 8 people got on and 4 got off. At the second stop 3 got on and 2 got off. How many people were left on the bus?

25) The Miami Dolphins scored 12 points in the first quarter, 7 points in the second quarter, 14 points in the third quarter, and 3 points in the fourth quarter. How many points did they score in the entire game?

26) Ms. Koshiama had a checking account balance of $5,167. If she wrote checks for $325, $578, and $290, how much money was left in her account?

27) The XYZ Corporation paid each of its 519 employees a bonus of $622. How much money did the company pay out in bonuses?

28) A total of 964 cartons of food was divided equally among four food pantries. How many cartons did each food pantry receive?

29)   76
$\quad\quad\underline{\times\ 6}$

30)   465
$\quad\quad\underline{\times\ 98}$

31)   6243
$\quad\quad\underline{\times\ 257}$

32) Multiply each of these numbers by 10, 100, and 1,000.

     a) 64   b) 4,000   c) 6   d) 70

33)   4128
    $\times\ 5000$

34) Simplify each of these.
    a) 4^3 b) 3^4 c) 10^0 d) 10^3

35) $8\overline{)4064}$ 36) $61\overline{)14387}$ 37) $60\overline{)72120}$

38) Divide each of these numbers by 1,000, 10,000, and 100,000.

    a) 3,500,000   b) 500,000   c) 434,000,000

39) A winning lottery prize of $14,500,000 is split 100 ways. How much is each person's share?

40) If there are 116 cans of tuna fish in a carton, how many cans are there in 2,900 cartons?

41) State the arithmetic properties each of these problems demonstrates.

    a) $3 + 6 = 6 + 3$ b) $5 + 0 = 5$

    c) $2(4 \cdot 3) = (2 \cdot 4)3$ d) $(2 + 3) + 4 = 2 + (4 + 3)$

42) Simplify the following.

    a) $4 + 2(8 - 4)$ b) $(3 \cdot 4^2) - 5^2$

    c) $(20 \div 5)^2 - 3^2$ d) $(3^4 - 9^2) \div (60 - 3^3)$

43) A band of explorers set off on a journey of 5,122 miles. After they covered 1,847 miles, how much further did they have to go?

44) One hundred ninety-six stockholders each held four thousand shares. How many shares did they hold all together?

1)
$$\begin{array}{r} 34 \\ \times\ 5 \\ \hline \end{array}$$

2)
$$\begin{array}{r} 192 \\ \times\ 7 \\ \hline \end{array}$$

3)
$$\begin{array}{r} 829 \\ \times\ 45 \\ \hline \end{array}$$

4)
$$\begin{array}{r} 209 \\ \times\ 687 \\ \hline \end{array}$$

5)
$$\begin{array}{r} 2{,}983 \\ \times\ 3{,}487 \\ \hline \end{array}$$

6)
$$\begin{array}{r} 10{,}982 \\ \times\ 2{,}981 \\ \hline \end{array}$$

7) Multiply each of these numbers by 10, 100, and 1,000.

a)   5   _____   _____   _____

b)   10   _____   _____   _____

c)   546   _____   _____   _____

d)   4,567   _____   _____   _____

8)
$$\begin{array}{r} 267 \\ \times\ 300 \\ \hline \end{array}$$

9)
$$\begin{array}{r} 2949 \\ \times\ 4000 \\ \hline \end{array}$$

10) Simplify each of the following:

a) 3^4 b) 4^3 c) 6^0 d) 4^4

11) $8\overline{)984}$ 12) $65\overline{)6825}$

13) $32\overline{)128256}$ 14) $136\overline{)3808}$

15) Divide each of the following numbers by 10, 100, and 1,000.

a) 52,000 _____ _____ _____

b) 1,000,000 _____ _____ _____

c) 720,000 _____ _____ _____

d) 1,000 _____ _____ _____

16) Find the average of the following test scores: 94, 68, 82, and 80. _____

17) State the arithmetic properties each of these problems demonstrates.

a) $3(4) = 4(3)$ b) $2(5 \cdot 7) = 5(7 \cdot 2)$

_____ _____

18) How much is $4 + 3(32 - 8 \div 2^3)$? 19) How much is $(2^0 - 1^{20}) \div 5^2$?

_____ _____

20) A bus with 55 passengers made two stops. At the first stop 10 people got on and 11 got off. At the second stop 3 got on and 21 got off. How many people were left on the bus?

_____

21) Miss Iorio had a checking account balance of $6,243. If she wrote checks for $235, $3,456, and $45, how much was left in her account?

_____

22) A total of 1,410 basketballs was shipped to six stores; if each store received an equal amount of basketballs, how many basketballs did each store receive?

_____

23) How many 8-ounce cups of water can be filled from a one gallon container? (There are 128 ounces in a gallon).

_____

24) If the heights of four brothers are 70 inches, 75 inches, 72 inches and 67 inches, what's the average height of the brothers in inches?

_____

25) A group of fifteen children went shopping, if they spent an average of $26 each on toys, how much did they spend all together?

_____

Chapter 2 • Test 2

1)   96
     × 4

2)   298
     × 873

3)   598
     × 67

4)   402
     × 521

5)   4,882
     × 6,798

6)   11,993
     × 4,588

7)   Multiply each of these numbers by 10, 100, and 1,000.

a)   6   _____   _____   _____

b)   76   _____   _____   _____

c)   748   _____   _____   _____

d)   5,870   _____   _____   _____

8)   398
   × 200

9)   2,938
   × 5,000

10) Simplify each of the following.

    a) 2^6 b) 7^0 c) 6^2 d) 5^3

11) $7\overline{)1218}$ 12) $26\overline{)15262}$

13) $65\overline{)65130}$ 14) $364\overline{)322868}$

15) Divide each of the following numbers by 10, 100, and 1,000.

    a)   13,000   _____   _____   _____

    b)   640,000   _____   _____   _____

    c)   1,000   _____   _____   _____

    d) 2,000,000   _____   _____   _____

16) Find the average of the following students' ages: 19, 22, 18, 21, 25.   _____

17) State the arithmetic properties each of these problems demonstrates.

    a) $2(0) = 0$ b) $3 + (4 + 5) = 4 + (3 + 5)$

    _____   _____

18) How much is $5 + 2(16 - 4 \div 2^2)$?

19) How much is $3^4 \div (9^2 - 3^4)$?

_____

_____

20) A bus with 30 passengers made two stops. At the first stop 6 people got on and 11 got off. At the second stop 5 got on and 3 got off. How many people were left on the bus?

_____

21) If a bin contains 11,760 M&M candies, how many packets containing 21 candies each can be formed from the bin?

_____

22) An inheritance of $46,701 is to be equally split among three sisters. If $2,000 is spent on legal fees, and $6,889 on taxes, how much will each sister inherit?

_____

23) A student's grades in this course are: 88, 92, 73, 52 and 0. What's the student's average in the course? How did the zero affect the grade?

_____

24) Mrs. Marek had a checking account balance of $7,910. She wrote checks for $67, $567, $890 and $2,334. She also made a deposit of $567. What's her new account balance?

_____

25) The weights in pounds of a group of children are: 55, 63, 29, 44, 77 and 56. Find the average weight.

_____

Chapter 3.   Proper Fractions

How many times a day do you hear ads on television—especially on the home shopping channels—offering you some pretty amazing products at just a *fraction* of what you would have to pay for them in a store? Of course you need to ask just what kind of fraction they're talking about. Is it $\frac{1}{2}$, $\frac{1}{3}$, $\frac{1}{4}$ or $\frac{9}{10}$?

We'll start with the fraction $\frac{1}{2}$. The top number is called the ***numerator*** and the bottom number is called the ***denominator***. So in the fraction $\frac{1}{2}$, the numerator is 1 and the denominator is 2. In the fraction $\frac{2}{3}$, the numerator is 2 and the denominator is 3.

In a ***proper fraction*** the denominator is always greater than the numerator. We already saw that $\frac{1}{2}$ and $\frac{1}{3}$ are proper fractions. How about $\frac{4}{5}$, $\frac{3}{8}$ and $\frac{19}{20}$? These, too, are proper fractions.

What do you think ***improper fractions*** look like? They look like these fractions: $\frac{2}{1}$, $\frac{17}{14}$ and $\frac{7}{5}$. So if the numerator is greater than the denominator, then it's an improper fraction.

What if the numerator and the denominator are equal (making the fraction equal to 1), as is the case with these fractions: $\frac{2}{2}$, $\frac{9}{9}$ and $\frac{20}{20}$? Are these proper or improper fractions? A while back someone decided that when the numerator and denominator are equal, we must call that an improper fraction.

What *is* important is to recognize the relationship between the numerator and the denominator. Let's take the improper fraction $\frac{4}{2}$. What are you supposed to do with it? Should we just leave it sitting there? Or maybe do a little division? Okay, you do a little division. Now what do you divide into what? You divide the 2 into the 4, which gives you 2.

So the relationship of the numerator to the denominator of a fraction is that you're supposed to divide the denominator (or bottom) of the fraction into the numerator (or top).

There's one more term we'd like to introduce, and then we can stop talking about fractions and start using them. The term is ***mixed number***, which consists of a whole number and a proper fraction. Examples would include numbers like $3\frac{3}{4}$, $1\frac{5}{8}$ and $4\frac{2}{3}$.

Do you really have to know all these terms? Not necessarily. Just remember *numerator* and *denominator*. If you can also remember *proper fraction*, *improper fraction* and *mixed number*, then you will have enriched your vocabulary, but you'll still have to get out of bed every morning, and you probably won't notice any major changes in the quality of your life.

When you have completed this chapter and the next, you will be able to convert improper fractions into mixed numbers and convert mixed numbers into improper fractions. You'll also be able to reduce, add, subtract, multiply, and divide proper fractions, improper fractions, and mixed numbers and simplify complex fractions.

In this chapter we'll work with proper fractions, and in the next, we'll work with improper fractions and mixed numbers. We'll begin with some pie charts which illustrate proper fractions.

3.1 Picturing Fractions

We'll start you off with a very easy one.

Example 1: What fraction of the pie is in the shaded area? Identify the numerator and the denominator.

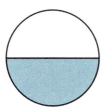

SOLUTION: The fraction is $\frac{1}{2}$ because one of two equal pieces is shaded. Its numerator is 1 and its denominator is 2.

Example 2: What fraction of the pie is in the shaded area? Identify the numerator and the denominator.

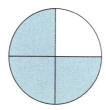

SOLUTION: The fraction is $\frac{3}{4}$. Its numerator is 3 and its denominator is 4.

Example 3: What fraction of this pie is in the shaded area? Identify the numerator and the denominator.

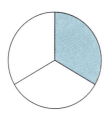

SOLUTION: The fraction is $\frac{1}{3}$. Its numerator is 1 and its denominator is 3.

Problem Set 3.1:

For each problem, answer these questions: (a) What fraction of the pie is in the shaded area? (b) Identify the numerator and denominator.

1)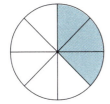

2)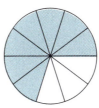

3)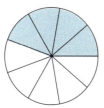

4)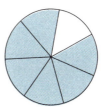

SOLUTIONS TO PROBLEM SET:

1) (a) $\frac{3}{8}$ (b) numerator is 3; denominator is 8

2) (a) $\frac{7}{10}$ (b) numerator is 7; denominator is 10

3) (a) $\frac{4}{9}$ (b) numerator is 4; denominator is 9

4) (a) $\frac{6}{7}$ (b) numerator is 6; denominator is 7

3.2   Fractions in Words and Numbers

Example 1: A class has 19 students, 8 of whom are boys. What fraction of the class is boys?

SOLUTION: The fraction of the class that is boys is $\frac{8}{19}$.

Example 2: There are 4 blue chairs and 5 red chairs in a room. What fraction of the chairs is red?

SOLUTION: The answer is $\frac{5}{9}$. First we find the total number of chairs in the room: $4 + 5 = 9$. Then we find the fraction of 9 that is red: $\frac{5}{9}$. While we're at it, what fraction of the chairs is blue? The answer is $\frac{4}{9}$.

Example 3: Paula has eleven beads. If 2 are green, 4 are orange, and 5 are black, what fraction of her beads are (a) green; (b) orange; and (c) black?

SOLUTION: (a) $\frac{2}{11}$ are green; (b) $\frac{4}{11}$ are orange; $\frac{5}{11}$ are black.

Example 4: José has 23 coins. If 5 are pennies, 6 are nickels, and 12 are dimes, what fraction of his coins are (a) pennies? (b) nickels? and (c) dimes?

SOLUTION: (a) $\frac{5}{23}$ are pennies; (b) $\frac{6}{23}$ are nickels; and (c) $\frac{12}{23}$ are dimes.

Example 5: Let's turn it up a notch. If Carol has 2 pennies, 5 nickels, and 10 dimes, what fraction of her coins are (a) pennies; (b) nickels; and (c) dimes?

SOLUTION: First we add up the coins to give us the fractions' denominator: $2 + 5 + 10 = 17$. (a) $\frac{2}{17}$ are pennies; (b) $\frac{5}{17}$ are nickels; and (c) $\frac{10}{17}$ are dimes.

Problem Set 3.2:

1) There are 10 balls in a box. If three of them are red, then what fraction of the balls is (a) red; (b) not red?

2) A kennel has 16 dogs and 13 cats. What fraction of the animals in the kennel is (a) dogs; (b) cats?

3) The same number of people live on each of the first, second, and third floors of a house. What fraction lives on the (a) first floor? (b) second floor? (c) third floor?

4) Of all the guests at a wedding, 19 were single, 44 were married, and 14 were divorced. What fraction of the guests were (a) never married? (b) married? and (c) divorced?

5) Martha, George, Bill, Hillary and Bob divided up a cake equally among themselves. What fraction of the cake did each person get?

6) If 10 out of 21 people at a party are male, what fractional part of the people at the party are (a) male, (b) female?

7) If a person spends 17 dollars out of 20 dollars on a pizza, what fractional part of their 20 dollars was spent on pizza?

8) If 25 students took a math test and 18 passed the test, (a) what fractional part of the class passed the test, and (b) what fractional part failed the test?

9) If 2 out of 3 dentists prefer their patients don't chew gum, what fractional part of the dentists prefer their patients don't chew gum?

10) If 1 out of every 100 math students loves to do word problems, what fractional part of the students loves to do word problems?

11) If in a class of 37 students, 23 are freshmen, what fractional part of the students are not freshmen?

12) If a person spends $2 out of every $9 they earn on food, what fractional part of their money is spent on food?

13) If 211 people finished a marathon and 212 people ran in the marathon, what fractional part of the runners (a) finished the marathon, (b) didn't finish the marathon?

14) If a family has 2 sons and 3 daughters, what fractional part of the children represents the sons?

15) If 2 out of 7 books are math books, what fractional part of the books are math books?

1)   (a) $\frac{3}{10}$ (b) $\frac{7}{10}$

2)   Total number of animals: $16 + 13 = 29$.
      (a) $\frac{16}{29}$ (b) $\frac{13}{29}$

3)   (a) $\frac{1}{3}$ (b) $\frac{1}{3}$ (c) $\frac{1}{3}$

4)   Total number of guests: $19 + 44 + 14 = 77$.
      (a) $\frac{19}{77}$ (b) $\frac{44}{77}$ (c) $\frac{14}{77}$

5) $\frac{1}{5}$

6)   (a) $\frac{10}{21}$ (b) $\frac{11}{21}$

7) $\frac{17}{20}$

8)   (a) $\frac{18}{25}$ (b) $\frac{7}{25}$

9) $\frac{2}{3}$ 10) $\frac{1}{100}$

11) $\frac{14}{37}$ 12) $\frac{2}{9}$

13)   (a) $\frac{210}{211}$ (b) $\frac{1}{211}$

14) $\frac{2}{5}$ 15) $\frac{2}{7}$

For each problem answer these questions: (a) What fraction of the pie is in the shaded area? (b) Identify the numerator. (c) Identify the denominator.

1)

2)

3)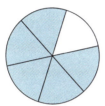

4)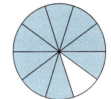

For each problem answer these questions: (a) What fraction of the pie is shaded? (b) Identify the numerator. (c) Identify the denominator.

5)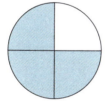

6)

7)

8)

9)

10)

Answer each of these questions:

11) There are 13 bottles in a row. If four are green, what fraction of the bottles is green?

12) There are eight puppies playing on the floor. If three of the puppies are brown, (a) what fraction of the puppies is brown? (b) What is the numerator of the fraction? (c) What is the denominator of the fraction?

13) A committee of 17 has 8 women. (a) What fraction of the committee's members is women? (b) What fraction is men?

14) A delivery man started the day with 29 cases of soda. If he dropped off 21 cases, (a) what fraction did he drop off? (b) What fraction did he not drop off?

15) A cash register had 12 one dollar bills, 9 five dollar bills, and 2 ten dollar bills. What fraction of the bills were (a) singles? (b) fives? and (c) tens?

16) Mrs. Yoshiba had 6 pennies, 8 nickels, 2 dimes, and 1 quarter. What fraction of her coins was in (a) pennies? (b) nickels? (c) dimes? and (d) quarters?

17) There are twelve horses in a barn. If five of the horses are white, what fraction of the horses is white?

18) There are eleven balls in a box. If nine of the balls are red, (a) what fraction of the balls is red? (b) What is the numerator of the fraction? (c) What is the denominator of the fraction?

19) There are six kittens sleeping in a basket. If one of the kittens is gray, (a) what fraction of the kittens is gray? (b) What is the numerator of the fraction? (c) What is the denominator of the fraction?

20) There are 10 pairs of shoes lined up in a closet. If seven pairs are black, then (a) what fraction of the pairs of shoes is black? (b) What fraction is not black?

21) The same number of apartments are on the first, second, third, and fourth floors of a building. What fraction of apartments is on the (a) first floor? (b) second floor? (c) third floor? (d) fourth floor?

22) A sales rep started the day with 23 samples. If she gave away 11, (a) what fraction did she give away? (b) What fraction did she not give away?

23) Carlos, Pei, Georges, Ingrid, Anna, and Marcello divided up a birthday cake equally among themselves. What fraction of the cake did each person get?

24) Lars had three red marbles, six white marbles, and five blue marbles. What fraction of his marbles was (a) red? (b) white? and (c) blue?

25) The Arizona Diamondbacks had 9 pitchers, 2 catchers, 6 outfielders, and 8 infielders on their team. What fraction of the team was (a) pitchers? (b) catchers? (c) outfielders? (d) infielders?

26) A collection of marbles consists of 3 blue, 7 red, and 5 green marbles. What fraction represents the red marbles?

27) If a class consisting of 21 students has 1 redhead, 4 blondes, and 16 brunettes, what fractional part of the class represents the number of (a) blondes, (b) redheads?

28) If 20 people started a math class and only 13 of them finished the class, what fractional part of the class finished the class?

29) If 7 people went out for dinner and 3 of them ordered dessert, what fractional part of the people had dessert?

30) If 30 children went to the playground and 11 of them went on the swings, what fractional part of the children went on the swings?

31) If 6,000,000 people bought a lottery ticket this week and one of them had a winning number, what fractional part of the buyers had the winning number?

32) If 12 people are at a meeting and 11 of them had a cup of coffee, what fractional part of the people had coffee?

33) If 300 people voted to abolish final exams and 761 people voted on this issue, what fractional part of the voters voted (a) to abolish finals, (b) not to abolish finals?

34) If 61 of the 231 cars driving past my house were speeding, what fractional part of the cars driving past my house were speeding?

35) If 4 out of the last 5 people I've hired were hard workers, what fractional part of the people I recently hired are hard workers?

3.3 Prime Numbers and Factoring

Before we can really work with fractions, we need to learn about prime numbers and factoring. It will take about one minute to learn what you'll need to know about prime numbers.

A *prime* number is any whole number greater than 1 whose only factors are 1 and itself. Some prime numbers are 2, 3, 5, 7, 11, 13, 17, 19,…. If a whole number greater than 1 is not prime, it is called *composite*. 12 is a composite number because it has factors of 1, 2, 3, 4, 6, and 12. When performing arithmetic operations with fractions, we sometimes need to write a number as a product of prime factors. This is called *prime factorization*. For example, the number 4 can be written as $4 = 2(2)$. The number 6 can be written as $6 = 2(3)$. When numbers with more factors need to be written in factored form, one way to easily do this is to use a factor tree. The number 1 is neither prime, nor composite.

3.3a Using the Factor Tree

Let's use a factor tree to write the number 20 as a product of primes. 20 breaks down in two possible ways: $20 = 2(10)$ or $20 = 4(5)$. It doesn't matter which way you start to break down 20 because we will get the same prime factorization when we're finished. Just remember, we're not finished factoring until we can't factor further. Now let's go back and finish the prime factorization of 20. $20 = 2(10)$ can go one more step and be written as $20 = 2(2)(5)$, because $10 = 2(5)$. $20 = 4(5)$ can go one more step and be written as $20 = 2(2)(5)$, because $4 = 2(2)$. In a tree form it would look this way:

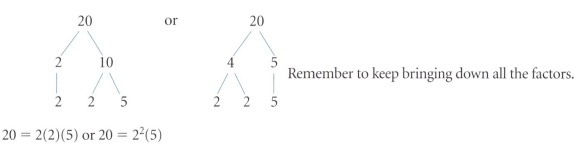

Remember to keep bringing down all the factors.

$20 = 2(2)(5)$ or $20 = 2^2(5)$

Example 1: Now let's find the prime factorization for 24.

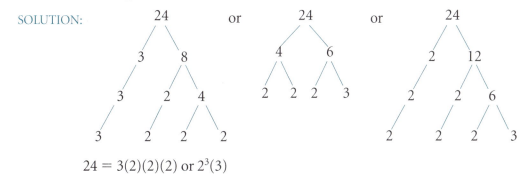

$24 = 3(2)(2)(2)$ or $2^3(3)$

Notice, we found the prime factorization in three ways, but we ended up with the same factors, so it doesn't matter how we begin factoring.

Example 2: Now let's find the prime factorization for 66.

SOLUTION:

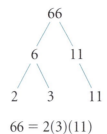 or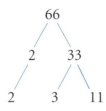

$$66 = 2(3)(11)$$

Problem Set 3.3:

Find the prime factors for each of these numbers.

1) 36   2) 128   3) 75   4) 39   5) 50

6) 336   7) 200   8) 98   9) 165   10) 37

SOLUTIONS TO PROBLEM SET:

1)

$$36 = (2)(3)(2)(3)$$
$$36 = 2^2(3^2)$$

2)

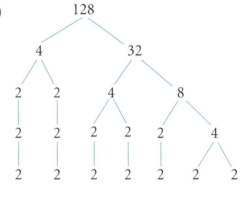

$$128 = 2^7$$

3)

$$75 = 5^2(3)$$
$$75 = (5)(5)(3)$$

4)

$$39 = 3(13)$$

5)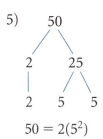

$$50 = 2(5^2)$$

6)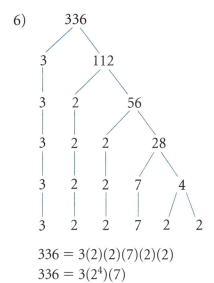

$$336 = 3(2)(2)(7)(2)(2)$$
$$336 = 3(2^4)(7)$$

7)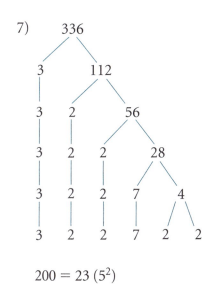

$$200 = 23 (5^2)$$

8)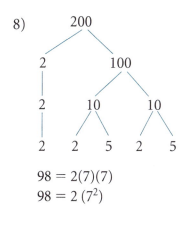

$$98 = 2(7)(7)$$
$$98 = 2 (7^2)$$

9)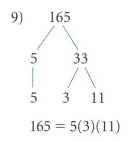

$$165 = 5(3)(11)$$

10)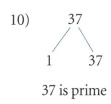

37 is prime

Find the prime factorization for the following numbers. If a number is prime, say so.

1) 340

2) 385

3) 154

4) 78

5) 200

6) 165

7) 45

8) 54

9) 62

10) 364

11) 99

12) 24

13) 135

14) 58

15) 231

16) 255

17) 16

18) 48

19) 80

20) 2002

21) 23

22) 71

23) 13

24) 31

3.3b Finding the Main Prime Factors

The numbers 2 and 3 happen to be prime numbers. We're going to be doing hundreds of problems in prime factoring, and we'll start out every problem by looking for prime factors of 2 and 3. If 2 is a prime factor of a number, that means that the number is divisible by 2. If 3 is a prime factor of a number, that means that number is divisible by 3. The number 6 is divisible by 2 and 3, so 2 and 3 are prime factors of 6. And the number 9 is divisible by 3, so 3 is a prime factor of 9.

When we're looking for the prime factors of a number, very often we're looking for the three main numbers: 2, 3, and 5.

Example 1: **What are the prime factors of 30?**

SOLUTION:

$$\begin{array}{r} 15 \\ 2\overline{)30} \\ -\ 2| \\ \hline 10 \\ -\ 10 \\ \hline \end{array} \qquad \begin{array}{r} 5 \\ 3\overline{)15} \\ -\ 15 \\ \hline \end{array}$$

$$30 = 2 \cdot 3 \cdot 5$$

Now we'll show you three easy facts that you can use to find these three prime factors.

Fact #1: All even numbers have the factor 2. In other words, a whole number is divisible by 2 if its last digit is 0, 2, 4, 6 or 8.

Fact #2: To find out if a number has a factor of 3, add all the digits and then see if that sum is divisible by 3. Try that with the number 417.

$4 + 1 + 7 = 12$. Since 12 is divisible by 3, then 417 is divisible by 3. If you don't believe us, we'll prove it:

$$\begin{array}{r} 139 \\ 3\overline{)417} \\ -\ 2| \\ \hline 11| \\ -\ 9| \\ \hline 27 \\ -\ 27 \\ \hline \end{array}$$

Fact #3: Any number ending in 0 or 5 has a primary factor of 5. You'll notice this when you count by 5s—5, 10, 15, 20, 25, 30....

As you work out more problems, see if you can apply these three facts to find prime factors of 2, 3, and 5. Some of these numbers may contain larger prime factors as well, such as 7, 11 or 13. But as you work out each problem, first take care of the prime factors of 2, 3 and 5, and then tackle any larger ones.

Example 2: **What is the prime factorization of 16?**

SOLUTION: Suppose you start out with 4×4. Is 4 a prime number? No, but 2 is. Since $4 = 2 \times 2$, we can write 4×4 as $2 \times 2 \times 2 \times 2$. So the prime factorization of 16 are $2 \times 2 \times 2 \times 2$, or $16 = 2^4$.

Example 3:   What is the prime factorization of 20?

SOLUTION: The prime factors of 20 are $2 \times 2 \times 5$. If you got this far – 4×5 – then you needed to find the prime factors of 4, which are 2×2. So we end up with $2 \times 2 \times 5$. $20 = 2^2(5)$

Problem Set 3.3b • 1:

Find the prime factorization for the following numbers.

1) 24   2) 18   3) 45   4) 27   5) 50

6) 35   7) 72   8) 120   9) 144   10) 150

SOLUTIONS TO PROBLEM SET:

1) $2 \cdot 2 \cdot 2 \cdot 3 = 2^3(3)$ 2) $2 \cdot 3 \cdot 3 = (2)3^2$ 3) $3 \cdot 3 \cdot 5 = 3^2(5)$

4) $3 \cdot 3 \cdot 3 = 3^3$ 5) $2 \cdot 5 \cdot 5 = 2(5^2)$ 6) $5 \cdot 7$

7) $2 \cdot 2 \cdot 2 \cdot 3 \cdot 3 = (2^3)3^2$ 8) $2 \cdot 2 \cdot 2 \cdot 3 \cdot 5 = 2^3(3)5$ 9) $2 \cdot 2 \cdot 2 \cdot 2 \cdot 3 \cdot 3 = (2^4)3^2$

10) $2 \cdot 3 \cdot 5 \cdot 5 = 2(3)5^2$

Let's stop for a minute to catch our breath. So far every number we've worked with had prime factors of 2, 3 or 5. Occasionally, however, we'll encounter larger prime factors, such as 7, 11, 13, 17, 19 and 23. Suppose you needed to find the prime factorization of the number 380. By just looking at it, would you know right off that it contained the prime factor of 19? Probably not. But, working step-by-step, you can discover all of the prime factors of 380, including 19. Start by dividing 380 by 2, and keep dividing until you find all of the prime factors.

$$
\begin{array}{r} 190 \\ 2\overline{)380} \\ -\ 2\vert \\ \hline 18 \\ -\ 18 \\ \hline \end{array}
\qquad
\begin{array}{r} 95 \\ 2\overline{)190} \\ -\ 18\vert \\ \hline 10 \\ -\ 10 \\ \hline \end{array}
\qquad
\begin{array}{r} 19 \\ 2\overline{)95} \\ -\ 5\vert \\ \hline 45 \\ -\ 45 \\ \hline \end{array}
$$

$$380 = 2 \cdot 2 \cdot 5 \cdot 19 \text{ or } 2^2(5)(19)$$

Example 4:   Find the prime factorization of 70.

SOLUTION:
$$
\begin{array}{r} 7 \\ 10\overline{)70} \\ -\ 70 \\ \hline \end{array}
\qquad
\begin{array}{r} 5 \\ 2\overline{)10} \\ -\ 10 \\ \hline \end{array}
$$
70 ends in a zero; therefore, we know ten is a factor of 70.

$$70 = (7)(5)(2)$$

Another way we could have found the prime factorization of 70 is to begin with 2 instead of 10. We'll still have the same factorization.

SOLUTION:
$$
\begin{array}{r} 35 \\ 2\overline{)70} \\ -\ 70 \\ \hline \end{array}
\qquad
\begin{array}{r} 7 \\ 5\overline{)35} \\ -\ 35 \\ \hline \end{array}
$$

$$70 = 2(7)(5)$$

Example 5:   Here's one more: Find the prime factorization of 177.

SOLUTION:

$$\begin{array}{r} 59 \\ 3\overline{)177} \\ -\ 171\ \downarrow \\ \hline 27 \\ -\ 27 \\ \hline \end{array}$$

Notice the steps we followed here. First, we saw that 177 was not divisible by 2. Since the sum of the digits—$1 + 7 + 7 = 15$—was divisible by 3, we divided 177 by 3. Then we saw that 59 was not divisible by 3—the sum of the digits of 59 is 14—so we considered trying 5. But 59, it turns out, is not divisible by 5, or by any other higher number. So 59 must be a prime number. That leaves us with our two prime factors of 3 and 59.

$$177 = (3)(59)$$

Problem Set 3.3b • 2:

Find the prime factorization for each of these numbers.

1) 90	2) 42	3) 100	4) 54	5) 55
6) 360	7) 195	8) 300	9) 174	10) 213

SOLUTIONS TO PROBLEM SET:

1) $2(3^2)(5)$ 2) $(2)(3)(7)$ 3) $2^2(5^2)$ 4) $2(3^3)$

5) $(5)(11)$ 6) $2^3(3^2)(5)$ 7) $(3)(5)(13)$ 8) $(2^2)(3)(5^2)$

9) $(2)(3)(29)$ 10) $(3)(71)$

How did you do? If you feel that you really know how to do these problems, go directly to the next section. *But if you need a little extra help, please read the accompanying box.*

Extra Help

We're going to go over the first three problems of the last problem set step-by-step.

1. To find the prime factorization of 90, we start by looking for any 2's. Since 90 is an even number (remember Fact #1?), we know there's a 2. So let's start out like this: 90 = (2)(45).

   Now we find the prime factors of 45. There are no more 2's, so we'll look for 3's. Applying Fact #2, we add the digits of 45, 4 + 5 = 9. Since 9 is divisible by 3, we know that 3 is a prime factor of 45. That gives us 90 = (2)(3)(15). Moving right along, we divide 15 by 3 and get 5. That gives us (2)(3)(3)(5)—which is our answer.

2. To find the prime factorization of 42, the first thing we look for is a prime factor of 2. 42 is an even number, so we get 42 = (2)(21). 21 has a prime factor of 3 (since 2 + 1 = 3), so now we have 42 = (2)(3)(7).

3. To find the prime factorization of 100, we first look for any 2's. First we get 100 = (2)(50). There's another 2: 100 = (2)(2)(25). Any 3's? No, because 2 + 5 = 7, which is not divisible by 3. Any 5's? Yes! Using Fact #3, we observe that 25 is divisible by 5 (since it ends in a 5). So we end up with 100 = (2)(2)(5)(5).

Find the prime factorization of each number. If a number is prime, say so.

1) 16

2) 30

3) 39

4) 24

5) 50

6) 77

7) 48

8) 94

9) 100

10) 91

11) 153

12) 426

13) 36

14) 27

15) 64

16) 51

17) 29

18) 63

19) 80

20) 99

21) 117

22) 152

23) 180

24) 143

25) 125

26) 210

27) 147

28) 195

29) 300

30) 207

31) 165

32) 450

33) 11

34) 41

35) 59

36) 73

3.3c Finding Prime Factors Using Continued Division

Actually there's nothing new in this section, but we'll show you that what you've been doing so far is really a process of continued division.

Example 1:   **What is the prime factorization of 8?**

SOLUTION: $8 = (2)(2)(2) = 2^3$

How did we get that answer? We did this:

$$
\begin{array}{r} 4 \\ 2\overline{)8} \\ -\underline{8} \end{array}
\qquad
\begin{array}{r} 2 \\ 2\overline{)4} \\ -\underline{4} \end{array}
$$

Example 2:   **Now find the prime factors of 100.**

SOLUTION:

$$
\begin{array}{r} 50 \\ 2\overline{)100} \\ -\underline{10\downarrow} \end{array}
\qquad
\begin{array}{r} 25 \\ 2\overline{)50} \\ -\underline{4\downarrow} \\ 10 \\ -\underline{10} \end{array}
\qquad
\begin{array}{r} 5 \\ 2\overline{)25} \\ -\underline{25} \end{array}
$$

$$100 = 2 \cdot 2 \cdot 5 \cdot 5 = 2^2(5^2)$$

Now here's a fun one for you to do.

Example 3:   **Find the prime factors of 4,400.**

SOLUTION:

$$
\begin{array}{r} 2200 \\ 2\overline{)4400} \\ -\underline{4\downarrow} \\ 4 \\ -\underline{4} \end{array}
\qquad
\begin{array}{r} 1100 \\ 2\overline{)2200} \\ -\underline{2\downarrow} \\ 2 \\ -\underline{2} \end{array}
\qquad
\begin{array}{r} 550 \\ 2\overline{)1100} \\ -\underline{10\downarrow} \\ 10 \\ -\underline{10} \end{array}
$$

$$
\begin{array}{r} 275 \\ 2\overline{)550} \\ -\underline{4\downarrow} \\ 15 \\ -\underline{14} \\ 10 \\ -\underline{10} \end{array}
\qquad
\begin{array}{r} 55 \\ 5\overline{)275} \\ -\underline{25\downarrow} \\ 25 \\ -\underline{25} \end{array}
\qquad
\begin{array}{r} 11 \\ 5\overline{)55} \\ -\underline{5\downarrow} \\ 5 \\ -\underline{5} \end{array}
$$

$$2{,}200 = 2 \cdot 2 \cdot 2 \cdot 2 \cdot 5 \cdot 5 \cdot 11 = 2^4(5^2)(11)$$

Example 4:   **Let's see if you can find the prime factors of 360.**

SOLUTION:

$$
\begin{array}{r} 180 \\ 2\overline{)360} \\ -\underline{2\downarrow} \\ 16 \\ -\underline{16} \end{array}
\qquad
\begin{array}{r} 90 \\ 2\overline{)180} \\ -\underline{18\downarrow} \end{array}
\qquad
\begin{array}{r} 45 \\ 2\overline{)90} \\ -\underline{8\downarrow} \\ 10 \\ -\underline{10} \end{array}
\qquad
\begin{array}{r} 15 \\ 3\overline{)45} \\ -\underline{3\downarrow} \\ 15 \\ -\underline{15} \end{array}
\qquad
\begin{array}{r} 5 \\ 3\overline{)15} \\ -\underline{15} \end{array}
$$

$$360 = 2 \cdot 2 \cdot 2 \cdot 3 \cdot 3 \cdot 5 = 2^3(3^2)(5)$$

Which method of factoring is best? Whichever one works for you. For the rest of this chapter and all through the next, you'll have many chances to use your factoring skills.

3.4 Reducing Fractions

In this section we will learn how to rewrite a fraction in a simpler form, thus making it more convenient to perform arithmetic operations with that fraction. For example, it would be much easier to add the fractions $\frac{105}{180}$ and $\frac{25}{60}$ if we rewrite them as $\frac{7}{12}$ and $\frac{5}{12}$. When two fractions have the same value but are written in different forms they are called *equivalent fractions*. $\frac{2}{4}$ and $\frac{1}{2}$ are equivalent fractions. Notice in the illustration below that the same part of the circle has been shaded for $\frac{2}{4}$ and $\frac{1}{2}$.

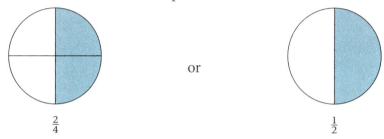

or

$$\frac{2}{4} \qquad\qquad \frac{1}{2}$$

Even though $\frac{2}{4}$ and $\frac{1}{2}$ are different fractions, they have the same value and therefore are equivalent fractions. Also notice that if both the numerator (the top of the fraction) and the denominator (the bottom of the fraction) of $\frac{2 \div 2}{4 \div 2}$ is divided by 2 we get $\frac{1}{2}$; $\frac{2}{4} = \frac{2 \div 2}{4 \div 2} = \frac{1}{2}$. Another example of equivalent fractions is $\frac{2}{5} = \frac{22}{55}$. Notice that if both the numerator and denominator of $\frac{2}{5}$ are multiplied by 11, we get $\frac{2(11)}{5(11)} = \frac{22}{55}$. Remember, if you multiply or divide the numerator of a fraction by a number, you must do the exact same thing to the denominator, or you'll change the value of the fraction. When the numerator and the denominator of a fraction doesn't have a common factor other than 1, the fraction is in *reduced form*, or *lowest terms*.

For example, $\frac{2}{6} = \frac{1}{3}$, $\frac{10}{12} = \frac{5}{6}$ and $\frac{2}{2} = 1$ are in reduced form.

The way to reduce a fraction is to eliminate or cancel the factors that are the same in the numerator and the denominator. These are called the *common factors*.

Example 1:   Reduce $\frac{330}{462}$

SOLUTION:

$$\frac{330}{462} = \frac{2(3)(5)(11)}{2(3)(7)(11)} = \frac{5}{7}$$

Notice we wrote 330 and 462 as a product of prime factors and then canceled the common factors. If you need to review how to write a number as a product of prime factors, go back to the last section and review prime factorization.

Example 2:   Reduce $\frac{81}{243}$

SOLUTION: $\dfrac{81}{243} = \dfrac{(3)(3)(3)(3)(1)}{(3)(3)(3)(3)(3)} = \dfrac{1}{3}$

Problem Set 3.4:

Now you try reducing these fractions to lowest terms.

1) $\underline{4}$ 2) $\underline{40}$ 3) $\underline{120}$ 4) $\underline{286}$

SOLUTIONS TO PROBLEM SET:

1) $\dfrac{4}{12} = \dfrac{\cancel{2}(\cancel{2})}{\cancel{2}(\cancel{2})(3)} = \dfrac{1}{3}$

2) $\dfrac{40}{48} = \dfrac{\cancel{2}(\cancel{2})(\cancel{2})(5)}{\cancel{2}(\cancel{2})(\cancel{2})(2)(3)} = \dfrac{5}{6}$

3) $\dfrac{120}{270} = \dfrac{2(2)(2)(\cancel{3})(\cancel{5})}{2(3)(3)(\cancel{3})(\cancel{5})} = \dfrac{4}{9}$

4) $\dfrac{286}{572} = \dfrac{\cancel{2}(\cancel{11})(\cancel{13})}{2(\cancel{2})(\cancel{11})(\cancel{13})} = \dfrac{1}{2}$

Reduce each of these fractions to its lowest terms.

1) $\dfrac{95}{100}$

2) $\dfrac{9}{12}$

3) $\dfrac{48}{144}$

4) $\dfrac{21}{30}$

5) $\dfrac{88}{99}$

6) $\dfrac{85}{100}$

7) $\dfrac{39}{52}$

8) $\dfrac{27}{81}$

9) $\dfrac{34}{102}$

10) $\dfrac{48}{120}$

11) $\dfrac{17}{49}$

12) $\dfrac{78}{130}$

13) $\dfrac{40}{50}$

14) $\dfrac{23}{58}$

15) $\dfrac{100}{240}$

16) $\dfrac{114}{121}$

17) $\dfrac{108}{135}$

18) $\dfrac{51}{85}$

19) $\dfrac{237}{555}$

20) $\dfrac{57}{190}$

21) $\dfrac{213}{300}$

22) $\dfrac{64}{256}$

23) $\dfrac{81}{405}$

24) $\dfrac{96}{288}$

25) $\dfrac{840}{910}$

26) $\dfrac{121}{154}$

27) $\dfrac{65}{286}$

28) $\dfrac{98}{350}$

29) $\dfrac{161}{483}$

30) $\dfrac{150}{625}$

3.5   Multiplying Proper Fractions

Example 1:   How much is one half of 4?

SOLUTION:

$$\frac{1}{2}\left(\frac{4}{1}\right)$$

Without working out this problem, we know half of 4 is 2, but we'll work it out anyway, so you know how to find the answer to other problems of this type.

Remember from Chapter 2, "of" means multiply.

$$\frac{1}{\cancel{2}_1}\left(\frac{\cancel{4}^2}{1}\right) = \frac{2}{1} = 2$$

When multiplying fractions we can reduce vertically or diagonally. *We can reduce diagonally ONLY in multiplication.*

Example 2:   How much is $\frac{1}{2}$ of $\frac{1}{4}$?

SOLUTION:

Before we multiply we'd like you to look at the following circle, which has been divided into four equal pieces.

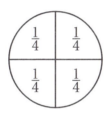

In the next circle, we divided one of the quarters of the circle in half.

If we look at the whole circle, which is now divided into eight equal pieces, we can see that the shaded area, which is half of one quarter, is one-eighth.

$$\frac{1}{2}\left(\frac{1}{4}\right) = \frac{1}{8}$$

Example 3: How much is $\dfrac{1}{5} \times \dfrac{1}{3}$?

SOLUTION: $\dfrac{1}{5} \times \dfrac{1}{3} = \dfrac{1 \times 1}{5 \times 3} = \dfrac{1}{15}$

Example 4: How much is $\dfrac{1}{7} \times \dfrac{3}{7}$?

SOLUTION: $\dfrac{1}{7} \times \dfrac{3}{7} = \dfrac{1 \times 3}{7 \times 7} = \dfrac{3}{49}$

Example 5: How much is $\dfrac{2}{5} \times \dfrac{1}{3}$?

SOLUTION: $\dfrac{2}{5} \times \dfrac{1}{3} = \dfrac{2 \times 1}{5 \times 3} = \dfrac{2}{15}$

Example 6: How much is one-eighth of a quarter? Remember, the word "of" means multiply.

SOLUTION: $\dfrac{1}{8} \times \dfrac{1}{4} = \dfrac{1}{32}$

Example 7: How much is one-third of one-eighth?

SOLUTION: $\dfrac{1}{3} \times \dfrac{1}{8} = \dfrac{1}{24}$

Example 8: Multiply $\dfrac{1}{6}$ by $\dfrac{1}{5}$.

SOLUTION: $\dfrac{1}{6} \times \dfrac{1}{5} = \dfrac{1}{30}$

Example 9: Multiply $\dfrac{3}{10}$ by $\dfrac{7}{11}$.

SOLUTION: $\dfrac{3}{10} \times \dfrac{7}{11} = \dfrac{21}{121}$

Problem Set 3.5:

1) Find one fifth of a quarter.

2) Find one half of one third.

3) Find one eighth of one half.

4) Find one quarter of one sixth.

5) Find one sixth of one third.

6) Find one fifth of one fifth.

7) How much is $\dfrac{1}{4} \times \dfrac{3}{5}$?

8) How much is $\dfrac{2}{9} \times \dfrac{1}{7}$?

9) Multiply $\dfrac{3}{11}$ by $\dfrac{1}{11}$.

10) Multiply $\dfrac{1}{4}$ by $\dfrac{5}{9}$.

SOLUTIONS TO PROBLEM SET:

1) $\dfrac{1}{5} \times \dfrac{1}{4} = \dfrac{1}{20}$

2) $\dfrac{1}{2} \times \dfrac{1}{3} = \dfrac{1}{6}$

3) $\dfrac{1}{8} \times \dfrac{1}{2} = \dfrac{1}{16}$

4) $\dfrac{1}{4} \times \dfrac{1}{6} = \dfrac{1}{24}$

5) $\dfrac{1}{6} \times \dfrac{1}{3} = \dfrac{1}{18}$

6) $\dfrac{1}{5} \times \dfrac{1}{5} = \dfrac{1}{25}$

7) $\dfrac{1}{4} \times \dfrac{3}{5} = \dfrac{3}{20}$

8) $\dfrac{2}{9} \times \dfrac{1}{7} = \dfrac{2}{63}$

9) $\dfrac{3}{11} \times \dfrac{1}{11} = \dfrac{3}{121}$

10) $\dfrac{1}{4} \times \dfrac{5}{9} = \dfrac{5}{36}$

1) Find one quarter of one third.

2) Find one half of one quarter.

3) How much is $\dfrac{1}{8} \times \dfrac{3}{5}$?

4) How much is $\dfrac{2}{5} \times \dfrac{1}{7}$?

5) Multiply $\dfrac{1}{11} \times \dfrac{7}{9}$.

6) Multiply $\dfrac{2}{13} \times \dfrac{4}{5}$.

7) Multiply $\dfrac{9}{10} \times \dfrac{3}{4}$.

8) Find two thirds of one fifth.

9) Find five sixths of one ninth.

10) Find one sixth of one seventh.

11) Multiply $\dfrac{3}{4}$ by $\dfrac{1}{2}$.

12) Multiply $\dfrac{8}{9}$ by $\dfrac{1}{5}$.

13) Find three-fourths of three-tenths.

14) Find five-eighths of three-elevenths.

15) Find one-half of three-sevenths.

16) Find one-third of one-half.

17) How much is $\dfrac{7}{8} \times \dfrac{3}{8}$?

18) How much is $\dfrac{2}{5} \times \dfrac{1}{9}$?

19) How much is $\dfrac{1}{10} \times \dfrac{3}{8}$?

20) Multiply $\dfrac{4}{11}$ by $\dfrac{2}{9}$.

21) Multiply $\dfrac{5}{8}$ by $\dfrac{1}{2}$.

22) Multiply $\dfrac{1}{6}$ by $\dfrac{5}{13}$.

23) How much is $\dfrac{1}{3} \times \dfrac{1}{4}$?

24) How much is $\dfrac{3}{7} \times \dfrac{2}{5}$?

3.6 Multiplying Fractions

3.6a Using Prime Factoring

So far we haven't had to reduce our answers to their lowest terms. Guess what we'll be doing next?

Example 1: How much is $\dfrac{3}{4} \times \dfrac{2}{9}$?

SOLUTION: $\dfrac{3}{4} \times \dfrac{2}{9} = \dfrac{\overset{1}{\cancel{3}}}{(2)(2)} \times \dfrac{\overset{1}{\cancel{2}}}{\underset{1}{(3)(3)}} = \dfrac{1}{6}$

You'll notice that simplifying or reducing the fractions made the multiplication a lot easier.

Example 2: Multiply $\dfrac{4}{5}$ by $\dfrac{5}{6}$.

SOLUTION: $\dfrac{4}{5} \times \dfrac{5}{6} = \dfrac{(2)(2)}{\underset{1}{\cancel{5}}} \times \dfrac{\overset{1}{\cancel{5}}}{(2)(3)} = \dfrac{2}{3}$

Example 3: How much is five-sevenths times nine-tenths?

SOLUTION: $\dfrac{5}{7} \times \dfrac{9}{10} = \dfrac{\overset{1}{\cancel{5}}}{7} \times \dfrac{(3)(3)}{\underset{1}{(2)\cancel{(5)}}} = \dfrac{9}{14}$

Example 4: How much is nine-tenths of five-twelfths?

SOLUTION: We'll answer this question using two different approaches. The first approach is to multiply first, then reduce.

$$\dfrac{9}{10} \times \dfrac{5}{12} = \dfrac{45}{120} = \dfrac{9}{24} = \dfrac{3}{8}$$

The second is to reduce first, then multiply.

$$\dfrac{\overset{3}{\cancel{9}}}{\underset{2}{\cancel{10}}} \times \dfrac{\overset{1}{\cancel{5}}}{\underset{4}{\cancel{12}}} = \dfrac{3}{8}$$

Based on this example, we hope you see that it's easier and faster to reduce first then multiply. If you reduce all the common factors before you multiply, you won't need to reduce your answer.

Problem Set 3.6a:

1) How much is $\frac{4}{5} \times \frac{5}{8}$?

2) Multiply $\frac{7}{10} \times \frac{5}{14}$.

3) How much is four-fifteenths of five-ninths?

4) How much is $\frac{7}{20}$ of $\frac{5}{8}$?

5) How much is $\frac{6}{7} \times \frac{14}{15}$?

6) How much is $\frac{5}{8}$ of $\frac{9}{10}$?

7) Multiply three-quarters times eight-ninths.

8) Multiply nine-fourteenths times seven-fifteenths.

9) What is three-twentieths of five-eighteenths?

10) Multiply $\frac{5}{12} \times \frac{8}{15}$.

SOLUTIONS TO PROBLEM SET:

1) $\frac{4}{5} \times \frac{5}{8} = \frac{(2)(2)}{5} \times \frac{5}{(2)(2)(2)} = \frac{5}{8}$

2) $\frac{7}{10} \times \frac{5}{14} = \frac{7}{(2)(5)} \times \frac{5}{(2)(7)} = \frac{1}{4}$

3) $\frac{4}{15} \times \frac{5}{9} = \frac{(2)(2)}{(3)(5)} \times \frac{5}{(3)(3)} = \frac{4}{27}$

4) $\frac{7}{20} \times \frac{5}{8} = \frac{7}{(2)(2)(5)} \times \frac{5}{(2)(2)(2)} = \frac{7}{32}$

5) $\frac{6}{7} \times \frac{14}{15} = \frac{(2)(3)}{7} \times \frac{(2)(7)}{(3)(5)} = \frac{4}{5}$

6) $\frac{5}{8} \times \frac{9}{10} = \frac{5}{(2)(2)(2)} \times \frac{(3)(3)}{(2)(5)} = \frac{9}{16}$

7) $\frac{3}{4} \times \frac{8}{9} = \frac{3}{(2)(2)} \times \frac{(2)(2)(2)}{(3)(3)} = \frac{2}{3}$

8) $\frac{9}{14} \times \frac{7}{15} = \frac{(3)(3)}{(2)(7)} \times \frac{7}{(3)(5)} = \frac{3}{10}$

9) $\frac{3}{20} \times \frac{5}{18} = \frac{5}{(2)(2)(3)} \times \frac{(3)(3)}{(2)(5)} = \frac{1}{24}$

10) $\frac{5}{12} \times \frac{8}{15} = \frac{5}{(2)(2)(3)} \times \frac{(2)(2)(9)}{(3)(5)} = \frac{2}{9}$

1) How much is $\dfrac{5}{8}$ of $\dfrac{12}{25}$?

2) How much is $\dfrac{9}{20}$ of $\dfrac{5}{18}$?

3) How much is $\dfrac{4}{9}$ of $\dfrac{15}{16}$?

4) How much is $\dfrac{13}{14}$ of $\dfrac{7}{26}$?

5) How much is $\dfrac{14}{15}$ of $\dfrac{5}{6}$?

6) Multiply $\dfrac{9}{20} \times \dfrac{8}{27}$.

7) Multiply $\dfrac{11}{20} \times \dfrac{4}{33}$.

8) Multiply $\dfrac{10}{21} \times \dfrac{7}{45}$.

9) Multiply $\dfrac{21}{50} \times \dfrac{15}{28}$.

10) What is four-ninths of three-eighths?

11) What is eleven-twelfths of eighteen thirty-thirds?

12) What is five-ninths of three-tenths?

13) How much is $\dfrac{12}{21} \times \dfrac{7}{8}$?

14) How much is $\dfrac{9}{25} \times \dfrac{15}{17}$?

15) How much is $\dfrac{27}{32} \times \dfrac{40}{63}$?

16) Multiply $\dfrac{9}{20} \times \dfrac{8}{15}$.

17) Multiply $\dfrac{9}{16} \times \dfrac{16}{21}$.

18) Multiply $\dfrac{5}{6} \times \dfrac{12}{25}$.

19) What is four-fifteenths of five-sixteenths?

20) What is ten twenty-firsts of seven-fifteenths?

21) What is four-elevenths of three-fortieths?

22) How much is $\dfrac{5}{6}$ of $\dfrac{12}{35}$?

23) How much is $\dfrac{7}{12}$ of $\dfrac{18}{35}$?

24) How much is $\dfrac{9}{10}$ of $\dfrac{15}{33}$?

3.6b Reducing Common Factors

Are you ready for a shortcut which will save you some work reducing fractions? All right, then, here it comes.

Let's start with this problem:

Example 1: $\frac{7}{12} \times \frac{8}{21}$

SOLUTION: We could use prime factoring to solve this:

$$\frac{7}{12} \times \frac{8}{21} = \frac{7}{(2)(2)(3)} \times \frac{(2)(2)(2)}{(3)(7)} = \frac{2}{9}$$

Now here's our shortcut solution:

$$\frac{7}{12} \times \frac{8}{21} = \frac{2}{9}$$

Let's go over what we just did step-by-step. First we reduced the common factor of 7 from the 7 and 21. Then we divided the 8 and 12 by 4, which left us with $\frac{2}{9}$.

You'll remember that you can divide the numerator and the denominator of a fraction by the same number. But aren't $\frac{7}{12} \times \frac{8}{21}$ really **two** separate fractions? Picky, picky. Once we've multiplied them, they become **one** fraction. So why wait? *Trust* us. You may divide the numerator of one fraction and the denominator of another fraction by the same number. You may do this whenever the two fractions are separated by a multiplication sign. You may reduce diagonally only in multiplication. You can always reduce the numerator and denominator of the same fraction by the same factor. You may always reduce up and down.

Question: May you reduce diagonally in addition, or in subtraction, or in division?

Answer: No, no, and no!

In case you missed the point: ***You may reduce diagonally ONLY in multiplication.***

Example 2:   How much is $\frac{5}{6} \times \frac{3}{4}$?

SOLUTION: $\frac{5}{6} \times \frac{3}{4} = \frac{5}{6} \times \frac{3}{4} = \frac{5}{8}$ You may also remove a common factor from the numerator and denominator of the same fraction.

In the last problem we performed a process called reducing common factors. We divided the 6 in $\frac{5}{6}$ by 3 and we divided the 3 in $\frac{3}{4}$ by 3. In other words, the 3 in the 6 and the 3 in the 3 reduced each other out. Try to reduce the common factors in the following problem:

Example 3: How much is $\frac{2}{3} \times \frac{1}{2}$?

SOLUTION: $\overset{1}{\cancel{\frac{2}{3}}} \times \frac{1}{\cancel{2}} = \frac{1}{3}$

Canceling out helps us reduce fractions to their lowest possible terms. It makes multiplication easier because it's much easier to work with smaller numbers.

Example 4: How much is $\frac{9}{16} \times \frac{8}{9}$?

SOLUTION: There are three ways we can simplify this:

$$1)\ \frac{9}{16} \times \frac{8}{9} = \frac{72}{144} = \frac{36}{72} = \frac{18}{36} = \frac{1}{2}$$

$$2)\ \frac{9}{16} \times \frac{8}{9} = \frac{(3)(3)}{(2)(2)(2)(2)} \times \frac{(2)(2)(2)}{(3)(3)} = \frac{1}{2}$$

$$3)\ \frac{\overset{1}{\cancel{9}}}{\underset{2}{\cancel{16}}} \times \frac{\overset{1}{\cancel{8}}}{\underset{1}{\cancel{9}}} = \frac{1}{2}$$

Reducing common factors certainly works out best here. But it's no more correct than either of the other ways we solved this problem. Use whichever method works best for you. We recommend you reduce before you multiply.

Problem Set 3.6b:

1) How much is four-fifths of one half?

2) Multiply $\frac{6}{7} \times \frac{14}{39}$.

3) How much is five-sixths of three tenths?

4) How much is eight-ninths of three quarters?

5) How much is four-fifths times five sixths?

6) Multiply $\frac{3}{8} \times \frac{5}{9}$.

7) Multiply $\frac{9}{20} \times \frac{2}{3}$.

8) How much is $\frac{8}{15} \times \frac{5}{16}$?

9) How much is $\frac{11}{18} \times \frac{9}{22}$?

10) Multiply $\frac{16}{25} \times \frac{5}{8}$.

SOLUTIONS TO PROBLEM SET:

1) $\overset{2}{\cancel{\frac{4}{5}}} \times \frac{1}{\underset{1}{\cancel{2}}} = \frac{2}{5}$

2) $\overset{2}{\cancel{\frac{6}{7}}} \times \overset{2}{\cancel{\frac{14}{39}}} = \frac{4}{13}$
   (with $\underset{1}{}$ under 7 and $\underset{13}{}$ under 39)

3) $\overset{1}{\underset{2}{\cancel{\frac{5}{6}}}} \times \overset{1}{\underset{2}{\cancel{\frac{3}{10}}}} = \frac{1}{4}$

4) $\overset{2}{\underset{3}{\cancel{\frac{8}{9}}}} \times \overset{1}{\underset{1}{\cancel{\frac{3}{4}}}} = \frac{2}{3}$

5) $\overset{2}{\underset{1}{\cancel{\frac{4}{5}}}} \times \overset{1}{\underset{3}{\cancel{\frac{5}{6}}}} = \frac{2}{3}$

6) $\frac{3}{8} \times \underset{3}{\cancel{\frac{5}{9}}} = \frac{5}{24}$

7) $\overset{3}{\underset{10}{\cancel{\frac{9}{20}}}} \times \overset{1}{\underset{1}{\cancel{\frac{2}{3}}}} = \frac{3}{10}$

8) $\underset{3}{\cancel{\frac{8}{15}}} \times \overset{1}{\underset{2}{\cancel{\frac{5}{16}}}} = \frac{1}{6}$

9) $\overset{1}{\underset{2}{\cancel{\frac{11}{18}}}} \times \overset{1}{\underset{2}{\cancel{\frac{9}{22}}}} = \frac{1}{4}$

10) $\overset{2}{\underset{5}{\cancel{\frac{16}{25}}}} \times \overset{1}{\underset{1}{\cancel{\frac{5}{8}}}} = \frac{2}{5}$

1) How much is five-eighths of four fifteenths?

2) How much is three-fifths of five ninths?

3) How much is two-sevenths of twenty-one twenty-sixths?

4) Multiply $\frac{7}{8} \times \frac{16}{21}$.

5) Multiply $\frac{9}{20} \times \frac{5}{18}$.

6) Multiply $\frac{16}{25} \times \frac{5}{12}$.

7) How much is $\frac{8}{15} \times \frac{3}{4}$?

8) How much is $\frac{15}{16} \times \frac{4}{5}$?

9) How much is $\frac{3}{4} \times \frac{8}{9}$?

10) Multiply $\frac{5}{8} \times \frac{24}{35}$.

11) Multiply $\frac{10}{13} \times \frac{39}{50}$.

12) Multiply $\frac{9}{22} \times \frac{11}{27}$.

13) How much is one-half of four-fifths?

14) How much is five-sixths of twelve twenty-fifths?

15) How much is seven-eighths of four twenty-firsts?

16) How much is fifteen-sixteenths of four-fifths?

17) How much is eight-ninths of twenty-seven thirty-seconds?

18) Multiply $\frac{3}{10} \times \frac{5}{6}$.

19) Multiply $\frac{13}{22} \times \frac{11}{52}$.

20) Multiply $\frac{35}{48} \times \frac{8}{21}$.

21) Multiply $\frac{5}{12} \times \frac{18}{25}$.

22) How much is $\frac{5}{8} \times \frac{16}{35}$?

23) How much is $\frac{10}{21} \times \frac{3}{3}$?

24) How much is $\frac{9}{16} \times \frac{4}{27}$?

3.6c Multiplication of Three Fractions

Example 1: How much is two-thirds of one-quarter?

SOLUTION: $\overset{1}{\underset{}{\frac{2}{3}}} \times \frac{1}{\underset{2}{4}} = \frac{1}{6}$

Example 2: How much is one quarter of two-thirds?

SOLUTION: $\frac{1}{\underset{2}{4}} \times \overset{1}{\frac{2}{3}} = \frac{1}{6}$

Did you notice what we just did? We multiplied the same fractions, but in a different order. So two-thirds of one quarter comes out the same as one quarter of two-thirds. When we multiply proper fractions, we get the same answer regardless of the order in which we place the numbers. We know this is true because we studied the Commutative Law of Multiplication earlier in this book.

Example 3: How much is $\frac{2}{5} \times \frac{5}{6} \times \frac{3}{4}$?

SOLUTION: $\overset{1}{\underset{1}{\frac{2}{5}}} \times \overset{1}{\underset{1}{\frac{5}{6}}} \times \overset{1}{\frac{3}{4}} = \frac{1}{4}$

You can see that we multiply three fractions just the way we multiply two fractions. Here's another one:

Example 4: Multiply $\frac{7}{8} \times \frac{3}{10} \times \frac{5}{6}$.

SOLUTION: $\frac{7}{8} \times \overset{1}{\underset{2}{\frac{3}{10}}} \times \overset{1}{\underset{2}{\frac{5}{6}}} = \frac{7}{32}$

Example 5: Multiply $\frac{5}{6} \times \frac{3}{10} \times \frac{7}{8}$.

SOLUTION: $\overset{1}{\underset{2}{\frac{5}{6}}} \times \overset{1}{\underset{2}{\frac{3}{10}}} \times \frac{7}{8} = \frac{7}{32}$

Once again we see that when we multiply fractions, we can change their order without changing the product.

Example 6: $\frac{2}{3}\left(\frac{1}{2} \times \frac{3}{4}\right)$

SOLUTION: $\frac{2}{3}\left(\frac{1}{2} \times \frac{3}{4}\right) = \overset{1}{\underset{2}{\frac{2}{3}}} \times \overset{1}{\underset{2}{\frac{3}{8}}} = \frac{1}{4}$

Example 7: $\left(\dfrac{3}{5} \times \dfrac{5}{6}\right)\dfrac{4}{5}$

SOLUTION: $\left(\dfrac{\overset{1}{3}}{\underset{1}{5}} \cdot \dfrac{\overset{1}{5}}{\underset{2}{6}}\right)\dfrac{4}{5} = \dfrac{1}{\underset{1}{2}} \cdot \dfrac{\overset{2}{4}}{5} = \dfrac{2}{5}$

Here's one more.

Example 8: $\dfrac{2}{9}\left(\dfrac{1}{5} \times \dfrac{10}{21}\right)$

SOLUTION: $\dfrac{2}{9}\left(\dfrac{1}{\underset{1}{5}} \times \dfrac{\overset{2}{10}}{21}\right) = \dfrac{2}{9}\left(\dfrac{2}{21}\right) = \dfrac{4}{189}$

Problem Set 3.6c:

1) Multiply $\dfrac{4}{5} \times \dfrac{5}{6} \times \dfrac{3}{8}$.

2) Multiply $\dfrac{7}{8} \times \dfrac{3}{10} \times \dfrac{5}{14}$.

3) $\dfrac{4}{7} \times \dfrac{2}{3} \times \dfrac{3}{14}$

4) $\dfrac{7}{12} \times \dfrac{8}{9} \times \dfrac{3}{14}$

5) $\dfrac{1}{2} \times \dfrac{5}{8} \times \dfrac{4}{5}$

6) $\dfrac{1}{3}\left(\dfrac{5}{6} \times \dfrac{3}{4}\right)$

7) $\dfrac{2}{5}\left(\dfrac{2}{7} \times \dfrac{1}{4}\right)$

8) $\dfrac{3}{10}\left(\dfrac{5}{12} \times \dfrac{3}{4}\right)$

9) $\left(\dfrac{6}{7} \times \dfrac{2}{3}\right)\dfrac{7}{8}$

10) $\left(\dfrac{11}{15} \times \dfrac{5}{8}\right)\dfrac{22}{27}$

SOLUTIONS TO PROBLEM SET:

1) $\dfrac{\overset{1}{4}}{\underset{1}{5}} \times \dfrac{\overset{1}{5}}{\underset{2}{6}} \times \dfrac{\overset{1}{3}}{\underset{2}{8}} = \dfrac{1}{4}$

2) $\dfrac{\overset{1}{7}}{8} \times \dfrac{3}{\underset{2}{10}} \times \dfrac{5}{\underset{2}{14}} = \dfrac{3}{32}$

3) $\dfrac{4}{7} \times \dfrac{\overset{1}{2}}{\underset{1}{3}} \times \dfrac{\overset{1}{3}}{\underset{7}{14}} = \dfrac{4}{49}$

4) $\dfrac{\overset{1}{7}}{\underset{3}{12}} \times \dfrac{\overset{8}{8}}{\underset{3}{9}} \times \dfrac{\overset{1}{3}}{\underset{1}{14}} = \dfrac{1}{9}$

5) $\dfrac{1}{2} \times \dfrac{\overset{1}{5}}{\underset{4}{8}} \times \dfrac{\overset{2}{4}}{\underset{1}{5}} = \dfrac{1}{4}$

6) $\dfrac{1}{3}\left(\dfrac{5}{\underset{2}{6}} \times \dfrac{\overset{1}{3}}{4}\right) = \dfrac{1}{3} \times \dfrac{5}{8} = \dfrac{5}{24}$

7) $\dfrac{2}{5}\left(\dfrac{\overset{1}{2}}{7} \times \dfrac{1}{\underset{2}{4}}\right) = \dfrac{2}{5} \times \dfrac{1}{14} = \dfrac{1}{35}$

8) $\dfrac{3}{10}\left(\dfrac{5}{\underset{4}{12}} \times \dfrac{\overset{1}{3}}{4}\right) = \dfrac{3}{10} \times \dfrac{5}{16} = \dfrac{3}{32}$

9) $\left(\dfrac{\overset{2}{6}}{7} \times \dfrac{2}{\underset{1}{3}}\right)\dfrac{7}{8} = \dfrac{\overset{1}{4}}{\underset{1}{7}} \times \dfrac{\overset{1}{7}}{\underset{2}{8}} = \dfrac{1}{2}$

10) $\left(\dfrac{11}{\underset{3}{15}} \times \dfrac{\overset{1}{5}}{8}\right)\dfrac{22}{27} = \dfrac{11}{\underset{12}{24}} \times \dfrac{\overset{11}{22}}{27} = \dfrac{121}{324}$

1) Multiply $\frac{1}{2} \times \frac{2}{3} \times \frac{3}{4}$.

2) Multiply $\frac{7}{8} \times \frac{9}{10} \times \frac{5}{12}$.

3) $\frac{5}{9} \times \frac{4}{5} \times \frac{3}{8}$

4) $\frac{2}{3}\left(\frac{1}{2} \times \frac{3}{5}\right)$

5) $\frac{2}{5}\left(\frac{15}{24} \times \frac{2}{3}\right)$

6) $\frac{3}{4}\left(\frac{7}{8} \times \frac{4}{21}\right)$

7) $\frac{9}{10}\left(\frac{5}{12} \times \frac{8}{9}\right)$

8) $\frac{3}{10}\left(\frac{1}{5} \times \frac{15}{16}\right)$

9) $\frac{5}{6}\left(\frac{1}{3} \times \frac{9}{10}\right)$

10) $\left(\frac{3}{4} \times \frac{6}{7}\right)\frac{14}{25}$

11) $\left(\frac{2}{7} \times \frac{3}{4}\right)\frac{14}{15}$

12) $\left(\frac{2}{9} \times \frac{3}{4}\right)\frac{9}{16}$

13) Multiply $\frac{1}{4} \times \frac{3}{8} \times \frac{12}{25}$.

14) Multiply $\frac{2}{3} \times \frac{4}{5} \times \frac{15}{48}$.

15) Multiply $\frac{1}{2} \times \frac{3}{5} \times \frac{5}{6}$.

16) Multiply $\frac{4}{9} \times \frac{3}{8} \times \frac{1}{3}$.

17) Multiply $\frac{3}{4} \times \frac{5}{6} \times \frac{8}{9}$.

18) $\frac{4}{5}\left(\frac{1}{2} \times \frac{6}{7}\right)$

19) $\frac{5}{8}\left(\frac{3}{4} \times \frac{2}{3}\right)$

20) $\frac{13}{15}\left(\frac{5}{26} \times \frac{2}{3}\right)$

21) $\frac{9}{10}\left(\frac{5}{6} \times \frac{1}{4}\right)$

22) $\frac{2}{5}\left(\frac{3}{7} \times \frac{7}{10}\right)$

23) $\left(\frac{5}{9} \times \frac{3}{4}\right)\frac{2}{5}$

24) $\left(\frac{5}{12} \times \frac{9}{16}\right)\frac{24}{35}$

3.7 Multiplying with Exponents

3.7a Squares

In Chapter 2 we talked about exponents, which are a shorthand way of indicating repeated multiplication. What are we supposed to do with this expression?

$$\left(\frac{2}{3}\right)^2$$

We need to square this fraction, or multiply it by itself:

$$\left(\frac{2}{3}\right)^2 = \frac{2}{3} \cdot \frac{2}{3} = \frac{4}{9}$$

Notice that we multiply the numerator by itself and then multiply the denominator by itself. See if you can do this one:

Example 1: $\left(\frac{4}{5}\right)^2$

SOLUTION: $\left(\frac{4}{5}\right)^2 = \frac{4}{5} \times \frac{4}{5} = \frac{16}{25}$

Here's one more:

Example 2: $\left(\frac{1}{7}\right)^2$

SOLUTION: $\left(\frac{1}{7}\right)^2 = \frac{1}{7} \times \frac{1}{7} = \frac{1}{49}$

Let's add a wrinkle. See what you can do with this one:

Example 3: $\frac{1}{5}\left(\frac{3}{8}\right)^2$

SOLUTION: $\frac{1}{5} \cdot \frac{3}{8} \cdot \frac{3}{8} = \frac{9}{320}$

Sometimes you can take a shortcut which can save you a lot of work. Let's consider this problem:

Example 4: $\left(\frac{9}{18}\right)^2$

SOLUTION: Without taking a shortcut, here's how we would simplify it:

$$\left(\frac{9}{18}\right)^1 = \frac{81}{324} = \frac{9}{36} = \frac{1}{4} \quad \text{or} \quad \left(\frac{9}{18}\right)^2 = \frac{\overset{1}{\cancel{9}}}{\underset{1}{\cancel{18}}} \cdot \frac{\overset{1}{\cancel{9}}}{\underset{1}{\cancel{18}}} = \frac{1}{4}$$

Here's the shortcut method:

$$\left(\frac{9}{18}\right)^2 = \left(\frac{1}{2}\right)^2 = \frac{1}{2} \times \frac{1}{2} = \frac{1}{4} \quad \text{Reduce first}$$

Always look to see if you can reduce the fraction inside the parentheses before you do the next operation. Here's another problem that lends itself to the shortcut method:

Example 5: $\left(\dfrac{5}{20}\right)^2 = \dfrac{25}{400} = \dfrac{5}{80} = \dfrac{1}{16}$

SOLUTION: $\left(\dfrac{5}{20}\right)^2 = \left(\dfrac{1}{4}\right)^2 = \dfrac{1}{4} \times \dfrac{1}{4} = \dfrac{1}{16}$

Example 6: $\left(\dfrac{5}{6}\right)^2 \dfrac{6}{25} = \,?$

SOLUTION: We'll use two methods to simplify this problem. The first method will multiply first, then reduce.

$$\left(\dfrac{5}{6}\right)^2 \dfrac{6}{25} = \left(\dfrac{5}{6} \cdot \dfrac{5}{6}\right) \dfrac{6}{25} = \dfrac{\overset{1}{\cancel{25}}}{\underset{6}{\cancel{36}}} \cdot \dfrac{\overset{1}{\cancel{6}}}{\underset{1}{\cancel{25}}} = \dfrac{1}{6}$$

The second method will reduce first, then multiply. We prefer the second method.

$$\dfrac{\overset{1}{\cancel{5}}}{6} \cdot \dfrac{\overset{1}{\cancel{5}}}{\underset{1}{\cancel{6}}} \cdot \dfrac{\overset{1}{\cancel{6}}}{\underset{\underset{1}{5}}{\cancel{25}}} = \dfrac{1}{6}$$

Problem Set 3.7a:

Work out the multiplication in each problem.

1) $\left(\dfrac{5}{8}\right)^2$

2) $\left(\dfrac{7}{12}\right)^2$

3) $\left(\dfrac{1}{9}\right)^2$

4) $\left(\dfrac{6}{13}\right)^2$

5) $\dfrac{2}{5}\left(\dfrac{5}{12}\right)^2$

6) $\dfrac{3}{4}\left(\dfrac{2}{9}\right)^2$

7) $\dfrac{1}{2}\left(\dfrac{6}{11}\right)^2$

8) $\dfrac{4}{9}\left(\dfrac{3}{10}\right)^2$

9) $\left(\dfrac{5}{9}\right)^2 \dfrac{3}{20}$

10) $\dfrac{5}{14}\left(\dfrac{7}{14}\right)^2$

11) $\left(\dfrac{2}{4}\right)^2 \left(\dfrac{4}{5}\right)$

12) $\dfrac{16}{17}\left(\dfrac{3}{12}\right)^2$

SOLUTIONS TO PROBLEM SET:

1) $\left(\dfrac{5}{8}\right)^2 = \dfrac{5}{8} \cdot \dfrac{5}{8} = \dfrac{25}{64}$

2) $\left(\dfrac{7}{12}\right)^2 = \dfrac{7}{12} \cdot \dfrac{7}{12} = \dfrac{49}{144}$

3) $\left(\dfrac{1}{9}\right)^2 = \dfrac{1}{9} \cdot \dfrac{1}{9} = \dfrac{1}{81}$

4) $\left(\dfrac{6}{13}\right)^2 = \dfrac{6}{13} \cdot \dfrac{6}{13} = \dfrac{36}{169}$

5) $\dfrac{2}{5}\left(\dfrac{5}{12}\right)^2 = \dfrac{\cancel{2}^{\,1}}{\cancel{5}_{\,1}} \cdot \dfrac{\cancel{5}^{\,1}}{\cancel{12}_{\,6}} \cdot \dfrac{5}{12} = \dfrac{5}{72}$

6) $\dfrac{3}{4}\left(\dfrac{2}{9}\right)^2 = \dfrac{\cancel{3}^{\,1}}{\cancel{4}_{\,2}} \cdot \dfrac{\cancel{2}^{\,1}}{\cancel{9}_{\,3}} \cdot \dfrac{2}{9} = \dfrac{1}{27}$

7) $\dfrac{1}{2}\left(\dfrac{6}{11}\right)^2 = \dfrac{1}{\cancel{2}_{\,1}} \cdot \dfrac{\cancel{6}^{\,3}}{11} \cdot \dfrac{6}{11} = \dfrac{18}{121}$

8) $\dfrac{4}{9}\left(\dfrac{3}{10}\right)^2 = \dfrac{\cancel{4}^{\,2}}{\cancel{9}_{\,3}}\cdot\dfrac{\cancel{3}^{\,1}}{\cancel{10}_{\,5}}\cdot\dfrac{\cancel{3}^{\,1}}{\cancel{10}_{\,5}} = \dfrac{1}{25}$

9) $\left(\dfrac{5}{9}\right)^2\dfrac{3}{20} = \dfrac{5}{\cancel{9}_{\,3}} \cdot \dfrac{5}{9} \cdot \dfrac{\cancel{3}^{\,1}}{\cancel{20}_{\,4}} = \dfrac{5}{108}$

10) $\dfrac{5}{14}\left(\dfrac{7}{14}\right)^2 = \dfrac{5}{14}\left(\dfrac{1}{2}\right)^2 = \dfrac{5}{14} \times \dfrac{1}{2} \times \dfrac{1}{2} = \dfrac{5}{56}$

11) $\left(\dfrac{2}{4}\right)^2\left(\dfrac{4}{5}\right) = \left(\dfrac{1}{2}\right)^2\left(\dfrac{4}{5}\right) = \left(\dfrac{1}{\cancel{2}_{\,1}}\right)\left(\dfrac{1}{\cancel{2}_{\,1}}\right)\left(\dfrac{\cancel{4}^{\,2}}{5}\right) = \dfrac{1}{5}$

12) $\dfrac{16}{17}\left(\dfrac{3}{12}\right)^2 = \dfrac{16}{17}\left(\dfrac{1}{4}\right)^2 = \dfrac{\cancel{16}^{\,4}}{17}\left(\dfrac{1}{\cancel{4}_{\,1}}\right)\left(\dfrac{1}{\cancel{4}_{\,1}}\right) = \dfrac{1}{17}$

Work out the multiplication in each problem.

1) $\left(\dfrac{2}{5}\right)^2$

2) $\left(\dfrac{4}{9}\right)^2$

3) $\left(\dfrac{5}{8}\right)^2$

4) $\left(\dfrac{1}{2}\right)^2$

5) $\dfrac{1}{4}\left(\dfrac{2}{7}\right)^2$

6) $\dfrac{3}{5}\left(\dfrac{2}{3}\right)^2$

7) $\dfrac{2}{7}\left(\dfrac{7}{10}\right)^2$

8) $\dfrac{3}{4}\left(\dfrac{6}{11}\right)^2$

9) $\left(\dfrac{5}{6}\right)^2\dfrac{1}{2}$

10) $\left(\dfrac{3}{10}\right)^2\dfrac{8}{9}$

11) $\left(\dfrac{4}{7}\right)^2\dfrac{7}{10}$

12) $\left(\dfrac{3}{4}\right)^2\dfrac{16}{27}$

13) $\left(\dfrac{1}{3}\right)^2$

14) $\left(\dfrac{2}{5}\right)^2$

15) $\left(\dfrac{5}{9}\right)^2$

16) $\left(\dfrac{2}{7}\right)^2$

17) $\dfrac{1}{2}\left(\dfrac{2}{3}\right)^2$

18) $\dfrac{3}{4}\left(\dfrac{2}{5}\right)^2$

19) $\dfrac{5}{8}\left(\dfrac{4}{9}\right)^2$

20) $\dfrac{1}{5}\left(\dfrac{1}{6}\right)^2$

21) $\dfrac{2}{9}\left(\dfrac{3}{4}\right)^2$

22) $\left(\dfrac{3}{10}\right)^2\dfrac{5}{6}$

23) $\left(\dfrac{5}{7}\right)^2\dfrac{14}{15}$

24) $\left(\dfrac{5}{6}\right)^2\dfrac{3}{10}$

25) $\dfrac{3}{5}\left(\dfrac{4}{6}\right)^2$

26) $\dfrac{2}{7}\left(\dfrac{7}{14}\right)^2$

27) $\left(\dfrac{2}{6}\right)^2\dfrac{36}{40}$

28) $\left(\dfrac{10}{15}\right)^2\dfrac{4}{6}$

3.7b Cubes

In the last section we squared fractions. That is, we multiplied fractions by themselves. Cubing a fraction means multiplying it by itself twice. For example, $\left(\frac{1}{2}\right)^3 = \frac{1}{2} \times \frac{1}{2} \times \frac{1}{2}$. So what is $\left(\frac{1}{2}\right)^3$ equal to?

SOLUTION: $\left(\frac{1}{2}\right)^3 = \frac{1}{2} \times \frac{1}{2} \times \frac{1}{2} = \frac{1}{8}$

Example 1: What is $\left(\frac{2}{3}\right)^3$ equal to?

SOLUTION: $\left(\frac{2}{3}\right)^3 = \frac{2}{3} \times \frac{2}{3} \times \frac{2}{3} = \frac{8}{27}$

Now do this problem:

Example 2: $\left(\frac{3}{5}\right)^3$

SOLUTION: $\left(\frac{3}{5}\right)^3 = \frac{3}{5} \times \frac{3}{5} \times \frac{3}{5} = \frac{27}{125}$

Let's add another wrinkle:

Example 3: What is $\frac{3}{10}\left(\frac{5}{6}\right)^3$ equal to?

SOLUTION: $\frac{3}{10}\left(\frac{5}{6}\right)^3 = \frac{3}{10} \times \left(\frac{5}{6} \times \frac{5}{6} \times \frac{5}{6}\right) = \overset{1}{\underset{2}{\cancel{\frac{3}{10}}}} \times \overset{25}{\underset{72}{\cancel{\frac{125}{216}}}} = \frac{25}{144}$

Example 4: What is $\left(\frac{3}{4}\right)^3 \frac{2}{9}$?

SOLUTION: $\left(\frac{3}{4}\right)^3 \frac{2}{9} = \left(\frac{3}{4} \times \frac{3}{4} \times \frac{3}{4}\right)\frac{2}{9} = \overset{3}{\underset{32}{\cancel{\frac{27}{64}}}} \times \overset{1}{\underset{1}{\cancel{\frac{2}{9}}}} = \frac{3}{32}$

Here's one more:

Example 5: What is $\frac{1}{2}\left(\frac{12}{16}\right)^3$?

SOLUTION: $\frac{1}{2}\left(\frac{12}{16}\right)^3 = \frac{1}{2}\left(\frac{3}{4}\right)^3 = \left(\frac{1}{2} \times \frac{3}{4} \times \frac{3}{4} \times \frac{3}{4}\right) = \frac{27}{128}$

Problem Set 3.7b:

Work out the multiplication in each problem.

1) $\left(\dfrac{1}{4}\right)^3$

2) $\left(\dfrac{5}{9}\right)^3$

3) $\left(\dfrac{24}{36}\right)^3$

4) $\left(\dfrac{3}{7}\right)^3$

5) $\dfrac{1}{2}\left(\dfrac{4}{9}\right)^3$

6) $\dfrac{7}{8}\left(\dfrac{2}{5}\right)^3$

7) $\dfrac{5}{7}\left(\dfrac{7}{10}\right)^3$

8) $\dfrac{3}{15}\left(\dfrac{5}{12}\right)^3$

9) $\left(\dfrac{2}{3}\right)^3\dfrac{9}{10}$

10) $\left(\dfrac{3}{10}\right)^3\dfrac{25}{36}$

SOLUTIONS TO PROBLEM SET:

1) $\left(\dfrac{1}{4}\right)^3 = \dfrac{1}{4} \times \dfrac{1}{4} \times \dfrac{1}{4} = \dfrac{1}{64}$

2) $\left(\dfrac{5}{9}\right)^3 = \dfrac{5}{9} \times \dfrac{5}{9} \times \dfrac{5}{9} = \dfrac{125}{729}$

3) $\left(\dfrac{24}{36}\right)^3 = \left(\dfrac{2}{3}\right)^3 = \dfrac{2}{3} \times \dfrac{2}{3} \times \dfrac{2}{3} = \dfrac{8}{27}$

4) $\left(\dfrac{3}{7}\right)^3 = \dfrac{3}{7} \times \dfrac{3}{7} \times \dfrac{3}{7} = \dfrac{27}{343}$

5) $\dfrac{1}{2}\left(\dfrac{4}{9}\right)^3 = \dfrac{1}{\cancel{2}_1} \cdot \dfrac{\cancel{4}^2}{9} \cdot \dfrac{4}{9} \cdot \dfrac{4}{9} = \dfrac{32}{729}$

6) $\dfrac{7}{8}\left(\dfrac{2}{5}\right)^3 = \dfrac{7}{\cancel{8}} \cdot \dfrac{\cancel{2}}{5} \cdot \dfrac{\cancel{2}}{5} \cdot \dfrac{\cancel{2}}{5} = \dfrac{7}{125}$

7) $\dfrac{5}{7}\left(\dfrac{7}{10}\right)^3 = \dfrac{5}{7} \times \dfrac{7}{10} \cdot \dfrac{7}{10} \cdot \dfrac{7}{10} = \dfrac{49}{200}$

8) $\dfrac{3}{15}\left(\dfrac{5}{12}\right)^3 = \dfrac{3}{15} \times \dfrac{5}{12} \times \dfrac{5}{12} \times \dfrac{5}{12} = \dfrac{25}{1728}$

9) $\left(\dfrac{2}{3}\right)^3 \dfrac{9}{10} = \dfrac{2}{3} \times \dfrac{2}{3} \times \dfrac{2}{3} \times \dfrac{9}{10} = \dfrac{4}{15}$

10) $\left(\dfrac{3}{10}\right)^3 \dfrac{25}{36} = \dfrac{3}{10} \cdot \dfrac{3}{10} \cdot \dfrac{3}{10} \cdot \dfrac{25}{36} = \dfrac{3}{160}$

Work out the multiplication in each problem.

1) $\left(\dfrac{1}{3}\right)^3$

2) $\left(\dfrac{4}{9}\right)^3$

3) $\left(\dfrac{3}{10}\right)^3$

4) $\left(\dfrac{20}{50}\right)^3$

5) $\left(\dfrac{1}{6}\right)^3$

6) $\dfrac{1}{2}\left(\dfrac{2}{3}\right)^3$

7) $\dfrac{3}{5}\left(\dfrac{5}{6}\right)^3$

8) $\dfrac{3}{4}\left(\dfrac{2}{9}\right)^3$

9) $\left(\dfrac{5}{6}\right)\left(\dfrac{3}{4}\right)^3$

10) $\dfrac{7}{10}\left(\dfrac{5}{7}\right)^3$

11) $\left(\dfrac{5}{12}\right)^3\dfrac{24}{35}$

12) $\left(\dfrac{9}{10}\right)^3\dfrac{100}{243}$

13) $\left(\dfrac{1}{5}\right)^3$

14) $\left(\dfrac{2}{9}\right)^3$

15) $\left(\dfrac{4}{5}\right)^3$

16) $\left(\dfrac{2}{11}\right)^3$

17) $\left(\dfrac{5}{10}\right)^3$

18) $\dfrac{4}{5}\left(\dfrac{5}{8}\right)^3$

19) $\dfrac{3}{4}\left(\dfrac{2}{3}\right)^3$

20) $\dfrac{1}{2}\left(\dfrac{4}{12}\right)^3$

21) $\dfrac{15}{28}\left(\dfrac{2}{5}\right)^3$

22) $\dfrac{27}{40}\left(\dfrac{2}{9}\right)^3$

23) $\left(\dfrac{3}{8}\right)^3\dfrac{64}{81}$

24) $\left(\dfrac{5}{6}\right)^3\dfrac{81}{125}$

3.7c   Roots of Fractions

Now that you have the hang of squaring and cubing a fraction, let's look at the inverse of raising a fraction to a power. In Chapter 2, we worked on roots of whole numbers. In this section we'll apply the same concept to roots of fractions.

Example 1:   Find the square root of $\frac{4}{25}$.

SOLUTION: This problem asks us, what fraction times itself is $\frac{4}{25}$. The square root of 4 is 2 and the square root of 25 is 5.

$$\sqrt{\frac{4}{25}} = \frac{\sqrt{4}}{\sqrt{25}} = \frac{2}{5}$$

Example 2:   Find $\sqrt{\frac{1}{121}}$.

SOLUTION: This problem asks us, what fraction times itself is $\frac{1}{121}$. The square root of 1 is 1 and the square root of 121 is 11.

$$\sqrt{\frac{1}{121}} = \frac{\sqrt{1}}{\sqrt{121}} = \frac{1}{11}$$

Example 3:   Find the cube root of $\frac{8}{27}$.

SOLUTION: The last two problems asked for the square root; this one asks for the cube root. Now we want to know what fraction times itself times itself is $\frac{8}{27}$.

$$\sqrt[3]{\frac{8}{27}} = \frac{\sqrt[3]{8}}{\sqrt[3]{27}} = \frac{2}{3}$$

The cube root of 8 is 2 and the cube root of 27 is 3.

Problem Set 3.7c:

Find the roots of the following problems.

1) $\sqrt{\frac{81}{100}}$

2) $\sqrt{\frac{64}{121}}$

3) $\sqrt[3]{\frac{1}{1000}}$

4) $\sqrt[3]{\frac{27}{125}}$

SOLUTIONS TO PROBLEM SET:

1) $\frac{9}{10}$ 2) $\frac{8}{11}$ 3) $\frac{1}{10}$ 4) $\frac{3}{5}$

Now let's look at some more challenging problems.

Example 4:   Find the fourth root of $\frac{16}{625}$.

SOLUTION: When working on problems with larger roots, it can be helpful to write the prime factorization of the numerator and denominator.

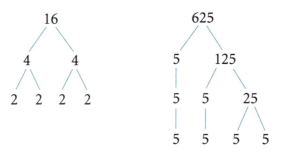

Now it's easy to see that $\sqrt[4]{\frac{16}{625}} = \frac{2}{5}$.

Example 5:   Find $\sqrt{\frac{100}{225}}$

SOLUTION: As is true in other problems we've done with fractions, it's usually best to reduce first.

$$\sqrt{\frac{100}{225}} = \sqrt{\frac{20}{45}} = \sqrt{\frac{4}{9}} = \frac{2}{3}$$

Simplify the following radicals.

1) $\sqrt{\dfrac{1}{4}}$

2) $\sqrt{\dfrac{1}{16}}$

3) $\sqrt{\dfrac{1}{9}}$

4) $\sqrt{\dfrac{1}{81}}$

5) $\sqrt{\dfrac{4}{25}}$

6) $\sqrt{\dfrac{9}{49}}$

7) $\sqrt{\dfrac{25}{36}}$

8) $\sqrt{\dfrac{81}{100}}$

9) $\sqrt{\dfrac{64}{121}}$

10) $\sqrt{\dfrac{169}{625}}$

11) $\sqrt[3]{\dfrac{1}{8}}$

12) $\sqrt[3]{\dfrac{1}{27}}$

13) $\sqrt[3]{\dfrac{27}{64}}$

14) $\sqrt[3]{\dfrac{64}{125}}$

15) $\sqrt[4]{\dfrac{1}{16}}$

16) $\sqrt[4]{\dfrac{16}{625}}$

17) $\sqrt{\dfrac{12}{75}}$

18) $\sqrt{\dfrac{8}{50}}$

19) $\sqrt{\dfrac{32}{98}}$

20) $\sqrt{\dfrac{50}{72}}$

21) $\sqrt[3]{\dfrac{16}{54}}$

22) $\sqrt[3]{\dfrac{54}{128}}$

3.8   Division with Fractions

Suppose I had a half a pie I wanted to share with two friends. And don't forget about me; I want a piece of the pie too. What I want to do is divide a half a pie into three equal pieces. The way to do that is to divide $\frac{1}{2}$ by 3. It would look like this: $\frac{1}{2} \div \frac{3}{1}$. The first thing I would have to do is multiply by the reciprocal of the divisor—which translates to English as flip the second fraction. The fraction before the ÷ is called the *dividend* and the fraction that's after the ÷ is called the *divisor*. These terms will be used from time to time in this course, and will be used often in the *Beginning Algebra* course that follows this course—so get used to hearing them. Division involving fractions actually becomes a multiplication problem; $\frac{1}{2} \div \frac{3}{1}$ becomes $\frac{1}{2} \times \frac{1}{3} = \frac{1}{6}$, so each piece of the pie is $\frac{1}{6}$ of the whole pie.

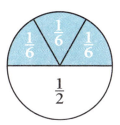

Notice that when we changed the division problem to a multiplication problem, the 3 became $\frac{1}{3}$ when we flipped it. Remember, any whole number can be written as itself over 1, so 3 can be written as $\frac{3}{1}$. The *reciprocal* of 3 is $\frac{1}{3}$. The reciprocal of any fraction is the fraction flipped upside down. The reciprocal for $\frac{2}{3}$ is $\frac{3}{2}$. The reciprocal for $\frac{2}{5}$ is $\frac{5}{2}$. Any number times its reciprocal is 1.

$$\left(\frac{\cancel{2}}{\cancel{3}}\right)\left(\frac{\cancel{3}}{\cancel{2}}\right) = 1 \qquad \left(\frac{\cancel{2}}{\cancel{5}}\right)\left(\frac{\cancel{5}}{\cancel{2}}\right) = 1$$

We think we owe you an explanation of why we can just say, flip the fraction that follows the ÷ and multiply.

We can write $\frac{1}{2} \div \frac{3}{1}$ in vertical form, $\dfrac{\frac{1}{2}}{\frac{3}{1}}$. When division of fractions is written in vertical form, it's called a complex fraction.

We'll begin by multiplying the numerator and the denominator of the fraction by $\frac{1}{3}$, which is the reciprocal of the denominator. (Remember any number except 0, divided by itself is 1). This will simplify the fraction by making the denominator of the fraction 1 (any number times its reciprocal is 1).

If we multiply the denominator by $\frac{1}{3}$, we also have to multiply the numerator by $\frac{1}{3}$. That way we're actually multiplying the fraction by 1. Remember that a number multiplied by 1 or divided by 1 doesn't change its value.

$$\dfrac{\frac{1}{2}}{\frac{3}{1}}\left(\dfrac{\frac{1}{3}}{\frac{1}{3}}\right) = \dfrac{\frac{1}{6}}{1} = \frac{1}{6}$$

Now we're going to show you four slightly different types of division problems. After each of them, we'll give you a couple of others to work out yourself.

Type 1:   **Example 1:** $\dfrac{2}{5} \div \dfrac{4}{5}$

SOLUTION: $\dfrac{\overset{1}{\cancel{2}}}{\underset{1}{\cancel{5}}}\left(\dfrac{\overset{1}{\cancel{5}}}{\underset{2}{\cancel{4}}}\right) = \dfrac{1}{2}$ When we divide, we have to flip the second fraction, so $\dfrac{2}{5} \div \dfrac{4}{5}$ becomes $\dfrac{2}{5}\left(\dfrac{5}{4}\right)$.

Next reduce the common factors: $\dfrac{\overset{1}{\cancel{2}}}{\underset{1}{\cancel{5}}}\left(\dfrac{\overset{1}{\cancel{5}}}{\underset{2}{\cancel{4}}}\right)$.

then multiply $\dfrac{\overset{1}{\cancel{2}}}{\underset{1}{\cancel{5}}}\left(\dfrac{\overset{1}{\cancel{5}}}{\underset{2}{\cancel{4}}}\right) = \dfrac{1}{2}$.

Example 2: $\dfrac{3}{4} \div \dfrac{5}{6}$

SOLUTION: $\dfrac{3}{\underset{2}{\cancel{4}}}\left(\dfrac{\overset{3}{\cancel{6}}}{5}\right) = \dfrac{9}{10}$ Multiply by the reciprocal of the division.

Example 3: $\dfrac{1}{8} \div \dfrac{7}{9}$

SOLUTION: $\dfrac{1}{8} \times \dfrac{9}{7} = \dfrac{9}{56}$ Multiply by the reciprocal of the division.

Type 2:   **Example 4:** $\dfrac{6}{15} \div \dfrac{15}{20}$

SOLUTION: $\dfrac{\overset{2}{\cancel{6}}}{\underset{3}{\cancel{15}}}\left(\dfrac{\overset{4}{\cancel{20}}}{\underset{5}{\cancel{15}}}\right) = \dfrac{8}{15}$ It's tempting to reduce the 15's in this problem, but we can reduce only diagonally in multiplication, NOT in addition, subtraction or division. So first write $\dfrac{6}{15} \div \dfrac{15}{20}$ as $\dfrac{6}{15}\left(\dfrac{20}{15}\right)$; then you can reduce diagonally and multiply:

$$\dfrac{\overset{2}{\cancel{6}}}{\underset{3}{\cancel{15}}}\left(\dfrac{\overset{4}{\cancel{20}}}{\underset{5}{\cancel{15}}}\right) = \dfrac{8}{15}$$

Example 5: $\dfrac{3}{10} \div \dfrac{5}{8}$

SOLUTION: $\dfrac{3}{\underset{5}{\cancel{10}}} \times \dfrac{\overset{4}{\cancel{8}}}{5} = \dfrac{12}{25}$

Example 6: $\dfrac{3}{8} \div \dfrac{1}{2}$

SOLUTION: $\dfrac{3}{\underset{4}{\cancel{8}}} \times \dfrac{\overset{1}{\cancel{2}}}{1} = \dfrac{3}{4}$

Type 3:   **Example 7:** $\frac{3}{9} \div \frac{20}{40}$

SOLUTION: $\frac{1}{3} \div \frac{1}{2}$ Notice that both fractions can be reduced: $\frac{3}{9} = \frac{1}{3}$ and $\frac{20}{40} = \frac{1}{2}$, before we divide. After we reduced the fractions, we multiplied to get the answer:

$\frac{1}{3}\left(\frac{2}{1}\right) = \frac{2}{3}$

$\frac{1}{3}\left(\frac{2}{1}\right) = \frac{2}{3}.$

Example 8: $\frac{10}{20} \div \frac{6}{8}$

SOLUTION: $\frac{1}{2} \div \frac{3}{4} = \frac{1}{2} \cdot \frac{\overset{2}{4}}{\underset{1}{3}} = \frac{2}{3}$

Example 9: $\frac{10}{15} \div \frac{16}{20}$

SOLUTION: $\frac{2}{3} \div \frac{4}{5} = \frac{2}{3} \cdot \frac{\overset{1}{5}}{\underset{2}{4}} = \frac{5}{6}$ We reduced first.

Type 4:   **Example 10:** $\frac{3}{20} \div \frac{2}{5} \div \frac{6}{12}$

SOLUTION: $\frac{\underset{4}{3}}{20}\left(\frac{\overset{1}{5}}{\underset{1}{2}}\right)\left(\frac{\overset{6}{12}}{\underset{1}{6}}\right) = \frac{3}{4}$

This problem involves division of three fractions. We can do this in one step by flipping the fractions that follow the $\div$ symbol, reducing, and multiplying.

Example 11: $\frac{1}{8} \div \frac{3}{4} \div \frac{2}{3}$

SOLUTION: $\frac{1}{8} \cdot \frac{\overset{2}{4}}{\underset{1}{3}} \cdot \frac{\overset{1}{3}}{\underset{1}{2}} = \frac{2}{8} = \frac{1}{4}$

Example 12: $\frac{2}{7} \div \frac{1}{2} \div \frac{13}{14}$

SOLUTION: $\frac{2}{7} \cdot \frac{2}{1} \cdot \frac{\overset{2}{14}}{13} = \frac{8}{13}$

Are you following all of this? If you are getting these practice problems right, go straight to the problem set that follows. But if you'd like some extra help, please work your way through the accompanying box.

Extra Help Dividing Fractions

The whole trick to dividing fractions is to convert each problem into a multiplication problem.

Let's start with this problem: $\frac{1}{5} \div \frac{1}{2} = ?$

We convert this into a multiplication problem by flipping the fraction following the division sign: $\frac{1}{5} \times \frac{2}{1}$. Then we simplify the problem: $\frac{1}{5} \times \frac{2}{1} = \frac{2}{5}$.

Next problem: $\frac{6}{20} \div \frac{20}{30}$.

First, let's reduce both fractions to their lowest terms: $\frac{3}{10} \div \frac{2}{3}$. Next we convert it into a multiplication problem by flipping the fraction after the division sign: $\frac{3}{10} \times \frac{3}{2}$. Then we carry out the multiplication: $\frac{3}{10} \times \frac{3}{2} = \frac{9}{20}$.

Here's another problem: $\frac{1}{6} \div \frac{2}{3} \div \frac{3}{4}$.

Here we need to flip *both* of the fractions that follow division signs, because division is performed twice: $\frac{1}{6} \times \frac{3}{2} \times \frac{4}{3}$.

And then we simplify the problem:
$$\frac{1}{6} \times \frac{3}{2} \times \frac{4}{3} = \frac{1}{3}$$

One last problem: $\frac{2}{9} \div \frac{1}{2} \div \frac{5}{8}$.

See if you can work it out yourself.

SOLUTION: $\frac{2}{9} \times \frac{2}{1} \times \frac{8}{5} = \frac{32}{45}$

Problem Set 3.8:

Perform the following division problems. Be sure to leave your answers in reduced form:

1) $\frac{1}{2} \div \frac{3}{5}$

2) $\frac{3}{4} \div \frac{4}{5}$

3) $\frac{5}{8} \div 4$

4) $\frac{2}{40} \div \frac{5}{25}$

5) $\frac{2}{3} \div 6$

6) $\frac{11}{50} \div \frac{3}{12} \div \frac{8}{9}$

7) $\frac{5}{8} \div 0$

8) $0 \div \frac{5}{8}$

9) $\frac{3}{5} \div \frac{25}{27}$

10) $\frac{2}{9} \div \frac{7}{14} \div \frac{3}{4}$

Let's take a closer look at problems 7 and 8. In problem 7 the divisor (the number that comes *after* the division sign), was 0. When we divide any fraction or whole number by 0, the quotient is undefined.

In problem 8, we have a dividend (the number that comes *before* the division sign) of 0. When 0 is divided by any fraction or whole number except 0, the quotient is 0. If you've forgotten about 0 and division, go back to Chapter 2, Section 3. Maybe the following example will help you to understand these concepts better.

- I have six bottles of beer to divide among 3 people. No problem; $\frac{6}{3} = 2$ bottles each.

- I have zero bottles of beer to divide among 3 people. No problem; $\frac{0}{3} = 0$ bottles each.

- I have six bottles of beer to divide among 0 people. The problem is not meaningful; therefore $\frac{6}{0}$ is undefined.

SOLUTIONS TO PROBLEM SET:

1) $\frac{1}{2} \div \frac{3}{5}$

$\frac{1}{2}\left(\frac{5}{3}\right) = \frac{5}{6}$

2) $\frac{3}{4} \div \frac{4}{5} = \frac{3}{4}\left(\frac{5}{4}\right) = \frac{15}{16}$

3) $\frac{5}{8} \div 4 = \frac{5}{8}\left(\frac{1}{4}\right) = \frac{5}{32}$

4) $\frac{2}{40} \div \frac{5}{25}$

$\frac{1}{20} \div \frac{1}{5}$

$\frac{1}{20}\left(\frac{\overset{1}{5}}{\underset{4}{1}}\right) = \frac{1}{4}$

5) $\frac{2}{3} \div 6 = \frac{\overset{1}{\cancel{2}}}{3}\left(\frac{1}{\underset{3}{\cancel{6}}}\right) = \frac{1}{9}$

6) $\frac{11}{50} \div \frac{3}{12} \div \frac{8}{9}$

$\frac{11}{50}\left(\frac{\overset{3}{\cancel{12}}}{\underset{1}{\cancel{3}}}\right)\left(\frac{\overset{3}{\cancel{9}}}{\underset{2}{\cancel{8}}}\right) = \frac{99}{100}$

7) $\frac{5}{8} \div 0$ is undefined

8) $0 \div \frac{5}{8} = 0$

9) $\frac{3}{5} \div \frac{25}{27}$

$\frac{3}{5}\left(\frac{27}{25}\right) = \frac{81}{125}$

10) $\frac{2}{9} \div \frac{7}{14} \div \frac{3}{4}$

$\frac{2}{9} \div \frac{1}{2} \div \frac{3}{4}$

$\frac{2}{9} \times \frac{2}{1} \times \frac{4}{3} = \frac{16}{27}$

Perform the following divisions. Write your answer in the lowest terms.

1) $\frac{2}{7} \div 3$

2) $\frac{25}{36} \div \frac{5}{6}$

3) $\frac{9}{16} \div \frac{3}{4}$

4) $\frac{1}{3} \div \frac{1}{2}$

5) $\frac{5}{8} \div \frac{15}{16}$

6) $\frac{27}{196} \div \frac{9}{392}$

7) $\frac{5}{32} \div \frac{10}{42}$

8) $4 \div \frac{1}{2}$

9) $0 \div \frac{11}{12}$

10) $\frac{11}{12} \div \frac{11}{24}$

11) $\frac{3}{8} \div \frac{9}{16}$

12) $\frac{10}{40} \div \frac{5}{8}$

13) $\frac{1}{4} \div 3$

14) $\frac{36}{48} \div \frac{5}{6}$

15) $\frac{9}{16} \div \frac{2}{3}$

16) $\frac{30}{40} \div 3$

17) $\frac{15}{20} \div 2 \div \frac{4}{5}$

18) $\frac{7}{8} \div 0$

19) $\frac{5}{15} \div \frac{1}{2} \div \frac{9}{10}$

20) $\frac{5}{9} \div \frac{3}{4} \div 2$

21) $\frac{81}{100} \div \frac{18}{20}$

22) $\frac{1}{12} \div \frac{5}{6} \div 2$

23) $\frac{5}{14} \div \frac{25}{29} \div 2$

24) $\frac{5}{16} \div \frac{500}{320}$

3.9　Addition and Subtraction of Fractions

3.9a　With Common Denominators

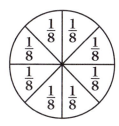

You'll notice that the circle above is divided into eight equal pieces. If we shade in $\frac{2}{8}$ and $\frac{1}{8}$ of the circle, $\frac{3}{8}$ of the circle has been shaded in: $\frac{2}{8} + \frac{1}{8} = \frac{3}{8}$

When adding the fractions, notice that we add the numerators—not the denominators. You need to remember to do two things when you add fractions that have a common denominator: add the numerators and reduce your answer to its lowest terms.

Example 1:　$\frac{1}{5} + \frac{3}{5}$

　　　　　SOLUTION:　　$\frac{1}{5} + \frac{3}{5} = \frac{4}{5}$

Example 2:　$\frac{1}{10} + \frac{3}{10}$

　　　　　SOLUTION:　　$\frac{1}{10} + \frac{5}{10} = \frac{6}{10} = \frac{3}{5}$

Example 3:　How much is $\frac{1}{15} + \frac{4}{15} + \frac{7}{15}$?

　　　　　SOLUTION:　　$\frac{1}{15} + \frac{4}{15} + \frac{7}{15} = \frac{12}{15} = \frac{4}{5}$

Are you ready to subtract fractions with common denominators?

Example 4: How much is $\frac{2}{7} - \frac{1}{7}$?

$$\frac{2}{7} \qquad - \qquad \frac{1}{7} \qquad = \qquad \frac{1}{7}$$

SOLUTION: The answer is $\frac{1}{7}$.

Example 5: $\frac{6}{14} - \frac{4}{14}$

SOLUTION: $\frac{6}{14} - \frac{4}{14} = \frac{2}{14} = \frac{1}{7}$

Example 6: $\frac{11}{20} - \frac{1}{20}$

SOLUTION: $\frac{11}{20} - \frac{1}{20} = \frac{10}{20} = \frac{1}{2}$

Problem Set 3.9a:

Add and simplify:

1) $\frac{1}{6} + \frac{3}{6}$

2) $\frac{5}{8} + \frac{1}{8}$

3) $\frac{2}{9} + \frac{4}{9}$

4) $\frac{2}{12} + \frac{8}{12}$

5) $\frac{2}{21} + \frac{4}{21} + \frac{8}{21}$

6) $\frac{4}{24} + \frac{5}{24} + \frac{9}{24}$

Subtract and simplify:

7) $\frac{7}{8} - \frac{3}{8}$

8) $\frac{5}{9} - \frac{2}{9}$

9) $\frac{3}{4} - \frac{1}{4}$

10) $\frac{9}{20} - \frac{3}{20}$

1) $\dfrac{1}{6} + \dfrac{3}{6} = \dfrac{4}{6} = \dfrac{2}{3}$

2) $\dfrac{5}{8} + \dfrac{1}{8} = \dfrac{6}{8} = \dfrac{3}{4}$

3) $\dfrac{2}{9} + \dfrac{4}{9} = \dfrac{6}{9} = \dfrac{2}{3}$

4) $\dfrac{2}{12} + \dfrac{8}{12} = \dfrac{10}{12} = \dfrac{5}{6}$

5) $\dfrac{2}{21} + \dfrac{4}{21} + \dfrac{8}{21} = \dfrac{14}{21} = \dfrac{2}{3}$

6) $\dfrac{4}{24} + \dfrac{5}{24} + \dfrac{9}{24} = \dfrac{18}{24} = \dfrac{3}{4}$

7) $\dfrac{7}{8} - \dfrac{3}{8} = \dfrac{4}{8} = \dfrac{1}{2}$

8) $\dfrac{5}{9} - \dfrac{2}{9} = \dfrac{3}{9} = \dfrac{1}{3}$

9) $\dfrac{3}{4} - \dfrac{1}{4} = \dfrac{2}{4} = \dfrac{1}{2}$

10) $\dfrac{9}{20} - \dfrac{3}{20} = \dfrac{6}{20} = \dfrac{3}{10}$

1) $\dfrac{4}{5} - \dfrac{1}{5}$

2) $\dfrac{3}{10} + \dfrac{2}{10}$

3) $\dfrac{5}{9} + \dfrac{1}{9}$

4) $\dfrac{7}{12} - \dfrac{1}{12}$

5) $\dfrac{1}{6} + \dfrac{4}{6}$

6) $\dfrac{14}{21} - \dfrac{7}{21}$

7) $\dfrac{17}{50} + \dfrac{8}{50}$

8) $\dfrac{4}{15} + \dfrac{2}{15}$

9) $\dfrac{1}{9} + \dfrac{2}{9} + \dfrac{3}{9}$

10) $\dfrac{17}{20} - \dfrac{7}{20}$

11) $\dfrac{2}{3} - \dfrac{1}{3}$

12) $\dfrac{1}{8} + \dfrac{2}{8} + \dfrac{3}{8}$

13) $\dfrac{15}{16} - \dfrac{5}{16}$

14) $\dfrac{14}{25} - \dfrac{3}{25}$

15) $\dfrac{9}{10} - \dfrac{1}{10}$

16) $\dfrac{19}{21} - \dfrac{5}{21}$

17) $\dfrac{4}{15} - \dfrac{6}{15}$

18) $\dfrac{1}{5} + \dfrac{1}{5}$

19) $\dfrac{9}{14} - \dfrac{2}{14}$

20) $\dfrac{1}{11} + \dfrac{3}{11} + \dfrac{5}{11}$

21) $\dfrac{19}{20} - \dfrac{4}{20}$

22) $\dfrac{3}{35} + \dfrac{2}{35} + \dfrac{2}{35}$

23) $\dfrac{3}{16} + \dfrac{5}{16}$

24) $\dfrac{2}{20} + \dfrac{5}{20} + \dfrac{7}{20}$

25) $\dfrac{1}{18} + \dfrac{8}{18}$

26) $\dfrac{2}{10} + \dfrac{3}{10}$

27) $\dfrac{2}{14} + \dfrac{5}{14}$

28) $\dfrac{1}{12} + \dfrac{2}{12} + \dfrac{3}{12}$

29) $\dfrac{7}{8} - \dfrac{3}{8}$

30) $\dfrac{5}{6} - \dfrac{1}{6}$

31) $\dfrac{1}{25} + \dfrac{3}{25} + \dfrac{16}{25}$

32) $\dfrac{11}{12} - \dfrac{3}{12}$

33) $\dfrac{2}{50} + \dfrac{9}{50} + \dfrac{14}{50}$

34) $\dfrac{8}{9} - \dfrac{2}{9}$

3.9b Addition and Subtraction of Fractions without Common Denominators

So far, we've added and subtracted fractions with common denominators. What do we do when the fractions we want to add or subtract don't have common denominators? In order to subtract these fractions, we must convert the fractions to the same denominator. It's always easiest to convert to the smallest or least common denominator (L.C.D.). The L.C.D. is the smallest number both denominators divide into with no remainder. For example, in the problem $\frac{1}{2} + \frac{1}{3}$, 6 is the smallest number both 2 and 3 divide into with no remainder, so we will use 6 as our L.C.D. for this problem. First we must use what we know about equivalent fractions to convert $\frac{1}{2}$ and $\frac{1}{3}$ to sixths. If we multiply the numerator and denominator of $\frac{1}{2}$ by 3 we get $\frac{1 \times 3}{2 \times 3} = \frac{3}{6}$. Now if we multiply the numerator and denominator of $\frac{1}{3}$ by 2 we get $\frac{1 \times 2}{3 \times 2} = \frac{2}{6}$. Now that the fractions have the same denominator it's easy to see that $\frac{3}{6} + \frac{2}{6} = \frac{5}{6}$.

Now let's look at some more examples of addition and subtraction with unlike denominators.

Example 1: $\dfrac{1}{10} + \dfrac{1}{25}$

SOLUTION:

$$\frac{5}{50} + \frac{2}{50} = \frac{7}{50}$$

The L.C.D. for 10 and 25 is 50. We convert $\frac{1}{10}$ to $\frac{1 \times 5}{10 \times 5} = \frac{5}{50}$ and $\frac{1}{25}$ to $\frac{1 \times 2}{25 \times 2} = \frac{2}{50}$, so $\frac{1}{10} + \frac{1}{25} = \frac{5}{50} + \frac{2}{50} = \frac{7}{50}$.

Example 2: $\dfrac{1}{3} + \dfrac{1}{15}$

SOLUTION:

$$\frac{5}{15} + \frac{1}{15} = \frac{6}{15} = \frac{2}{5}$$

The L.C.D. for 3 and 15 is 15. We only need to convert the $\frac{1}{3}$ to $\frac{5}{15}$ because the $\frac{1}{15}$ is already written as fifteenths, so $\frac{1}{3} + \frac{1}{15} = \frac{5}{15} + \frac{1}{15} = \frac{6}{15} = \frac{2}{5}$. Remember to reduce your answer.

Example 3: Add $\dfrac{1}{4}$ and $\dfrac{2}{3}$:

SOLUTION:

$$\frac{1 \times 3}{4 \times 3} + \frac{2 \times 4}{3 \times 4} = \frac{3}{12} + \frac{8}{12} = \frac{11}{12}$$

Example 4: Add $\dfrac{2}{5}$ and $\dfrac{1}{4}$.

SOLUTION:

$$\frac{2 \times 4}{5 \times 4} + \frac{1 \times 5}{4 \times 5} = \frac{8}{20} + \frac{5}{20} = \frac{13}{20}$$

Problem Set 3.9b • 1:

Add the fractions in each problem:

1) $\frac{1}{3} + \frac{1}{5}$

2) $\frac{3}{8} + \frac{1}{2}$

3) $\frac{3}{10} + \frac{2}{5}$

4) $\frac{1}{6} + \frac{3}{5}$

5) $\frac{1}{5} + \frac{2}{15}$

6) $\frac{2}{7} + \frac{5}{21}$

7) $\frac{1}{3} + \frac{4}{9}$

8) $\frac{1}{5} + \frac{3}{10}$

9) $\frac{1}{4} + \frac{3}{5}$

10) $\frac{1}{6} + \frac{5}{8}$

SOLUTIONS TO PROBLEM SET:

1) $\frac{1 \times 5}{3 \times 5} + \frac{1 \times 3}{5 \times 3} = \frac{5}{15} + \frac{3}{15} = \frac{8}{15}$

2) $\frac{3}{8} + \frac{1 \times 4}{2 \times 4} = \frac{3}{8} + \frac{4}{8} = \frac{7}{8}$

3) $\frac{3}{10} + \frac{2 \times 2}{5 \times 2} = \frac{3}{10} + \frac{4}{10} = \frac{7}{10}$

4) $\frac{1 \times 5}{6 \times 5} + \frac{3 \times 6}{5 \times 6} = \frac{5}{30} + \frac{18}{30} = \frac{23}{30}$

5) $\frac{1 \times 3}{5 \times 3} + \frac{2}{15} = \frac{3}{15} + \frac{2}{15} = \frac{5}{15} = \frac{1}{3}$

6) $\frac{2 \times 3}{7 \times 3} + \frac{5}{21} = \frac{6}{21} + \frac{5}{21} = \frac{11}{21}$

7) $\frac{1 \times 3}{3 \times 3} + \frac{4}{9} = \frac{3}{9} + \frac{4}{9} = \frac{7}{9}$

8) $\frac{1 \times 2}{5 \times 2} + \frac{3}{10} = \frac{2}{10} + \frac{3}{10} = \frac{5}{10} = \frac{1}{2}$

9) $\frac{1 \times 5}{4 \times 5} + \frac{3 \times 4}{5 \times 4} = \frac{5}{20} + \frac{12}{20} = \frac{17}{20}$

10) $\frac{1 \times 4}{6 \times 4} + \frac{5 \times 3}{8 \times 3} = \frac{4}{24} + \frac{15}{24} = \frac{19}{24}$

So far we've been adding two fractions that have different denominators. Let's see if you can add three fractions with different denominators.

Example 5: $\frac{1}{4} + \frac{1}{3} + \frac{1}{6}$

SOLUTION: $\frac{1 \times 3}{4 \times 3} + \frac{1 \times 4}{3 \times 4} + \frac{1 \times 2}{6 \times 2} = \frac{3}{12} + \frac{4}{12} + \frac{2}{12} = \frac{9}{12} = \frac{3}{4}$

Example 6: $\frac{2}{5} + \frac{1}{10} + \frac{3}{20}$

SOLUTION: $\frac{2 \times 4}{5 \times 4} + \frac{1 \times 2}{10 \times 2} + \frac{3}{20} = \frac{8}{20} + \frac{2}{20} + \frac{3}{20} = \frac{13}{20}$

Example 7: $\frac{1}{3} + \frac{1}{6} + \frac{4}{12}$

SOLUTION: $\frac{1 \times 2}{3 \times 2} + \frac{1}{6} + \frac{2}{6} = \frac{2}{6} + \frac{1}{6} + \frac{2}{6} = \frac{5}{6}$

We took a shortcut. We reduced $\frac{4}{12}$ to $\frac{2}{6}$. That way we were able to make our common denominator 6, instead of 12. That saved us a little time. Do you **have** to take shortcuts like that? No—but as you do more and more problems, you'll probably end up taking plenty of shortcuts.

Problem Set 3.9b • 2:

Add the fractions in each problem:

1) $\frac{1}{3} + \frac{1}{6} + \frac{2}{9}$

2) $\frac{1}{5} + \frac{3}{10} + \frac{6}{15}$

3) $\frac{1}{4} + \frac{3}{8} + \frac{2}{16}$

4) $\frac{1}{2} + \frac{1}{4} + \frac{1}{8}$

5) $\frac{2}{10} + \frac{3}{15} + \frac{1}{5}$

6) $\frac{1}{7} + \frac{1}{14} + \frac{2}{28}$

7) $\frac{1}{6} + \frac{1}{4} + \frac{5}{12}$

8) $\frac{3}{10} + \frac{2}{15} + \frac{13}{30}$

9) $\frac{2}{8} + \frac{1}{4} + \frac{3}{12}$

10) $\frac{1}{3} + \frac{2}{9} + \frac{6}{18}$

SOLUTIONS TO PROBLEM SET:

1) $\frac{1 \times 6}{3 \times 6} + \frac{1 \times 3}{6 \times 3} + \frac{2 \times 2}{9 \times 2} = \frac{6}{18} + \frac{3}{18} + \frac{4}{18} = \frac{13}{18}$

2) $\frac{1}{5} + \frac{3}{10} + \frac{2}{5} = \frac{3}{5} + \frac{3}{10} = \frac{3 \times 2}{5 \times 2} + \frac{3}{10} = \frac{6}{10} + \frac{3}{10} = \frac{9}{10}$

3) $\frac{1 \times 2}{4 \times 2} + \frac{3}{8} + \frac{1}{8} = \frac{2}{8} + \frac{3}{8} + \frac{1}{8} = \frac{6}{8} = \frac{3}{4}$

4) $\frac{1 \times 4}{2 \times 4} + \frac{1 \times 2}{4 \times 2} + \frac{1}{8} = \frac{4}{8} + \frac{2}{8} + \frac{1}{8} = \frac{7}{8}$

5) $\frac{2}{10} + \frac{3}{15} + \frac{1}{5} = \frac{1}{5} + \frac{1}{5} + \frac{1}{5} = \frac{3}{5}$

6) $\frac{1 \times 2}{7 \times 2} + \frac{1}{14} + \frac{1}{14} = \frac{2}{14} + \frac{1}{14} + \frac{1}{14} = \frac{4}{14} = \frac{2}{7}$

7) $\frac{1 \times 2}{6 \times 2} + \frac{1 \times 3}{4 \times 3} + \frac{5}{12} = \frac{2}{12} + \frac{3}{12} + \frac{5}{12} = \frac{10}{12} = \frac{5}{6}$

8) $\dfrac{3 \times 3}{10 \times 3} + \dfrac{2 \times 2}{15 \times 2} + \dfrac{13}{30} = \dfrac{9}{30} + \dfrac{4}{30} + \dfrac{13}{30} = \dfrac{26}{30} = \dfrac{13}{15}$

9) $\dfrac{1}{4} + \dfrac{1}{4} + \dfrac{1}{4} = \dfrac{3}{4}$

10) $\dfrac{1 \times 3}{3 \times 3} + \dfrac{2}{9} + \dfrac{3}{9} = \dfrac{3}{9} + \dfrac{2}{9} + \dfrac{3}{9} = \dfrac{8}{9}$

Sometimes it's not easy to find the Least Common Denominator (L.C.D.). When this happens, we will use a method that uses prime factorization. We'll find the L.C.D. for the following example by using prime factorization.

Example 8: $\dfrac{1}{18} + \dfrac{1}{21} + \dfrac{1}{14}$

SOLUTION:

First we write each denominator as a product of its primes.

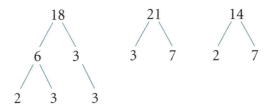

Notice the factors are 2, 3, and 7. To find the L.C.D., we take the greatest number of times a 2, 3 or 7 appears in any *ONE* of the denominators (*NOT* the total number of times a 2, 3 or 7 appears altogether). The greatest number of times a 2 appears in one of these denominators is once, a 7 is once and a 3 is twice so the L.C.D. is $2(3)(3)(7) = 126$.

Example 9:   Find the L.C.D. for $\dfrac{1}{8} + \dfrac{1}{10} + \dfrac{1}{25}$

SOLUTION:

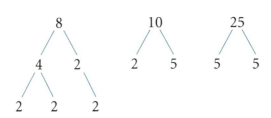

We have factors of 2 and 5. The greatest number of times a 2 appears in any one of these denominators is three times. The greatest number of times a 5 appears in any one of these denominators is twice, so the L.C.D. is $(2)(2)(2)(5)(5) = 200$.

Example 10: Find the L.C.D. for $\dfrac{1}{36} + \dfrac{1}{50}$

SOLUTION:

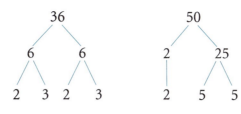

$$\text{L.C.D.} = 2(2)(3)(3)(5)(5) = 900$$

Example 11: Find the L.C.D. for $\dfrac{4}{36} + \dfrac{20}{40}$.

SOLUTION: Whenever possible, first reduce the fractions before you start looking for an L.C.D. It usually makes the problem much easier.

$\dfrac{4}{36} = \dfrac{1}{9}$, $\dfrac{20}{40} = \dfrac{1}{2}$; now the problem is $\dfrac{1}{9} + \dfrac{1}{2}$, and it's easy to see the L.C.D. is 18.

Example 12: Find the L.C.D. for $\dfrac{5}{6} + \dfrac{5}{100} + \dfrac{7}{147}$.

SOLUTION: First we'll reduce: $\dfrac{5}{100} = \dfrac{1}{20}$, $\dfrac{4}{147} = \dfrac{1}{21}$.

Now the problem looks like this: $\dfrac{5}{6} + \dfrac{1}{20} + \dfrac{1}{21}$.

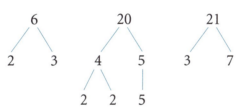

The L.C.D. is: $2(2)(3)(5)(7) = 420$

Remember always to reduce first. Now you try to find the L.C.D. for the following problem set.

Problem Set 3.9b • 3:

Find the L.C.D. for the following problems.

1) $\dfrac{1}{20} + \dfrac{3}{4}$

2) $\dfrac{1}{30} + \dfrac{4}{10}$

3) $\dfrac{1}{32} + \dfrac{4}{40}$

4) $\dfrac{1}{6} + \dfrac{1}{15}$

5) $\dfrac{10}{20} - \dfrac{1}{18}$

6) $\dfrac{2}{12} + \dfrac{3}{15} + \dfrac{5}{25}$

7) $\dfrac{5}{12} - \dfrac{7}{18}$ 8) $\dfrac{5}{18} + \dfrac{3}{14}$

9) $\dfrac{8}{25} - \dfrac{3}{20}$ 10) $\dfrac{3}{10} + \dfrac{5}{12} + \dfrac{1}{6}$

11) $\dfrac{10}{33} + \dfrac{3}{22} - \dfrac{13}{78}$ 12) $\dfrac{2}{21} + \dfrac{6}{108} + \dfrac{11}{105}$

SOLUTIONS TO PROBLEM SET:

1) 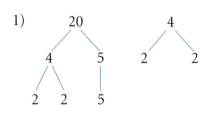   LCD = 2(2)(5) = 20

2) $\dfrac{4}{10} = \dfrac{2}{5}$ LCD = 3(2)(5) = 30

3) $\dfrac{4}{40} = \dfrac{1}{10}$ LCD = 2(2)(2)(2)(2)(5) = 160

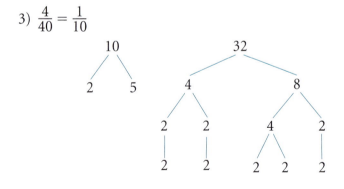

4)   LCD = 2(3)(5) = 30

5) $\frac{10}{20} = \frac{1}{2}$

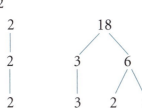

LCD = (3)(2)(3) = 18

6) $\frac{2}{12} = \frac{1}{6}, \frac{3}{15} = \frac{1}{5}, \frac{5}{25} = \frac{1}{5}$

LCD = 30

7)

LCD = (3)(3)(2)(2) = 36

8)

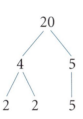

LCD = (3)(3)(2)(7) = 126

9)

LCD = (5)(5)(2)(2) = 100

10)

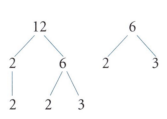

LCD = 2(2)(3)(5)

11) $\frac{13}{78} = \frac{1}{6}$

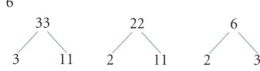

LCD = 3(11)(2) = 66

12) $\dfrac{6}{108} = \dfrac{1}{18}$ $LCD = 3(3)(7)(2)(5) = 630$

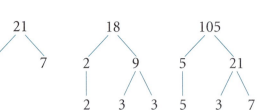

Now that you know how to find an L.C.D., we can do addition and subtraction of fractions. To perform an addition or subtraction problem, we follow this simple step-by-step procedure.

Step 1:   Reduce all fractions.

Step 2:   Find the L.C.D.

Step 3:   Convert all fractions to an equivalent fraction with the L.C.D.

Step 4:   Add or subtract the numerators NOT the denominators.

Step 5:   Reduce your answer.

Now let's go back to Example 8 from earlier in this section and finish the problem. We found the L.C.D. to be 126.

Example 8: $\dfrac{1}{18} + \dfrac{1}{21} + \dfrac{1}{14}$

SOLUTION: $\dfrac{1}{18} = \dfrac{7}{126}$ First we must convert each fraction to the L.C.D.

$+\dfrac{1}{21} = \dfrac{6}{126}$ $\dfrac{1}{18}\left(\dfrac{7}{7}\right) = \dfrac{7}{126}, \dfrac{1}{21}\left(\dfrac{6}{6}\right) = \dfrac{6}{126}, \dfrac{1}{14}\left(\dfrac{9}{9}\right) = \dfrac{9}{126}$

$+\dfrac{1}{14} = \dfrac{9}{126}$ Next add the fractions and reduce our answer to lowest terms.

$\overline{\phantom{+\dfrac{1}{14}}\;\dfrac{22}{126}} = \dfrac{11}{63}$

Now let's go back to Example 9 and finish the problem.

Example 9: $\dfrac{1}{8} + \dfrac{1}{10} + \dfrac{1}{25}$

SOLUTION: $\dfrac{1}{8} = \dfrac{25}{200}$ The L.C.D. is 200. Let's convert the fractions to the common denominator.

$+\dfrac{1}{10} = \dfrac{20}{200}$ $\dfrac{1}{8}\left(\dfrac{25}{25}\right) = \dfrac{25}{200}, \dfrac{1}{10}\left(\dfrac{20}{20}\right) = \dfrac{20}{200}, \dfrac{1}{25}\left(\dfrac{8}{8}\right) = \dfrac{8}{200}$

$+\dfrac{1}{25} = \dfrac{8}{200}$ Now add the fractions.

$\overline{\phantom{+\dfrac{1}{25}}\;\dfrac{53}{200}}$

Now let's go back to Example 10 and finish the problem.

Example 10: $\dfrac{1}{36} + \dfrac{1}{50}$

SOLUTION:

$$\begin{aligned} \frac{1}{36} &= \frac{25}{900} \\ +\frac{1}{50} &= \frac{18}{900} \\ \hline &\quad\frac{43}{900} \end{aligned}$$

First let's convert the fractions to the L.C.D. of 900.

$$\frac{1}{36}\left(\frac{25}{25}\right) = \frac{25}{900}, \frac{1}{50}\left(\frac{18}{18}\right) = \frac{18}{900}$$

Now add:

$$\frac{25}{900} + \frac{18}{900} = \frac{43}{900}$$

Now let's finish Example 11.

Example 11: $\dfrac{4}{36} + \dfrac{20}{40}$

SOLUTION:

$$\frac{4}{36} + \frac{20}{40} = \frac{1}{9} + \frac{1}{2}$$

$$\begin{aligned} \frac{1}{9} &= \frac{2}{18} \\ +\frac{1}{2} &= \frac{9}{18} \\ \hline &\quad\frac{11}{18} \end{aligned}$$

Convert the fraction to the L.C.D. of 18.

$$\frac{1}{9}\left(\frac{2}{2}\right) = \frac{2}{18} \qquad \frac{1}{2}\left(\frac{9}{9}\right) = \frac{9}{18}$$

Now add the fractions.

$$\frac{2}{18} + \frac{9}{18} = \frac{11}{18}$$

Problem Set 3.9b • 4:

Perform the following operations. Write your answers in reduced form.

1) $\dfrac{1}{20} + \dfrac{3}{4}$

2) $\dfrac{1}{30} + \dfrac{4}{10}$

3) $\dfrac{1}{32} + \dfrac{4}{40}$

4) $\dfrac{1}{6} + \dfrac{1}{5}$

5) $\dfrac{10}{20} - \dfrac{1}{18}$

6) $\dfrac{2}{12} + \dfrac{3}{15} + \dfrac{5}{25}$

7) $\dfrac{5}{12} - \dfrac{7}{18}$

8) $\dfrac{5}{18} + \dfrac{3}{14}$

9) $\dfrac{8}{25} - \dfrac{3}{20}$

10) $\dfrac{3}{10} + \dfrac{5}{12} + \dfrac{1}{6}$

SOLUTIONS TO PROBLEM SET:

1) $\dfrac{1}{20} + \dfrac{3}{4}$ LCD = 20

$$\frac{1}{20} + \frac{15}{20} = \frac{16}{20} = \frac{4}{5}$$

2) $\dfrac{1}{30} + \dfrac{4}{10}$ LCD = 30

$$\frac{1}{30} + \frac{2}{5}$$

$$\frac{1}{30} + \frac{12}{30} = \frac{13}{30}$$

3) $\dfrac{1}{32} + \dfrac{4}{40}$ LCD = 160

$\dfrac{1}{32} + \dfrac{1}{10}$

$\dfrac{1}{32} = \dfrac{5}{160}$

$+ \dfrac{1}{10} = \dfrac{16}{160}$

$\overline{\hspace{2cm}}$

$\dfrac{21}{160}$

4) $\dfrac{1}{6} + \dfrac{1}{5}$ LCD = 30

$\dfrac{1}{6} = \dfrac{5}{30}$

$+ \dfrac{1}{15} = \dfrac{2}{30}$

$\overline{\hspace{2cm}}$

$\dfrac{7}{30}$

5) $\dfrac{10}{20} - \dfrac{1}{18}$ LCD = 18

$\dfrac{1}{2} - \dfrac{1}{18}$

$\dfrac{1}{2} = \dfrac{9}{18}$

$- \dfrac{1}{18}$

$\overline{\hspace{2cm}}$

$\dfrac{8}{18} = \dfrac{4}{9}$

6) $\dfrac{2}{12} + \dfrac{3}{15} + \dfrac{5}{25}$ LCD = 30

$\dfrac{1}{6} + \dfrac{1}{5} + \dfrac{1}{5}$

$\dfrac{5}{30} + \dfrac{6}{30} + \dfrac{6}{30} = \dfrac{17}{30}$

7) $\dfrac{5}{12} - \dfrac{7}{18}$ LCD = 36

$\dfrac{5}{12} = \dfrac{15}{36}$

$\dfrac{7}{18} = \dfrac{14}{36}$

$\overline{\hspace{2cm}}$

$\dfrac{1}{36}$

8) $\dfrac{5}{18} + \dfrac{3}{14}$ LCD = 126

$\dfrac{5}{18} = \dfrac{35}{126}$

$\dfrac{3}{14} = \dfrac{27}{126}$

$\overline{\hspace{2cm}}$

$\dfrac{62}{126} = \dfrac{31}{63}$

9) $\dfrac{8}{25} - \dfrac{3}{20}$ LCD = 100

$\dfrac{8}{25} = \dfrac{32}{100}$

$\dfrac{3}{20} = \dfrac{12}{100}$

$\overline{\hspace{2cm}}$

$\dfrac{20}{100} = \dfrac{1}{5}$

10) $\dfrac{3}{10} + \dfrac{5}{12} + \dfrac{1}{6}$ LCD = 60

$\dfrac{3}{10} = \dfrac{18}{60}$

$\dfrac{5}{12} = \dfrac{25}{60}$

$+ \dfrac{1}{6} = \dfrac{10}{60}$

$\overline{\hspace{2cm}}$

$\dfrac{53}{60}$

Perform the following operations. Put your answer in reduced form.

1) $\dfrac{2}{12} + \dfrac{1}{10}$

2) $\dfrac{3}{20} - \dfrac{1}{24}$

3) $\dfrac{6}{15} + \dfrac{7}{49}$

4) $\dfrac{13}{16} + \dfrac{1}{24}$

5) $\dfrac{2}{15} + \dfrac{1}{12} + \dfrac{4}{25}$

6) $\dfrac{6}{16} - \dfrac{1}{24}$

7) $\dfrac{2}{9} + \dfrac{3}{14} + \dfrac{1}{21}$

8) $\dfrac{6}{60} + \dfrac{3}{30} + \dfrac{4}{40}$

9) $\dfrac{5}{12} + \dfrac{1}{16} + \dfrac{1}{9}$

10) $\dfrac{11}{33} + \dfrac{2}{10} + \dfrac{6}{50}$

11) $\dfrac{19}{42} + \dfrac{13}{70}$

12) $\dfrac{25}{84} + \dfrac{41}{90}$

13) $\dfrac{3}{70} + \dfrac{5}{28}$

14) $\dfrac{17}{66} + \dfrac{7}{44}$

15) $\dfrac{13}{104} + \dfrac{13}{78}$

16) $\dfrac{15}{75} + \dfrac{15}{90}$

17) $\dfrac{1}{2} + \dfrac{1}{3} + \dfrac{1}{8}$

18) $\dfrac{1}{2} + \dfrac{1}{3} + \dfrac{1}{4}$

19) $\dfrac{2}{11} + \dfrac{3}{22} + \dfrac{4}{10}$

20) $\dfrac{3}{12} + \dfrac{4}{16} + \dfrac{1}{4}$

So far we've added and subtracted fractions with common denominators, and we've added fractions without common denominators. Now we'll subtract fractions without common denominators.

Example 12: How much is $\frac{1}{3} - \frac{1}{4}$?

SOLUTION: $\frac{1 \times 4}{3 \times 4} - \frac{1 \times 3}{4 \times 3} = \frac{4}{12} - \frac{3}{12} = \frac{1}{12}$

Example 13: $\frac{3}{5} - \frac{3}{10}$

SOLUTION: $\frac{3 \times 2}{5 \times 2} - \frac{3}{10} = \frac{6}{10} - \frac{3}{10} = \frac{3}{10}$

Example 14: $\frac{5}{6} - \frac{2}{3}$

SOLUTION: $\frac{5}{6} - \frac{2 \times 2}{3 \times 2} = \frac{5}{6} - \frac{4}{6} = \frac{1}{6}$

Problem Set 3.9b • 5:

Subtract and simplify:

1) $\frac{3}{5} - \frac{2}{10}$

2) $\frac{7}{8} - \frac{1}{4}$

3) $\frac{4}{9} - \frac{1}{3}$

4) $\frac{5}{6} - \frac{2}{9}$

5) $\frac{2}{11} - \frac{1}{8}$

6) $\frac{5}{6} - \frac{2}{5}$

7) $\frac{5}{8} - \frac{5}{12}$

8) $\frac{5}{6} - \frac{3}{16}$

9) $\frac{7}{10} - \frac{9}{15}$

10) $\frac{2}{3} - \frac{4}{9}$

SOLUTIONS TO PROBLEM SET:

1) $\frac{3}{5} - \frac{1}{5} = \frac{2}{5}$

2) $\frac{7}{8} - \frac{1 \times 2}{4 \times 2} = \frac{7}{8} - \frac{2}{8} = \frac{5}{8}$

3) $\frac{4}{9} - \frac{1 \times 3}{3 \times 3} = \frac{4}{9} - \frac{3}{9} = \frac{1}{9}$

4) $\frac{5 \times 3}{6 \times 3} - \frac{2 \times 2}{9 \times 2} = \frac{15}{18} - \frac{4}{18} = \frac{11}{18}$

5) $\frac{2 \times 8}{11 \times 8} - \frac{1 \times 11}{8 \times 11} = \frac{16}{88} - \frac{11}{88} = \frac{5}{88}$

6) $\frac{5 \times 5}{9 \times 5} - \frac{2 \times 9}{5 \times 9} = \frac{25}{45} - \frac{18}{45} = \frac{7}{45}$

7) $\dfrac{5 \times 3}{8 \times 3} - \dfrac{5 \times 2}{12 \times 2} = \dfrac{15}{24} - \dfrac{10}{24} = \dfrac{5}{24}$

8) $\dfrac{5 \times 8}{6 \times 8} - \dfrac{3 \times 3}{16 \times 3} = \dfrac{40}{48} - \dfrac{9}{48} = \dfrac{31}{48}$

9) $\dfrac{7}{10} - \dfrac{3}{5} = \dfrac{7}{10} - \dfrac{6}{10} = \dfrac{1}{10}$

10) $\dfrac{2 \times 3}{3 \times 3} - \dfrac{4}{9} = \dfrac{6}{9} - \dfrac{4}{9} = \dfrac{2}{9}$

Subtract and simplify the following:

1) $\dfrac{1}{3} - \dfrac{1}{5}$

2) $\dfrac{2}{3} - \dfrac{3}{9}$

3) $\dfrac{4}{6} - \dfrac{1}{3}$

4) $\dfrac{3}{4} - \dfrac{5}{8}$

5) $\dfrac{6}{7} - \dfrac{3}{5}$

6) $\dfrac{1}{2} - \dfrac{2}{5}$

7) $\dfrac{4}{5} - \dfrac{3}{10}$

8) $\dfrac{7}{10} - \dfrac{3}{5}$

9) $\dfrac{5}{9} - \dfrac{1}{6}$

10) $\dfrac{9}{15} - \dfrac{1}{5}$

11) $\dfrac{4}{11} - \dfrac{2}{7}$

12) $\dfrac{5}{7} - \dfrac{3}{14}$

13) $\dfrac{9}{10} - \dfrac{2}{5}$

14) $\dfrac{7}{8} - \dfrac{1}{4}$

15) $\dfrac{1}{2} - \dfrac{5}{12}$

16) $\dfrac{13}{16} - \dfrac{2}{4}$

17) $\dfrac{2}{3} - \dfrac{2}{5}$

18) $\dfrac{14}{21} - \dfrac{1}{3}$

19) $\dfrac{9}{10} - \dfrac{3}{5}$

20) $\dfrac{4}{5} - \dfrac{3}{6}$

21) $\dfrac{5}{8} - \dfrac{1}{4}$

22) $\dfrac{5}{13} - \dfrac{1}{3}$

23) $\frac{12}{16} - \frac{1}{4}$

24) $\frac{13}{15} - \frac{7}{10}$

25) $\frac{9}{12} - \frac{1}{4}$

26) $\frac{5}{8} - \frac{2}{5}$

27) $\frac{3}{14} - \frac{1}{7}$

28) $\frac{11}{12} - \frac{5}{24}$

29) $\frac{7}{9} - \frac{1}{6}$

30) $\frac{8}{9} - \frac{2}{3}$

31) $\frac{4}{9} - \frac{3}{11}$

32) $\frac{5}{11} - \frac{9}{33}$

33) $\frac{9}{10} - \frac{1}{5} - \frac{1}{3}$

34) $\frac{11}{12} - \frac{1}{4} - \frac{3}{16}$

35) $\frac{69}{70} - \frac{3}{28}$

36) $\frac{55}{66} - \frac{4}{44}$

37) $\frac{55}{104} - \frac{11}{78}$

38) $\frac{74}{75} - \frac{11}{90}$

39) $\frac{40}{120} - \frac{3}{60}$

40) $\frac{77}{200} - \frac{7}{150}$

41) $\frac{27}{28} - \frac{7}{14} - \frac{2}{12}$

42) $\frac{60}{64} - \frac{10}{30} - \frac{5}{25}$

3.10 Order of Operations Applied to Fractions

This section applies the order of operations to fractions. Remember the order of operations is:

Step 1:   Inside groupings first.

Step 2:   Exponents

Step 3:   Multiplication/Division (left to right)

Step 4:   Addition/Subtraction (left to right)

In the next five examples, we will apply the order of operations to problems involving fractions, and then we will provide you with some practice problems for you to try.

Example 1: $\dfrac{1}{2} + \dfrac{1}{8}\left(\dfrac{1}{10} + \dfrac{3}{10}\right)$

SOLUTION: First you must do inside the parentheses:

$$\dfrac{1}{10} + \dfrac{3}{10} = \dfrac{4}{10}$$

$\dfrac{1}{2} + \dfrac{1}{\cancel{8}_2}\left(\dfrac{\cancel{4}^1}{10}\right)$ Next you multiply:

$$\dfrac{1}{\cancel{8}_2}\left(\dfrac{\cancel{4}^1}{10}\right) = \dfrac{1}{20}$$

$\dfrac{1}{2} + \dfrac{1}{20}$ Finally you use 20 as a common denominator to add:

$\dfrac{10}{20} + \dfrac{1}{20} = \dfrac{11}{20}$ $\dfrac{1}{2} = \dfrac{10}{20}$ $\dfrac{10}{20} + \dfrac{1}{20} = \dfrac{11}{20}$

Example 2: $\dfrac{6}{25} + \left(\dfrac{3}{5}\right)^2$

SOLUTION: You must do the exponent before the addition:

$$\left(\dfrac{3}{5}\right)^2 = \left(\dfrac{3}{5}\right)\left(\dfrac{3}{5}\right) = \dfrac{9}{25}$$

$\dfrac{6}{25} + \dfrac{9}{25}$ Now add the fractions:

$$\dfrac{6}{25} + \dfrac{9}{25} = \dfrac{15}{25} = \dfrac{3}{5}$$

Always remember to reduce your answer.

Example 3: $15 + \dfrac{13}{14} \div \dfrac{13}{42}$

SOLUTION: Do the division first.

$15 + \dfrac{\cancel{13}^1}{\cancel{14}_1}\left(\dfrac{\cancel{42}^3}{\cancel{13}_1}\right)$ $\dfrac{13}{14} \div \dfrac{13}{42} = \dfrac{\cancel{13}^1}{\cancel{14}_1}\left(\dfrac{\cancel{42}^3}{\cancel{13}_1}\right) = 3$

Now do the addition.

$15 + 3 = 18$ $15 + 3 = 18$

Example 4: $\left(\frac{1}{4}\right)^3 \div \left(\frac{1}{4}\right)^2$

SOLUTION: Remember to do the exponents first.

$$\left(\frac{1}{4}\right)^3 = \left(\frac{1}{4}\right)\left(\frac{1}{4}\right)\left(\frac{1}{4}\right) = \frac{1}{64}$$

$$\left(\frac{1}{4}\right)^2 = \left(\frac{1}{4}\right)\left(\frac{1}{4}\right) = \frac{1}{16}$$

$$\frac{1}{64} \div \frac{1}{16}$$

Now divide. Remember to flip the second fraction, not the first.

$$\frac{1}{\overset{}{\underset{4}{64}}}\left(\frac{\overset{1}{16}}{1}\right) \qquad \frac{1}{\overset{}{\underset{4}{64}}}\left(\frac{\overset{1}{16}}{1}\right) = \frac{1}{4}$$

$$\frac{1}{4}$$

Example 5: $100 \div \left(\frac{5}{7}\right)^2 + 200 \div \left(\frac{2}{3}\right)^2$

SOLUTION: The order in which we must do this problem is the exponents first.

$$\left(\frac{5}{7}\right)^2 = \left(\frac{5}{7}\right)\left(\frac{5}{7}\right) = \frac{25}{49}$$

$$100 \div \left(\frac{25}{49}\right) + 200 \div \left(\frac{4}{9}\right) \qquad \left(\frac{2}{3}\right)^2 = \left(\frac{2}{3}\right)\left(\frac{2}{3}\right) = \frac{4}{9}$$

Then the multiplications:

$$\frac{\overset{4}{100}}{1}\left(\frac{49}{\underset{1}{25}}\right) + \frac{\overset{50}{200}}{1}\left(\frac{9}{\underset{1}{4}}\right) \qquad \frac{\overset{4}{100}}{1}\left(\frac{49}{\underset{1}{25}}\right) = 196 \qquad \frac{\overset{50}{200}}{1}\left(\frac{9}{\underset{1}{4}}\right) = 450$$

Last is the addition:

$$196 + 450 = 646 \qquad \begin{array}{r} 196 \\ + 450 \\ \hline 646 \end{array}$$

Problem Set 3.10:

1) $\frac{1}{10} + \left(\frac{2}{3}\right) \div \frac{8}{9}$

2) $\left(\frac{1}{2} + \frac{1}{5}\right)\left(\frac{5}{6} - \frac{1}{3}\right)$

3) $\frac{4}{5} + \left(\frac{1}{5}\right)^2$

4) $\frac{2}{3} \div \left(\frac{6}{7}\right)^2$

5) $\frac{3}{4} - \left(\frac{2}{3}\right)^3$

6) $\left(\frac{7}{8}\right)^2 \left(\frac{1}{2}\right)^3$

7) $\frac{5}{8}\left(\frac{1}{4}\right) - \frac{1}{8}$

8) $\left(\frac{1}{3}\right)^2 \div \left(\frac{1}{2}\right)^2$

9) $\frac{2}{5} \div \left(\frac{3}{5} \div \frac{4}{5}\right)$

10) $\dfrac{\frac{1}{16} - \left(\frac{1}{4}\right)^2}{\frac{2}{3}}$

SOLUTIONS TO PROBLEM SET:

1) $\dfrac{1}{10} + \left(\dfrac{2}{3}\right) \div \dfrac{8}{9} = \dfrac{1}{10} + \left(\dfrac{2}{3}\right) \times \dfrac{9}{8} = \dfrac{1}{10} + \dfrac{3}{4} = \dfrac{1 \times 2}{10 \times 2} - \dfrac{3 \times 5}{4 \times 5} = \dfrac{2}{10} + \dfrac{15}{20} = \dfrac{17}{20}$

2) $\left(\dfrac{1}{2} + \dfrac{1}{5}\right)\left(\dfrac{5}{6} - \dfrac{1}{3}\right) = \left(\dfrac{1 \times 5}{2 \times 5} + \dfrac{1 \times 2}{5 \times 2}\right)\left(\dfrac{5}{6} - \dfrac{1 \times 2}{3 \times 2}\right) = \left(\dfrac{5}{10} + \dfrac{2}{10}\right)\left(\dfrac{5}{6} - \dfrac{2}{6}\right) = \dfrac{7}{10}\left(\dfrac{3}{6}\right) =$

$\dfrac{7}{10} \times \dfrac{1}{2} = \dfrac{7}{20}$

3) $\dfrac{4}{5} + \left(\dfrac{1}{5}\right)^2 = \dfrac{4}{5} + \left(\dfrac{1}{5} \times \dfrac{1}{5}\right) = \dfrac{4}{5} + \dfrac{1}{25} = \dfrac{4 \times 5}{5 \times 5} + \dfrac{1}{25} = \dfrac{20}{25} + \dfrac{1}{25} = \dfrac{21}{25}$

4) $\dfrac{2}{3} \div \left(\dfrac{6}{7}\right)^2 = \dfrac{2}{3} \div \left(\dfrac{6}{7} \times \dfrac{6}{7}\right) = \dfrac{2}{3} \div \dfrac{36}{49} = \dfrac{2}{3} \times \dfrac{49}{36} = \dfrac{49}{54}$

5) $\dfrac{3}{4} - \left(\dfrac{2}{3}\right)^3 = \dfrac{3}{4} - \left(\dfrac{2}{3} \cdot \dfrac{2}{3} \cdot \dfrac{2}{3}\right) = \dfrac{3}{4} - \dfrac{8}{27} = \dfrac{3 \times 27}{4 \times 27} - \dfrac{8 \times 4}{27 \times 4} = \dfrac{81}{108} - \dfrac{32}{108} = \dfrac{49}{108}$

6) $\left(\dfrac{7}{8}\right)^2 \left(\dfrac{1}{2}\right)^3 = \left(\dfrac{7}{8} \cdot \dfrac{7}{8}\right)\left(\dfrac{1}{2} \cdot \dfrac{1}{2} \cdot \dfrac{1}{2}\right) = \dfrac{49}{64} \cdot \dfrac{1}{8} = \dfrac{49}{512}$

7) $\dfrac{5}{8}\left(\dfrac{1}{4}\right) - \dfrac{1}{8} = \dfrac{5}{32} - \dfrac{1}{8} = \dfrac{5}{32} - \dfrac{1 \times 4}{8 \times 4} = \dfrac{5}{32} - \dfrac{4}{32} = \dfrac{1}{32}$

8) $\left(\dfrac{1}{3}\right)^2 \div \left(\dfrac{1}{2}\right)^2 = \left(\dfrac{1}{3} \times \dfrac{1}{3}\right) \div \left(\dfrac{1}{2} \times \dfrac{1}{2}\right) = \dfrac{1}{9} \div \dfrac{1}{4} = \dfrac{1}{9} \times \dfrac{4}{1} = \dfrac{4}{9}$

9) $\dfrac{2}{5} \div \left(\dfrac{3}{5} \div \dfrac{4}{5}\right) = \dfrac{2}{5} \div \left(\dfrac{3}{5} \times \dfrac{5}{4}\right) = \dfrac{2}{5} \div \dfrac{3}{4} = \dfrac{2}{5} \times \dfrac{4}{3} = \dfrac{8}{15}$

10) $\dfrac{\dfrac{1}{16} - \left(\dfrac{1}{4}\right)^2}{\dfrac{2}{3}} = \dfrac{\dfrac{1}{16} - \left(\dfrac{1}{4} \times \dfrac{1}{4}\right)}{\dfrac{2}{3}} = \dfrac{\dfrac{1}{16} - \dfrac{1}{16}}{\dfrac{2}{3}} = \dfrac{0}{\dfrac{2}{3}} = 0$

Simplify the following:

1) $\frac{3}{5} + \left(\frac{1}{4}\right) \div \frac{7}{8}$

2) $\frac{7}{8} - \left(\frac{1}{3}\right) \div \frac{5}{6}$

3) $\left(\frac{1}{3} + \frac{2}{5}\right)\left(\frac{7}{10} - \frac{2}{5}\right)$

4) $\frac{1}{8} + \left(\frac{3}{4}\right)\left(\frac{3}{10} - \frac{1}{5}\right)$

5) $\frac{2}{3} - \left(\frac{1}{3}\right)^2$

6) $\left(\frac{1}{6} + \frac{3}{4}\right)\left(\frac{5}{8} - \frac{1}{4}\right)$

7) $\frac{5}{12} \div \left(\frac{3}{4}\right)^2$

8) $\frac{2}{9} - \left(\frac{1}{3}\right)^2$

9) $\frac{13}{15} - \left(\frac{1}{2}\right)^3$

10) $\frac{5}{16} \div \left(\frac{5}{8}\right)^2$

11) $\left(\frac{4}{5}\right)^2 \left(\frac{1}{5}\right)^3$

12) $\frac{3}{14} \div \left(\frac{6}{7}\right)^2 + \frac{1}{8}$

13) $\frac{5}{8}\left(\frac{1}{4}\right) - \frac{1}{8}$

14) $\frac{7}{9} - \left(\frac{1}{3}\right)^3$

15) $\left(\frac{2}{3}\right)^2 \div \left(\frac{5}{6}\right)^2$

16) $\left(\frac{3}{4}\right)^2 \left(\frac{1}{4}\right)^3$

17) $\frac{1}{5} \div \left(\frac{1}{5} \div \frac{3}{5}\right)$

18) $\frac{5}{6}\left(\frac{2}{3}\right) - \frac{1}{6}$

19) $\dfrac{\frac{5}{8} - \left(\frac{1}{4}\right)^2}{\frac{3}{4}}$

20) $\left(\frac{4}{9}\right)^2 \div \left(\frac{2}{3}\right)^2$

21) $\frac{3}{10} \div \left(\frac{3}{5} \div \frac{4}{5}\right)$

22) $\dfrac{\frac{15}{16} - \left(\frac{3}{4}\right)^2}{\frac{3}{4}}$

3.11 Applications for Fractions

Now it's time to apply the arithmetic operations we just learned to word problems. When we start to solve a word problem, first we have to figure out which operation we need to perform to solve the word problem. As we solve the following examples, we'll explain how we knew which operation to perform. Let's look at some problems; then you can try some.

Example 1: A recipe calls for $\frac{1}{2}$ cup of sugar and $\frac{1}{4}$ cup of oil. What is the total amount of sugar and oil used in the recipe?

SOLUTION: Here we need two things taken together, the sugar and the oil. Whenever we need two things taken together, we add. The total amount of sugar and oil used is.

$$\frac{1}{2} + \frac{1}{4} = \frac{2}{4} + \frac{1}{4} = \frac{3}{4} \text{ cup}$$

Example 2: My favorite cookie recipe calls for $\frac{1}{4}$ cup of orange juice. I want to triple this recipe because all my friends want some of my cookies. How much orange juice do I need to triple this recipe?

SOLUTION: In this example I want 3 times the recipe, which means I must multiply $\frac{1}{4}\left(\frac{3}{1}\right) = \frac{3}{4}$ cup.

Example 3: One half of a cake is shared by three people. What fraction of the cake is each person's share?

SOLUTION: We need to ask ourselves, then, how much is $\frac{1}{2}$ divided by 3? So we need to divide:

$$\frac{1}{2} \div 3 = \frac{1}{2} \div \frac{3}{1} = \frac{1}{2} \times \frac{1}{3} = \frac{1}{6}$$

Example 4: Find the difference between the opening price of a stock $\left(\frac{13}{16}\right)$ and the closing price $\left(\frac{5}{8}\right)$.

SOLUTION: This is a subtraction problem because we need to find the difference between a larger number and a smaller number.

$$\frac{13}{16} - \frac{5}{8} = \frac{13}{16} - \frac{5 \times 2}{8 \times 2} = \frac{13}{16} - \frac{10}{16} = \frac{3}{16}$$

Students are always asking, how do you know when to add, subtract, multiply or divide? In **Example 1**, we had a recipe that called for $\frac{1}{2}$ cup of sugar *and* $\frac{1}{4}$ cup of oil. Then we're asked, what is the total amount of sugar and oil used in the recipe? The "and" tells us to add. The word "total" also tells us to add.

Example 2: My favorite cookie recipe calls for $\frac{1}{4}$ cup of orange juice. I want to *triple* this recipe. If you *triple* any number, you multiply that number by 3. When you double or triple a number, you multiply that number by 2 or by 3, respectively.

Example 3: One half of a cake is *shared* by three people. What fraction of the cake is each person's share? If three people are *sharing* half a cake, they're dividing that cake into three parts, so we divide.

Example 4: Find the *difference* between the opening price of a stock $\left(\frac{13}{16}\right)$ and the closing price $\left(\frac{5}{8}\right)$. To find the *difference* between two fractions, start with the first fraction and subtract the second fraction. When you find a *difference*, you must subtract.

Problem Set 3.11:

1) If you read half of a book, and then another quarter of that book, what fraction of the book would you have read in all?

2) Mary bought a half of a cheese cake. If she ate one third of it, what fraction of a whole cheese cake did she eat?

3) If my friend and I are painting a room, how much of the room still needs to be painted if I have already painted $\frac{7}{10}$ of the room and my friend has already painted $\frac{1}{4}$ of the room?

4) A quarter is what fraction of a dollar? Reduce this fraction to its lowest terms.

5) Luigi answered 16 of 20 questions correctly on a test. What fractional part did he answer correctly? Reduce your answer to its lowest terms.

SOLUTIONS TO PROBLEM SET:

1) The key word here is "***and***," which tells us to add $\frac{1}{2}$ and $\frac{1}{4}$.

$$\frac{1}{2} + \frac{1}{4} = \frac{1 \times 2}{2 \times 2} + \frac{1}{4} = \frac{2}{4} + \frac{1}{4} = \frac{3}{4}$$

2) The key question can be restated: How much is $\frac{1}{3}$ of $\frac{1}{2}$? The key word here is "***of***," which tells us to multiply $\frac{1}{2} \times \frac{1}{3}$.

$$\frac{1}{2} \times \frac{1}{3} = \frac{1}{6}$$

3) Let's add the fractions to see what part of the room we have finished painting:

$$\frac{7}{10} + \frac{1}{4} = \frac{14}{20} + \frac{5}{20} = \frac{19}{20}$$

Now let's subtract $\frac{19}{20}$ from 1 (which represents the entire room) to see what part of the room still needs to be painted:

$$1 - \frac{19}{20} = \frac{20}{20} - \frac{19}{20} = \frac{1}{20}$$

So $\frac{1}{20}$ of the room still needs to be painted.

That was really an addition problem and a subtraction problem combined.

4) A quarter is 25 cents and a dollar is 100 cents. What fraction, or part of a dollar is 25 cents? This question is answered by setting up the fraction $\frac{25}{100}$, and reducing the fraction to its lowest terms:

$$\frac{25}{100} = \frac{\overset{1}{5} \cdot \overset{1}{5}}{5 \cdot 20} = \frac{1}{4}$$

5) This is the same type of question as question 9. What part, or fraction of the 20 questions did Luigi get right? He got 16 right out of 20, or $\frac{16}{20}$, which we reduce to its lowest terms.

$$\frac{\overset{4}{16}}{\underset{5}{20}} = \frac{4}{5}$$

1) If you paint one quarter of a room, and then paint another fifth, what fraction of the room would you have painted in all?

2) A dime is what fraction of a dollar? Reduce this fraction to its lowest terms.

3) Two-thirds of a cake is divided equally among five people. How much of the cake does each person get?

4) An inheritance is shared equally by five people. One of the people gives one third of her share to her daughter. What fraction of the inheritance does her daughter get?

5) Sarah answered 40 of 50 questions correctly on a test. What fraction did she answer correctly? Reduce your answer to its lowest terms.

6) What fractional part of a day is eight hours?

7) John starts the day with $\frac{3}{4}$ of a yard of cloth. If he gets $\frac{1}{3}$ of a yard from Sue and gives Mike $\frac{3}{8}$ of a yard, how much cloth does John have left?

8) Bill gets 72 correct answers on a 100-question test. What fractional part did he answer correctly? Reduce your answer to its lowest terms.

9) Charlene owns $\frac{11}{12}$ of an acre of land. If she sells off $\frac{1}{3}$ of an acre, what fractional part of her land does she have left?

10) A nickel is what fraction of a dollar? Reduce this fraction to its lowest terms.

Chapter 3 • Review

1) $\left(\frac{1}{4}\right)^3 =$

2) How much is $\frac{2}{7} + \frac{3}{7}$?

3) Reduce to $\frac{25}{30}$ to its lowest terms.

4) Find the prime factors of 480.

5) $\frac{1}{6} \div \frac{2}{3}$

6) $\frac{3}{8} \times \frac{5}{9}$

7) $\frac{1}{2} - \frac{1}{3}$

8) $\left(\frac{4}{7}\right)^2 \frac{7}{12}$

9) A committee of 21 has 14 women. What fraction of the committee's members are (a) women? (b) men?

For problems 10 and 11, answer these questions: (a) What fraction of the pie is in the shaded area? (b) Identify the numerator and the denominator.

10)

11)

12) A pie is shared equally by Ben, Cathy, Juan, and Carla. What fraction of the pie does each person receive?

13) What fraction of a day is 4 hours?

14) $\dfrac{2}{15} + \dfrac{3}{5}$

15) $\dfrac{3}{8} \div \dfrac{3}{4}$

16) $\dfrac{1}{2} + \dfrac{1}{4} + \dfrac{1}{8}$

17) $\dfrac{4}{5} - \dfrac{3}{4}$

18) $\left(\dfrac{1}{5}\right)^2 \div \left(\dfrac{3}{10}\right)^2$

19) How much is $\dfrac{2}{5}$ of $\dfrac{2}{3}$ of $\dfrac{1}{2}$?

20) Find $\dfrac{1}{9} \times \dfrac{3}{5} \times \dfrac{5}{6}$.

21) $\dfrac{7}{8} - \dfrac{3}{4}$

22) $\dfrac{1}{6} + \dfrac{1}{4} + \dfrac{3}{8}$

23) How much is $\dfrac{2}{5}$ divided by $\dfrac{4}{11}$?

24) $\dfrac{9}{10} - \dfrac{3}{10}$

25) A class has 12 men and 16 women. What fraction of the class is (a) men? (b) women?

26) Find $\dfrac{1}{3} \times \dfrac{4}{5} \times \dfrac{5}{8}$.

27) How much is $\dfrac{4}{9}$ of $\dfrac{12}{13}$?

28) Find the prime factors of 300.

29) John bought half a pineapple upside down cake. If he ate one quarter of it, what fraction of a whole pineapple upside down cake did he eat?

30) $\left(\frac{1}{4}\right)^2 + \left(\frac{1}{2}\right)^3$

31) Reduce $\frac{105}{945}$ to its lowest terms.

32) Find the prime factors of 1,728.

33) $\frac{5}{9}\left(\frac{2}{5}\right)^3$

34) How much is four-fifths of five-twelfths?

35) Jane has 9 pennies, 6 nickels, and 10 dimes. What fraction of the coins are (a) pennies? (b) nickels? (c) dimes?

36) $\frac{1}{4}\left(\frac{4}{5}\right)^2 - \left(\frac{1}{5}\right)^2$

37) $\left(\frac{1}{8}\right)^2\left(\frac{2}{3}\right)^3$

38) $\frac{16}{25} = \left(\frac{4}{5}\right)^2$

39) $\frac{2}{3} - \frac{3}{5} + \frac{1}{25}$

40) $\frac{2}{3} \div \frac{3}{4}$

41) $\dfrac{2}{\frac{1}{8} - \left(\frac{1}{2}\right)^3}$

42) $\frac{3}{8} \div \frac{4}{5} \div \frac{9}{10}$

43) $\left(\frac{1}{3} - \frac{2}{9}\right) + \frac{2}{81} \div \left(\frac{3}{9}\right)^2$

44) $\frac{1}{2}\left(\frac{2}{7} \div \frac{6}{7}\right)^2 + \frac{2}{3}$

45) $\left(\frac{2}{3}\right)^3 \div \left(\frac{2}{3}\right)^2 \div \frac{2}{3}$

46) $\frac{4}{16} \div \left(\frac{3}{4}\right)^2$

47) $\sqrt{\dfrac{100}{144}}$

48) $\sqrt[3]{\dfrac{27}{1000}}$

49) $\sqrt{\dfrac{121}{196}}$

50) $\sqrt[3]{\dfrac{16}{54}}$

1) $\frac{2}{3}\left(\frac{1}{4}\right)$

2) $\frac{2}{3} + \frac{1}{4}$

3) $\frac{1}{4} \div \frac{2}{3}$

4) $\frac{2}{3} - \frac{1}{4}$

5) $\frac{2}{5} + \frac{1}{4} \div \frac{9}{16}$

6) One week you walked one-tenth of a mile on Monday, two-fifths of a mile on Wednesday, and one-fifteenth of a mile on Saturday. How far did you walk that week?

7) $\left(\frac{2}{3}\right)^2$

8) $\left(\frac{4}{5}\right)^3$

9) Reduce $\frac{126}{504}$

10) Reduce $\frac{105}{3675}$

11) $\left(\frac{2}{3} - \frac{1}{4}\right) \div \left(\frac{1}{8} + \frac{1}{3}\right)$

12) $\frac{5}{10} + \frac{2}{10}\left(\frac{5}{8}\right)$

13) $\frac{2}{5} - \left(\frac{2}{5}\right)^2$

14) $\left(\frac{2}{5}\right)^2 \div \left(\frac{2}{5}\right)$

15) $\left(\frac{6}{7}\right) - \left(\frac{6}{7}\right)^3$

16) $\frac{3}{4} - \frac{1}{2} + \frac{2}{3} = 3$

17) $\left[\frac{1}{4} - \left(\frac{1}{2}\right)^4\right] \div \left(\frac{3}{4}\right)^2$

18) $\frac{3}{5} + \frac{1}{10}\left(4\right)$

19) Find the sum of one-twelfth, five-fiftieths and two-ninths.

20) $\left(\frac{3}{4}\right)^2 + \left(\frac{1}{2}\right)^4$

21) A class has 17 boys and 34 girls. What fraction of the class are (a) boys? (b) girls?

22) How much is $\frac{5}{9}$ divided by $\frac{5}{8}$?

23) Find the prime factors of 200.

24) Sarah bought half a strawberry shortcake. If she ate one third of it, what fraction of a whole strawberry shortcake did she eat?

25) A cake is shared equally by Juanita, Sven, Lars, Dolph, and Greta. What fraction of the cake does each person receive?

26) What fraction of a day is 6 hours?

27) A team of 36 has 27 boys. What fraction of the team's members are (a) boys? (b) girls?

28) How much is $\frac{3}{5}$ of $\frac{3}{4}$?

29) How much is $\frac{1}{3}$ of $\frac{5}{8}$ of $\frac{2}{5}$?

30) $\sqrt{\frac{8}{72}}$

31) $\sqrt[3]{\frac{64}{216}}$

Chapter 3 • Test 2

1) $\frac{1}{6}\left(\frac{3}{5}\right)$

2) $\frac{1}{6} + \frac{3}{5}$

3) $\frac{1}{6} \div \frac{3}{5}$

4) $\frac{2}{3} - \frac{1}{4}$

5) $\frac{4}{15} + \frac{1}{8} \div \frac{45}{64}$

6) One week you walked one-forth of a mile on Monday, one-fifth of a mile on Wednesday, and one-twentieth of a mile on Saturday. How far did you walk that week?

7) $\left(\frac{7}{8}\right)^2$

8) $\left(\frac{4}{5}\right)^3$

9) Reduce $\frac{420}{525}$

10) Reduce $\frac{143}{195}$

11) $\left(\frac{1}{4} - \frac{1}{6}\right) \div \left(\frac{1}{2} + \frac{5}{12}\right)$

12) $\frac{3}{12} + \frac{1}{12}\left(\frac{6}{8}\right)$

Chapter 3 • Test 2

13) $\frac{3}{7} - \left(\frac{3}{7}\right)^2$

14) $\left(\frac{3}{8}\right)^2 \div \left(\frac{3}{8}\right)$

15) $\left(\frac{6}{12}\right) - \left(\frac{6}{12}\right)^3$

16) $\frac{1}{6} - \frac{1}{8} + \frac{3}{9} \div 9$

17) $\left[\frac{1}{4} - \left(\frac{1}{2}\right)^4\right] \div \left(\frac{3}{4}\right)^2$

18) $\frac{2}{25} + \frac{1}{15}\left(5\right)$

19) Find the sum of one-ninth, five-eighteenths and two-twentieths.

20) $\left(\frac{3}{4}\right)^2 + \left(\frac{1}{2}\right)^4$

21) Georgia has 4 pennies, 7 nickels, and 12 dimes. What fraction of her money are (a) pennies? (b) nickels? (c) dimes?

22) How much is $\frac{1}{3}$ of $\frac{3}{8}$ of $\frac{3}{4}$?

23) A team of 30 has 18 girls. What fraction of the team's members are (a) girls? (b) boys?

24) Find the prime factorization of 1680.

25) How much is three-fifths of five-eighths?

26) Marge, Roger, Phyllis, Raoul, Kirstin and Melissa share a pie. What fraction of the pie does each person get?

27) What fraction of the entire month is the third week of September?

28) How much is $\frac{7}{8}$ of $\frac{9}{14}$?

29) Samantha works an 8-hour shift that begins at midnight. What fraction of her shift is completed at 6 a.m.?

30) $\sqrt{\dfrac{64}{169}}$ 31) $\sqrt[3]{\dfrac{128}{250}}$

Chapter 4. Improper Fractions, Complex Fractions, and Mixed Numbers

In the last chapter, which is, by far, the longest in this book, you learned how to add, subtract, multiply, and divide proper fractions. As you know, proper fractions have values of less than one, because their numerators are smaller than their denominators.

Now we'll be working with improper fractions like $\frac{6}{6}$, $\frac{4}{3}$, $\frac{9}{5}$, and $\frac{7}{2}$. You'll notice that each of these fractions has a value of at least one, because their numerators are larger than, or equal to, their denominators.

We'll also be working with mixed numbers. What's a mixed number? It's a whole number and a proper fraction combined. Examples are $5\frac{1}{4}$, $2\frac{5}{8}$, and $19\frac{1}{2}$. $5\frac{1}{4}$ means $5 + \frac{1}{4}$, $2\frac{5}{8}$ means $2 + \frac{5}{8}$, and $19\frac{1}{2}$ means $19 + \frac{1}{2}$.

Later in the chapter, we'll work with complex fractions like

$$\frac{\frac{1}{2}}{\frac{3}{4}}, \quad \frac{\frac{1}{2} + \frac{1}{3}}{\frac{1}{5} - \frac{1}{15}} \quad \text{and} \quad \frac{2\frac{1}{8} + 3\frac{1}{3}}{1\frac{1}{6} + 5\frac{1}{4}}$$

Don't worry: these are a lot easier to work out than they look.

We'll begin with the easiest topic in the chapter. So what are we waiting for? Let's get started!

4.1 Converting Improper Fractions to Mixed Numbers and Mixed Numbers to Improper Fractions

To be able to work out many of the problems in this chapter, you'll need to convert mixed numbers into improper fractions. And you'll need to convert improper fractions into mixed numbers to get your final answers.

4.1a Converting Improper Fractions to Mixed Numbers

Figure 4.1

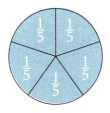 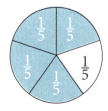

Look at Figure 4.1. We see two pies that have been divided into 10 equal parts, one $\frac{1}{5}$ each. If we shade in 9 of the parts, that leaves us with $1\frac{4}{5}$ pies. We can see that $1\frac{4}{5} = 1 + \frac{4}{5} = \frac{5}{5} + \frac{4}{5} = \frac{9}{5}$.

Remember that *an improper fraction is a fraction whose denominator is less than, or equal to, its numerator.* $\frac{9}{5}$, $\frac{11}{4}$, $\frac{6}{3}$, and $\frac{2}{2}$ are examples of improper fractions. Let's look at the improper fraction $\frac{9}{5}$. Remember, a fraction is just another way of writing a division problem. It's very easy to convert an improper fraction to a mixed number; all you have to do is divide the numerator by the denominator:

$$
\begin{array}{r}
1 \\
5\overline{)9} \\
-5 \\
\hline
4
\end{array}
$$

The remainder goes over the denominator in the fractional part, so $\frac{9}{5}$ is $1\frac{4}{5}$, which is called a mixed number.

Example 1:   **Convert $\frac{11}{4}$ to a mixed number.**

SOLUTION: To convert $\frac{11}{4}$ to a mixed number, just divide 11 by 4 and put the remainder over the divisor:
$$\frac{11}{4} = 2\frac{3}{4}$$

Example 2:   **Convert $\frac{6}{3}$ to a mixed number.**

SOLUTION: When we divide 6 by 3 there is no remainder, so this improper fraction represents the whole number 2. The same thing is true for $\frac{2}{2}$. When we divide 2 by 2 there is no remainder, so we get the whole number 1.

Sometimes we can reduce an improper fraction before converting it to a mixed number. Here's how that's done with $\frac{20}{12}$.

Example 3:   **Convert $\frac{20}{12}$ to a mixed number.**

SOLUTION: $\frac{20}{12} = \frac{5}{3} = 1\frac{2}{3}$

Problem Set 4.1a:

Convert the following improper fractions to mixed numbers.

1) $\frac{7}{2}$

2) $\frac{5}{3}$

3) $\frac{16}{12}$

4) $\frac{4}{4}$

SOLUTIONS TO PROBLEM SET:

1) $5 \div 3 = \frac{5}{3} = 1\frac{2}{3}$

2) $\frac{7}{2} = 3\frac{1}{2}$

3) $5 \div 3 = \frac{5}{3} = 1\frac{2}{3}$

4) $4 \div 4 = 1$

4.1b. Converting Mixed Numbers to Improper Fractions

Figure 4.2

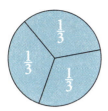  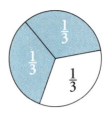

Look at Figure 4.2. We see three pies that have been divided into thirds. $2\frac{2}{3}$ or 8 pieces of the pies have been shaded. So $2\frac{2}{3} = \frac{8}{3}$.

Now let's reverse what we just did and convert mixed numbers to improper fractions. Instead of dividing and subtracting, we will do the opposite; we will multiply and add. The first example we did was to convert $\frac{9}{5}$ to $1\frac{4}{5}$. Now let's convert the mixed number $1\frac{4}{5}$ to the improper fraction $\frac{9}{5}$. The way to convert a mixed number to an improper fraction is to multiply the denominator times the whole number, then add the numerator, and put that number over the denominator. Let's do this to $1\frac{4}{5}$. First multiply the denominator of 5 times the whole number 1 and add the numerator 4; $5 \times 1 + 4 = 5 + 4 = 9$, put 9 over the denominator of 5 to get the improper fraction $\frac{9}{5}$.

Example 1: **Convert $2\frac{3}{4}$ to an improper fraction.**

SOLUTION: Multiply 4×2 and then add 3. This gives us 11. Put the 11 over the denominator of 4 and we have the improper fraction $\frac{11}{4}$. Let's do one more, and then you try the following problem set.

Example 2: **Convert $2\frac{1}{5}$ to an improper fraction.**

SOLUTION: Multiply 5×2 and add 1; then put the number 11 over the denominator 5 to get the improper fraction $\frac{11}{5}$.

Problem Set 4.1b:

Convert the following mixed numbers to improper fractions.

1) $3\frac{1}{3}$

2) $3\frac{1}{4}$

3) $5\frac{4}{5}$

4) $8\frac{11}{12}$

SOLUTIONS TO PROBLEM SET:

1) $3\frac{1}{3} = \frac{3(3) + 1}{3} = \frac{10}{3}$

2) $3\frac{1}{4} = \frac{4(3) + 1}{4} = \frac{13}{4}$

3) $5\frac{4}{5} = \frac{5(5) + 4}{5} = \frac{29}{5}$

4) $8\frac{11}{12} = \frac{12(8) + 11}{12} = \frac{107}{12}$

Convert the following improper fractions to mixed numbers.

1) $\frac{12}{11}$

2) $\frac{60}{11}$

3) $\frac{17}{12}$

4) $\frac{55}{3}$

5) $\frac{27}{2}$

6) $\frac{12}{3}$

7) $\frac{33}{5}$

8) $\frac{5}{2}$

9) $\frac{121}{11}$

10) $\frac{20}{14}$

11) $\frac{16}{14}$

12) $\frac{28}{15}$

13) $\frac{3}{2}$

14) $\frac{12}{5}$

15) $\frac{80}{8}$

16) $\frac{40}{40}$

17) $\frac{71}{8}$

18) $\frac{16}{9}$

Convert the following mixed numbers to improper fractions.

19) $2\frac{4}{7}$

20) $6\frac{1}{2}$

21) $6\frac{1}{5}$

22) $4\frac{30}{60}$

23) $18\frac{3}{12}$

24) $12\frac{2}{5}$

25) $8\frac{11}{13}$

26) $8\frac{9}{10}$

27) $4\frac{3}{5}$

28) $7\frac{2}{3}$

29) $10\frac{2}{5}$

30) $11\frac{11}{12}$

31) $3\frac{6}{18}$

32) $5\frac{3}{4}$

33) $9\frac{2}{9}$

34) $10\frac{6}{7}$

35) $21\frac{2}{11}$

36) $3\frac{2}{3}$

37) $5\frac{3}{8}$

38) $90\frac{10}{40}$

39) $20\frac{1}{2}$

4.2 Multiplication and Division of Mixed Numbers

There's only one new rule to remember when working on a multiplication or division problem involving mixed numbers: *You must convert to improper fractions before you multiply or divide.* If you remember this one rule, you can't go wrong. Let's look at some problems. The rule for multiplying improper fractions is the same as the rule for proper fractions.

Example 1: $1\frac{1}{2} \times 1\frac{1}{3}$

SOLUTION:

$$\overset{1}{\underset{1}{\cancel{\frac{3}{2}}}}\left(\overset{2}{\underset{1}{\cancel{\frac{4}{3}}}}\right) = 2$$

The first thing we did was to convert the mixed numbers to improper fractions, then multiply. Remember to reduce the common factors before you multiply.

Example 2: $1\frac{1}{2} \div 1\frac{1}{3}$

SOLUTION:

$$\frac{3}{2} \div \frac{4}{3}$$

$$\frac{3}{2}\left(\frac{3}{4}\right) = \frac{9}{8} = 1\frac{1}{8}$$

Convert to improper fractions first; then multiply by the reciprocal of the divisor (flip the second fraction) because this is a division problem; then multiply. Write your answer as a mixed number.

Example 3: $\left(3\frac{1}{3}\right)^3$

SOLUTION:

$$\left(\frac{10}{3}\right)^3 = \frac{1000}{27} = 37\frac{1}{27}$$

The exponent means multiply, so we must convert $3\frac{1}{3}$ to $\frac{10}{3}$ and then apply the exponent. Write your answer as a mixed number.

Example 4: $\frac{4}{15} \div \left(1\frac{1}{5} \cdot 2\frac{1}{12}\right)$

SOLUTION:

$$\frac{4}{15} \div \left(\overset{1}{\underset{1}{\cancel{\frac{6}{5}}}} \times \overset{5}{\underset{2}{\cancel{\frac{25}{12}}}}\right) = \frac{4}{15} \div \frac{5}{2}$$

$$= \frac{4}{15} \cdot \frac{2}{5} = \frac{8}{75}$$

Always do the work inside the parenthesis first, converting to improper fractions before you do the multiplication. Then divide.

Problem Set 4.2:

1) $\left(2\frac{2}{3}\right)^2$

2) $1\frac{1}{5}\left(2\frac{1}{12}\right)$

3) $1\frac{1}{7}\left(1\frac{5}{16} \div 1\frac{1}{2}\right)$

4) $1\frac{3}{4} \div 5\frac{3}{5}$

5) $1\frac{7}{8} \div 1\frac{1}{4}$

6) $1\frac{5}{7}\left(2\frac{5}{8} \div 4\frac{1}{2}\right)$

SOLUTIONS TO PROBLEM SET:

1) $\left(\frac{8}{3}\right)^2 = \left(\frac{8}{3}\right)\left(\frac{8}{3}\right) = \frac{64}{9} = 7\frac{1}{9}$

2) $\frac{\overset{1}{\cancel{6}}}{\underset{1}{\cancel{5}}}\left(\frac{\overset{5}{\cancel{25}}}{\underset{2}{\cancel{12}}}\right) = \frac{5}{2} = 2\frac{1}{2}$

3) $\frac{8}{7}\left(\frac{21}{16} \div \frac{3}{2}\right) = \frac{8}{7}\left(\frac{\overset{7}{\cancel{21}}}{\underset{8}{\cancel{16}}} \bullet \frac{\overset{1}{\cancel{2}}}{\underset{1}{\cancel{3}}}\right) = \frac{\overset{1}{\cancel{8}}}{\cancel{7}}\left(\frac{\cancel{7}}{\underset{1}{\cancel{8}}}\right) = 1$

4) $\frac{7}{4} \div \frac{28}{5} = \frac{\overset{1}{\cancel{7}}}{4} \times \frac{5}{\underset{4}{\cancel{28}}} = \frac{5}{16}$

5) $\frac{15}{8} \div \frac{5}{4} = \frac{\overset{3}{\cancel{15}}}{\underset{2}{\cancel{8}}} \times \frac{\overset{1}{\cancel{4}}}{\underset{1}{\cancel{5}}} = \frac{3}{2} = 1\frac{1}{2}$

6) $\frac{12}{7}\left(\frac{21}{8} \div \frac{9}{2}\right) = \frac{12}{7}\left(\frac{\overset{7}{\cancel{21}}}{\underset{4}{\cancel{8}}} \times \frac{\overset{1}{\cancel{2}}}{\underset{3}{\cancel{9}}}\right) = \frac{\overset{1}{\cancel{12}}}{\underset{1}{\cancel{7}}} \times \frac{\overset{1}{\cancel{7}}}{\underset{1}{\cancel{12}}} = \frac{1}{1} = 1$

Perform the following operations. Be sure your answer is in proper form.

1) $1\frac{1}{5}\left(2\frac{1}{12}\right)$

2) $\left(2\frac{6}{10}\right)^2$

3) $1\frac{2}{10} \div 1\frac{1}{2}$

4) $2\frac{6}{7}\left(9\frac{4}{5}\right)$

5) $\left(1\frac{1}{2}\right)^3$

6) $1\frac{1}{5} \div 2\frac{2}{5}$

7) $1\frac{1}{7} \div 2\frac{2}{7}$

8) $1\frac{5}{9} \div 1\frac{1}{6}$

9) $3\frac{1}{2} \cdot \frac{5}{14}$

10) $\left(1\frac{4}{16}\right)^3$

11) $4 \div 13\frac{1}{3}$

12) $1\frac{4}{11}\left(9\frac{3}{22}\right)\left(1\frac{7}{15}\right)$

13) $3\frac{3}{4} \div \left(2\frac{1}{2}\right)^2$

14) $12 \div 2\frac{2}{3}$

15) $1\frac{1}{5} \div 1\frac{2}{7}$

16) $\left(\frac{8}{9}\right)\left(1\frac{1}{2}\right)^2$

17) $1\frac{1}{8}\left(3\frac{1}{5}\right) \div 4\frac{4}{5}$

18) $3\frac{9}{24} \div 1\frac{1}{6}\left(1\frac{1}{2}\right)^2$

19) $2\frac{1}{2}\left(\frac{1}{2} \div 2\frac{1}{2}\right)$

20) $1\frac{5}{8} \div 2\frac{7}{16}\left(1\frac{1}{2}\right)$

21) $\frac{3}{4} \div 1\frac{4}{11}$

22) $2\frac{1}{4} \div 12$

23) $7\frac{1}{5} \cdot \frac{5}{9}$

CHALLENGE PROBLEMS

24) $14\frac{2}{3}\left(1\frac{8}{25} \times 33\frac{1}{3}\right)$

25) $\left[1\frac{11}{12}\left(1\frac{1}{23}\right)\right](4)$

26) $\left(1\frac{1}{8} \times 7\frac{1}{9}\right) \div \left(\frac{1}{3} \div 2\frac{2}{3}\right)$

27) $\left(2\frac{3}{4} \times 3\frac{1}{3}\right) \div \left(\frac{3}{4}\right)^2$

28) $1\frac{4}{7}\left(5\frac{3}{5} \cdot 2\right)\left(1\frac{1}{11}\right)$

4.3 Addition and Subtraction of Mixed Numbers

When adding and subtracting mixed numbers, we usually will not use improper fractions. All the basic rules for addition and subtraction of fractions apply to mixed numbers. We must find the L.C.D. and convert to it before we add or subtract. Don't forget to write your answer in proper form. Let's look at a few problems together.

Example 1: $1\frac{1}{2} + 3\frac{1}{5}$

SOLUTION:

$$1\frac{1}{2} = 1\frac{5}{10}$$
$$+ 3\frac{1}{5} = 3\frac{2}{10}$$
$$\overline{\quad\quad 4\frac{7}{10}}$$

Convert the fractions to the common denominator of 10; then add the fractions and the whole numbers.

Example 2: $3\frac{11}{12} + 4\frac{5}{6}$

SOLUTION:

$$3\frac{11}{12} = 3\frac{11}{12}$$
$$+ 4\frac{5}{6} = 4\frac{10}{12}$$
$$\overline{\quad\quad 7\frac{21}{12}}$$

$$7\frac{21}{12} = 8\frac{9}{12} = 8\frac{3}{4}$$

Convert the fractions to the common denominator of 12; then add the fractions and the whole numbers. When we add the fractions, we get $\frac{21}{12}$ which needs to be written as a mixed number. Now add 7 and $1\frac{9}{12}$ to get $8\frac{9}{12}$. Always reduce your answer: $8\frac{9}{12} = 8\frac{3}{4}$

Example 3: $16\frac{14}{15} + 12\frac{9}{10}$

SOLUTION:

$$16\frac{14}{15} = 16\frac{28}{30}$$
$$+12\frac{9}{10} = 12\frac{27}{30}$$
$$\overline{\quad\quad 28\frac{55}{30}}$$

$$28\frac{55}{30} = 28 + 1\frac{5}{6} = 29\frac{5}{6}$$

Convert the fractions to the common denominator of 30; then add the fractions and the whole numbers. Convert $\frac{55}{30}$ to $1\frac{25}{30} = 1\frac{5}{6}$; then add 28 and $1\frac{5}{6}$, reduce.

Before we try subtraction, do the following problem set.

Problem Set 4.3:

1) $\begin{array}{r} 6\frac{2}{3} \\ +\, 2\frac{2}{5} \\ \hline \end{array}$

2) $9\frac{1}{3} + 5\frac{9}{10}$

3) $10\frac{2}{3} + 3\frac{6}{7} + 4\frac{3}{21}$

4) $\begin{array}{r} 8\frac{3}{4} \\ +\, 3\frac{2}{3} \\ \hline \end{array}$

5) $5\frac{5}{6} + 2\frac{3}{7} + 1\frac{1}{3}$

6) $4\frac{1}{3} + 2\frac{1}{6} + 1\frac{3}{4}$

1)
$$6\tfrac{2}{3} = 6\tfrac{10}{15}$$
$$+\,2\tfrac{2}{5} = 2\tfrac{6}{15}$$
$$8\tfrac{16}{15} = 8 + 1\tfrac{1}{15} = 9\tfrac{1}{15}$$

2)
$$9\tfrac{1}{3} = 9\tfrac{10}{30}$$
$$+\,5\tfrac{9}{10} = 5\tfrac{27}{30}$$
$$14\tfrac{37}{30} = 14 + 1\tfrac{7}{30} = 15\tfrac{7}{30}$$

3)
$$10\tfrac{2}{3} = 10\tfrac{14}{21}$$
$$+\,3\tfrac{6}{7} = 3\tfrac{18}{21}$$
$$+\,4\tfrac{3}{21} = 4\tfrac{3}{21}$$
$$17\tfrac{35}{21} = 17 + 1\tfrac{14}{21} = 18\tfrac{2}{3}$$

4)
$$8\tfrac{3}{4} = 8\tfrac{9}{12}$$
$$+\,3\tfrac{2}{3} = 3\tfrac{8}{12}$$
$$11\tfrac{17}{12} = 11 + 1\tfrac{5}{12} = 12\tfrac{5}{12}$$

5)
$$5\tfrac{5}{6} = 5\tfrac{35}{42}$$
$$+\,2\tfrac{3}{7} = 2\tfrac{18}{42}$$
$$+\,1\tfrac{1}{3} = 1\tfrac{14}{42}$$
$$8\tfrac{67}{42} = 8 + 1\tfrac{25}{42} = 9\tfrac{25}{42}$$

6)
$$4\tfrac{1}{3} = 4\tfrac{4}{12}$$
$$+\,2\tfrac{1}{6} = 2\tfrac{2}{12}$$
$$+\,1\tfrac{3}{4} = 1\tfrac{9}{12}$$
$$7\tfrac{15}{12} = 7 + 1\tfrac{3}{12} = 8\tfrac{1}{4}$$

Now that you have mastered addition with mixed numbers, let's try subtraction with mixed numbers.

Example 4: $2\tfrac{2}{3} - 1\tfrac{1}{4}$

SOLUTION:
$$2\tfrac{2}{3} = 2\tfrac{8}{12}$$
$$-\,1\tfrac{1}{4} = 1\tfrac{3}{12}$$
$$1\tfrac{5}{12}$$

First convert the fractions to the L.C.D. of 12; then subtract the fractions and the whole numbers.

Now we're going to look at an example that most students find confusing. We'll do this example in two ways. You choose whichever method you prefer. The first method uses mixed numbers and borrowing; the second uses improper fractions.

Example 5: $5\tfrac{1}{8} - 2\tfrac{5}{24}$

SOLUTION:

$$\overset{4}{\cancel{5}}\tfrac{1}{8} = 4\tfrac{3}{24} + \tfrac{24}{24} = 4\tfrac{27}{24}$$
$$-\,2\tfrac{5}{24} \qquad\qquad\qquad -\,2\tfrac{5}{24}$$
$$\rule{3cm}{0.4pt}$$
$$2\tfrac{22}{24} = 2\tfrac{11}{12}$$

Convert the fractions to the common denominator of 24. The next step is to subtract. But we can't subtract $\tfrac{5}{24}$ from $\tfrac{3}{24}$, so we need to borrow 1 from the 5. Next we have to add the one we borrowed to $\tfrac{3}{24}$. When we add the 1, we have to add it as $\tfrac{24}{24}$, so we get $\tfrac{3}{24} + \tfrac{24}{24} = \tfrac{27}{24}$. Now we can subtract the fractions and whole numbers and reduce our answer.

Let's do Example 5 by the other method involving improper fractions.

Example 5: $5\frac{1}{8} - 2\frac{5}{24}$

SOLUTION:
$$5\frac{1}{8} = \frac{41}{8} = \frac{123}{24}$$
$$-2\frac{5}{24} = \qquad \frac{53}{24}$$
$$\overline{\qquad \frac{70}{24} = 2\frac{22}{24} = 2\frac{11}{12}}$$

First convert the mixed numbers to improper fractions; then convert to the L.C.D. of 24 and subtract. Write your answer as a mixed number and reduce.

Example 6: $12\frac{7}{12} - 6\frac{7}{8}$

SOLUTION:
$$\overset{11}{\cancel{12}}\frac{7}{12} = 11\frac{14}{24} + \frac{24}{24} = 11\frac{38}{24}$$
$$-6\frac{7}{8} = \qquad\qquad\qquad -6\frac{21}{24}$$
$$\overline{\qquad\qquad\qquad\qquad\quad 5\frac{17}{24}}$$

First convert to the L.C.D. of 24, then borrow 1 and add the 1 you borrowed to $\frac{14}{24}$ as $\frac{24}{24}$ to get $\frac{38}{24}$. Now subtract the fractions and the whole numbers to get $5\frac{17}{24}$.

Example 7: $10\frac{2}{3} - 7\frac{3}{4}$

SOLUTION:
$$10\frac{2}{3} = \frac{32}{3} = \frac{128}{12}$$
$$-7\frac{3}{4} = \frac{31}{4} = \frac{93}{12}$$
$$\overline{\qquad\qquad\quad \frac{35}{12} = 2\frac{11}{12}}$$

First convert to improper fractions, then to the L.C.D. of 12. Now subtract and convert to mixed number form.

Sometimes a problem calls for addition *and* subtraction. Usually it's easier to do the addition first, and then the subtraction.

Example 8: $8\frac{5}{8} - 5\frac{1}{2} + 3\frac{3}{4}$

SOLUTION:
$$8\frac{5}{8} = 8\frac{5}{8}$$
$$+3\frac{3}{4} = 3\frac{6}{8}$$
$$\overline{\qquad 11\frac{11}{8} = 11 + 1\frac{3}{8} = 12\frac{3}{8}}$$

First add $8\frac{5}{8} + 3\frac{3}{4}$.

$$\overset{11}{\cancel{12}}\frac{3}{8} + \frac{8}{8} = 11\frac{11}{8}$$
$$-5\frac{1}{2} \qquad\qquad -5\frac{4}{8}$$
$$\overline{\qquad\qquad\qquad 6\frac{7}{8}}$$

Then subtract $5\frac{1}{2}$ from $12\frac{3}{8}$.

So far, all our examples showed a mixed number minus a mixed number. Let's look at an example involving a whole number minus a mixed number. We'll look at two approaches to performing this subtraction.

Example 9: $5 - 2\frac{1}{2}$

SOLUTION:

First Approach:

$$\begin{array}{r} \overset{4}{\cancel{5}}\frac{2}{2} \\ -2\frac{1}{2} \\ \hline 2\frac{1}{2} \end{array}$$

We'll borrow 1 from the 5 and write the 1 as $\frac{2}{2}$. Now it's easy to subtract.

Second Approach:

$$\begin{array}{r} \frac{5}{1} = \frac{10}{2} \\ -2\frac{1}{2} = -\frac{5}{2} \\ \hline \frac{5}{2} = 2\frac{1}{2} \end{array}$$

We'll write 5 as the improper fraction $\frac{10}{2}$ and $2\frac{1}{2}$ as $\frac{5}{2}$, then subtract. Write your answer as a mixed number. This way we don't have to borrow.

Remember, you can always check a subtraction problem by adding:

$$\begin{array}{r} 2\frac{1}{2} \\ +2\frac{1}{2} \\ \hline 4\frac{2}{2} = 4 + 1 = 5 \end{array}$$

Perform the following operations and put your answer in proper form.

1) $3\frac{2}{9} + 2\frac{1}{5}$

2) $12\frac{1}{2} + 3\frac{2}{3}$

3) $4\frac{4}{7}$
$+3\frac{2}{3}$

4) $6\frac{11}{12} - 4\frac{1}{24}$

5) $11\frac{9}{10}$
$-4\frac{6}{7}$

6) $5 - 3\frac{3}{8}$

7) $10\frac{7}{8} + 9\frac{1}{6}$

8) $16\frac{3}{5}$
$-14\frac{14}{15}$

9) $20\frac{2}{15}$
$-10\frac{7}{20}$

10) $8\frac{7}{8}$
$-2\frac{2}{3}$

11) $14\frac{3}{10}$
$+25\frac{4}{25}$

12) $15\frac{2}{5} - 11\frac{6}{7}$

13) $12\frac{1}{2} + 4\frac{1}{2}$

14) $11\frac{1}{2} - 2\frac{2}{3}$

15) $31\frac{1}{26}$
$-13\frac{12}{13}$

16) $12\frac{3}{4} + 6\frac{4}{5} + 3\frac{9}{10}$

17) $6\frac{2}{5} + 3\frac{4}{15} + 2\frac{5}{6}$

18) 10
$-3\frac{2}{3}$

19) $3\frac{5}{6} + 4\frac{11}{14} + 2\frac{11}{21}$

20) $9\frac{9}{9}$
$+13\frac{2}{21}$

21) $7\frac{7}{10} + 6\frac{11}{20} + 3\frac{14}{15}$

22) $6\frac{2}{6} + 5\frac{3}{9} + 4\frac{2}{30}$

23) $1\frac{1}{2} + 1\frac{2}{3} + 1\frac{4}{15}$

24) $11\frac{3}{4}$
$-2\frac{11}{14}$

25) $12\frac{1}{2}$
$+10\frac{1}{2}$

26) $18\frac{1}{10}$
$+19\frac{11}{12}$

27) $12 - 6\frac{7}{8}$

28) $30 - 8\frac{4}{5}$

CHALLENGE PROBLEMS

29) $4\frac{2}{3} + 1\frac{5}{6} - 5\frac{1}{6}$

30) $10\frac{1}{8} + 8\frac{3}{4} - 5\frac{1}{2}$

31) $15\frac{5}{8} - 9\frac{1}{3} + 4\frac{5}{6}$

32) $12\frac{3}{5} - 6\frac{2}{3} + 8\frac{2}{15}$

4.4   Applications

By now you should have a pretty good idea of when to add, subtract, multiply or divide. In this section you'll be combining that skill with what you've learned about improper fractions and mixed numbers.

Example 1:   Senator Stacey Long, a candidate for the Republican presidential nomination, was assured of $17\frac{1}{2}$ votes from the Connecticut delegation, $42\frac{1}{4}$ votes from the Pennsylvania delegation, and $6\frac{3}{4}$ votes from the West Virginia delegation. What was the total number of votes Senator Long was assured by these three delegations?

SOLUTION:
$$17\frac{1}{2} = 17\frac{2}{4}$$
$$42\frac{1}{4} = 42\frac{1}{4}$$
$$6\frac{3}{4} = 6\frac{3}{4}$$
$$65\frac{6}{4} = 65 + 1\frac{2}{4} = 65 + 1\frac{1}{2} = 66\frac{1}{2} \text{ votes}$$

Example 2:   If we're sewing tops that require $1\frac{2}{3}$ yards of material for one top, and we have 48 yards of material, how many tops can we make?

SOLUTION:

We have to divide because the 48 yards of material will be cut into smaller parts.
$$48 \div 1\frac{2}{3} = \frac{48}{1} \div \frac{5}{3} = \frac{48}{1} \cdot \frac{3}{5} = \frac{144}{5} = 28\frac{4}{5} \text{ tops}$$

Example 3:   Juan Carlos earns $58\frac{3}{4}$ pesos an hour. How much would he earn for $10\frac{1}{2}$ hours work?

SOLUTION:
$$58\frac{3}{4} \cdot 10\frac{1}{2} = \frac{235}{4} \cdot \frac{21}{2} = \frac{4935}{8} = 616\frac{7}{8} \text{ pesos}$$

Example 4:   Yesterday a share of Dell Computer opened at $53\frac{1}{8}$ and fell by $2\frac{3}{4}$ during the day. What was its price at the end of the day?

SOLUTION:
$$\overset{52}{5}3\frac{1}{8} + \frac{8}{8} = 52\frac{9}{8}$$
$$\underline{-2\frac{3}{4} \qquad\qquad -2\frac{6}{8}}$$
$$50\frac{3}{8}$$

Problem Set 4.4:

1) Beverly O'Connor owned $4\frac{7}{8}$ shares of a French company whose stock is priced at $15\frac{3}{4}$ francs. How many francs is her stock worth?

2) How many pieces of pipe that are $2\frac{1}{2}$ feet long must be laid together to make a pipe 80 feet long?

3) A bottle holds $1\frac{4}{5}$ pints of liquid. How many bottles would it take to hold 90 pints of liquid?

4) John worked $8\frac{3}{4}$ hours, but was docked $1\frac{1}{2}$ hours pay for forgetting to punch in after lunch. How many hours pay did he receive?

5) In the triple jump Ann jumped $9\frac{1}{2}$ feet, $8\frac{5}{8}$ feet, and $7\frac{2}{3}$ feet. How far did she jump all together?

6) If one dress can be sewn from $2\frac{2}{3}$ yards of material, how many dresses can be sewn from 48 yards?

SOLUTIONS TO PROBLEM SET:

1) $\left(4\frac{7}{8}\right)\left(15\frac{3}{4}\right) = \frac{39}{8} \cdot \frac{63}{4} = \frac{2457}{32} = 76\frac{25}{32}$ francs

2) $80 \div 2\frac{1}{2} = 80 \div \frac{5}{2} = \frac{\overset{16}{\cancel{80}}}{1}\left(\frac{2}{\cancel{5}}\right) = 32$ pieces of pipe

3) $90 \div 1\frac{4}{5} = 90 \div \frac{9}{5} = \frac{\overset{10}{\cancel{90}}}{1}\left(\frac{5}{\cancel{9}}\right) = 50$ bottles

4) $8\frac{3}{4} - 1\frac{1}{2} = 8\frac{3}{4} - 1\frac{2}{4} = 7\frac{1}{4}$ hours

5) $9\frac{1}{2} + 8\frac{5}{8} + 7\frac{2}{3} = 9\frac{12}{24} + 8\frac{15}{24} + 7\frac{16}{24} = 24\frac{43}{24} = 24 + 1\frac{19}{24} = 25\frac{19}{24}$ feet

6) $48 \div 2\frac{2}{3} = 48 \div \frac{8}{3} = \frac{\overset{6}{\cancel{48}}}{1}\left(\frac{3}{\cancel{8}}\right) = 18$ dresses

1) A woman started digging a hole on Tuesday, when she dug $2\frac{2}{3}$ feet. On Wednesday she dug $1\frac{3}{4}$ feet, and on Thursday she dug $2\frac{1}{2}$ feet. How far down had she dug by Thursday?

2) If a glass of punch contained $7\frac{1}{2}$ ounces and if $28\frac{1}{3}$ cups of punch were consumed at a party, how many ounces of punch were consumed?

3) Sam Hong walked $2\frac{1}{4}$ miles before breakfast, $3\frac{7}{8}$ miles before lunch, $2\frac{1}{2}$ miles before supper, and $3\frac{3}{4}$ miles after supper. How far did he walk that day?

4) A box that contained $4\frac{3}{4}$ pounds of candy was opened at a party. If people ate $2\frac{7}{8}$ pounds, how much candy was left?

5) The weight of water is $62\frac{1}{2}$ lb. per cubic foot. Find the weight of $10\frac{1}{5}$ cubic feet of water.

6) A recipe calls for $2\frac{2}{3}$ cups of flour. How much flour would be needed to triple the recipe?

7) The average amount of rainfall in Hackettstown is $3\frac{1}{6}$ inches per month. Based on this information, how much rainfall should we expect over an 18 month period?

8) A recipe calls for $1\frac{3}{4}$ cups of sugar. How much sugar is needed to make half this recipe?

9) A piece of material $4\frac{1}{4}$ yards long is to be used to make shirts. If each shirt uses $\frac{1}{4}$ of a yard, how many shirts can be made from the material?

10) If it rained $10\frac{1}{4}$ inches in $2\frac{1}{4}$ days, on the average, how many inches did it rain per day?

11) If the average amount of bird seed consumed by a parrot per day is $5\frac{1}{3}$ ounces, how many days will $106\frac{2}{3}$ ounces of seed feed the parrot?

CHALLENGE PROBLEMS

12) Jessica left home at 9 a.m. for a long run. Running in the same direction, she covered $4\frac{1}{8}$ miles in the first half hour, $3\frac{9}{10}$ miles in the second half hour, and $3\frac{5}{6}$ miles in the third half hour. Turning back toward home, she ran $3\frac{2}{3}$ miles in the last half hour. After two hours of running, how far was she from home?

13) Suppose you wanted to build a patio in a yard that was $47\frac{1}{2}$ feet long and $33\frac{4}{5}$ feet wide. If you left an unpaved border of 2 feet around the patio, how many cubic feet of concrete would you need if the patio is to be $1\frac{1}{4}$ feet high?

4.5   Complex Fractions

A *complex fraction* is a fraction that contains fractions within it.

For example, $\dfrac{\frac{1}{2}}{\frac{3}{4}}$, $\dfrac{\frac{2}{3}+\frac{3}{5}}{\frac{4}{5}-\frac{2}{7}}$, $\dfrac{\frac{7}{8}}{\frac{2}{5}+\frac{1}{10}}$ and $\dfrac{\frac{9}{10}-\frac{2}{5}}{\frac{1}{2}}$ are all complex fractions.

They can be rewritten in horizontal form because the fraction bar means divide.

When it's written in horizontal form like this: $\left(\frac{1}{10}+\frac{1}{5}\right)\div\left(\frac{5}{6}-\frac{1}{15}\right)$, it looks like a regular division problem from Section 3.9.

When it's written in the form $\dfrac{\frac{1}{10}+\frac{1}{5}}{\frac{5}{6}-\frac{1}{15}}$, it's called a complex fraction because there are fractions within the fraction.

Also notice, these are two different ways of writing the same thing: $\dfrac{\frac{1}{10}+\frac{1}{5}}{\frac{5}{6}-\frac{1}{15}} = \left(\frac{1}{10}+\frac{1}{5}\right)\div\left(\frac{5}{6}-\frac{1}{15}\right)$

So a complex fraction is just a vertical way of showing a division problem involving fractions. In this section we'll learn two ways to simplify a complex fraction.

Example 1:   **Let's take a look at another complex fraction and simplify it together.**

$$\dfrac{\frac{3}{5}-\frac{2}{15}}{\frac{2}{3}-\frac{1}{15}}$$

SOLUTION: We'll use the first method here. Then immediately after the next problem set we'll redo this problem by the second method.

$$\dfrac{\frac{3}{5}-\frac{2}{15}}{\frac{2}{3}-\frac{1}{15}}$$

Let's simplify this complex fraction by putting it in horizontal form: Remember your order of operations and do the work inside the parentheses first before you divide.

$\left(\frac{3}{5}-\frac{2}{15}\right)\div\left(\frac{2}{3}-\frac{1}{15}\right)$ First find the L.C.D.s for the fractions inside the parenthesis.

$\left(\frac{9}{15}-\frac{2}{15}\right)\div\left(\frac{10}{15}-\frac{1}{15}\right)$ Convert $\frac{3}{5}$ to $\frac{9}{15}$ and subtract $\frac{9}{15}-\frac{2}{15}$ to get $\frac{7}{15}$.

$\frac{7}{15}\div\frac{9}{15}$ Convert the $\frac{2}{3}$ to $\frac{10}{15}$ and subtract $\frac{10}{15}-\frac{1}{15}$ to get $\frac{9}{15}$.

$\frac{7}{\cancel{15}}\cdot\dfrac{\overset{1}{\cancel{15}}}{9} = \frac{7}{9}$ Now multiply by the reciprocal of the divisor (flip the second fraction). Don't forget to reduce the common factors (cancel).

So far all the problems we've done have involved proper fractions. Let's really have fun and try some problems with mixed numbers. We'll do the same routine, but be sure to remember, *we must convert to improper fractions before we multiply or divide.*

Let's try one together.

$$\frac{1\frac{1}{2} + 2\frac{1}{4}}{1\frac{1}{3} + 1\frac{5}{6}}$$

$\left(1\frac{1}{2} + 2\frac{1}{4}\right) \div \left(1\frac{1}{3} + 1\frac{5}{6}\right)$ First let's write this example in horizontal form.

$\left(1\frac{2}{4} + 2\frac{1}{4}\right) \div \left(1\frac{2}{6} + 1\frac{5}{6}\right)$ Next, let's convert to the L.C.D.'s.

$3\frac{3}{4} \div 2\frac{7}{6}$ Now add the whole numbers and the fractions.

$\frac{15}{4} \div \frac{19}{6}$ Convert to improper fractions before you divide.

$\frac{15}{\overset{}{\underset{2}{4}}}\left(\frac{\overset{3}{6}}{19}\right) = \frac{45}{38} = 1\frac{7}{38}$ Divide and reduce the common factor of 2 out of the 4 and the 6. Write your answer in proper form.

Let's do one more together and then you can try the problem set that follows.

SOLUTION: $$\frac{5\frac{1}{3} + 2\frac{3}{4}}{4\frac{5}{9} - 1\frac{1}{3}}$$

$\left(5\frac{1}{3} + 2\frac{3}{4}\right) \div \left(4\frac{5}{9} - 1\frac{1}{3}\right)$ First let's write this example in horizontal form.

$\left(5\frac{4}{12} + 2\frac{9}{12}\right) \div \left(4\frac{5}{9} - 1\frac{3}{9}\right)$ Now, convert to the common denominators because we are adding and subtracting.

$7\frac{13}{12} \div 3\frac{2}{9}$ Add the whole numbers and the fractions.

$\frac{97}{12} \div \frac{29}{9}$ Convert to improper fractions before you divide.

$\frac{97}{\underset{4}{12}} \cdot \frac{\overset{3}{9}}{29} = \frac{291}{116} = 2\frac{59}{116}$ Write your answer as a mixed number.

Problem Set 4.5:

Simplify the following complex fractions. Put your answer in proper form.

1) $\dfrac{\frac{2}{15}}{\frac{4}{5}}$ 2) $\dfrac{2\frac{2}{3}}{\frac{2}{3}}$ 3) $\dfrac{1\frac{1}{3} + 2\frac{3}{5}}{2}$

4) $\dfrac{2\frac{2}{3} - 1\frac{1}{6}}{1\frac{3}{8} + 1\frac{1}{2}}$ 5) $\dfrac{1\frac{1}{3} + \frac{3}{4}}{\frac{7}{12} + \frac{5}{6}}$ 6) $\dfrac{1\frac{1}{5} + \frac{2}{3}}{1\frac{1}{9} + 1\frac{1}{15}}$

1) $\frac{2}{15} \div \frac{4}{5} = \frac{2}{15} \times \frac{5}{4} = \frac{1}{6}$

2) $2\frac{2}{3} \div \frac{2}{3} = \frac{8}{3} \times \frac{3}{2} = 4$

3) $\left(1\frac{1}{3} + 2\frac{3}{5}\right) \div \frac{2}{1} = \left(\frac{4}{3} + \frac{13}{5}\right) \cdot \frac{1}{2} = \left(\frac{20}{15} + \frac{39}{15}\right) \cdot \frac{1}{2} = \frac{59}{15} \cdot \frac{1}{2} = \frac{59}{30} = 1\frac{29}{30}$

4) $\left(2\frac{2}{3} - 1\frac{1}{6}\right) \div \left(1\frac{3}{8} + 1\frac{1}{2}\right) = \left(2\frac{4}{6} - 1\frac{1}{6}\right) \div \left(1\frac{3}{8} + 1\frac{4}{8}\right) = 1\frac{1}{2} \div 2\frac{7}{8} = \frac{3}{2} \cdot \frac{8}{23} = \frac{12}{23}$

5) $\left(1\frac{1}{3} + \frac{3}{4}\right) \div \left(\frac{7}{12} + \frac{5}{6}\right) = \left(1\frac{4}{12} + \frac{9}{12}\right) \div \left(\frac{7}{12} + \frac{10}{12}\right) = 1\frac{13}{12} \div \frac{17}{12} = \frac{25}{12} \cdot \frac{12}{17} = \frac{25}{17} = 1\frac{8}{17}$

6) $\left(1\frac{1}{5} + \frac{2}{3}\right) \div \left(1\frac{1}{9} + 1\frac{1}{15}\right) = \left(1\frac{3}{15} + \frac{10}{15}\right) \div \left(1\frac{5}{45} + 1\frac{3}{45}\right) = \left(1\frac{13}{15}\right) \div \left(2\frac{8}{45}\right)$

$= \frac{28}{15} \div \frac{98}{45} = \frac{28}{15} \times \frac{45}{98} = \frac{6}{7}$

By now you should be pretty good with complex fractions, so you're ready to see the other method for simplifying complex fractions. Let's take a look at how to simplify a problem we did earlier.

Example 3: $\dfrac{\frac{3}{5} - \frac{2}{15}}{\frac{2}{3} - \frac{1}{15}}$

SOLUTION:

$\dfrac{\frac{15}{1}\left(\frac{3}{5}\right) - \left(\frac{2}{15}\right)\frac{15}{1}}{\frac{15}{1}\left(\frac{2}{3}\right) - \left(\frac{1}{15}\right)\frac{15}{1}}$

Step 1: Find the L.C.D. (least common denominator) for ALL the fractions in the numerator and the denominator and multiply EVERY term by the L.C.D. of 15.

$\dfrac{\overset{3}{\cancel{15}}\left(\frac{3}{\cancel{5}}\right) - \left(\frac{2}{\cancel{15}}\right)\overset{1}{\cancel{15}}}{\underset{1}{\cancel{15}}\left(\frac{2}{\cancel{3}}\right) - \left(\frac{1}{\cancel{15}}\right)\overset{1}{\cancel{15}}}$

Step 2: Reduce the common factors.

$\dfrac{3(3) - 2(1)}{5(2) - 1(1)}$

Step 3: Multiply.

$\dfrac{9 - 2}{10 - 1}$

Step 4: Subtract and reduce.
(Be sure to leave your answer in proper form.)

$\dfrac{7}{9}$

Let's do another one together; then you can try some.

Example 4:
$$\dfrac{1\frac{1}{2} + 2\frac{1}{4}}{1\frac{1}{3} + 1\frac{5}{6}}$$

SOLUTION:

$$\dfrac{\frac{3}{2} + \frac{9}{4}}{\frac{4}{3} + \frac{11}{6}}$$

Convert to improper fractions and find the L.C.D., 12, for ALL the fractions.

$$\dfrac{\overset{6}{\underset{1}{\cancel{12}}}\left(\dfrac{3}{\underset{1}{\cancel{2}}}\right) + \left(\dfrac{9}{\underset{1}{\cancel{4}}}\right)\overset{3}{\underset{1}{\cancel{12}}}}{\overset{4}{\underset{1}{\cancel{12}}}\left(\dfrac{4}{\underset{1}{\cancel{3}}}\right) + \left(\dfrac{11}{\underset{1}{\cancel{6}}}\right)\overset{2}{\underset{1}{\cancel{12}}}}$$

Multiply all the terms by the L.C.D. of 12 and reduce.

$$\dfrac{6(3) + 9(3)}{4(4) + 11(2)} = \dfrac{18 + 27}{16 + 22}$$

Multiply.

$$\dfrac{45}{38}$$

Add.

$$1\frac{7}{38}$$

Write your answer as a mixed number.

You give this one a try:

Example 5:
$$\dfrac{5\frac{1}{3} + 2\frac{3}{4}}{4\frac{5}{9} - 1\frac{1}{3}}$$

SOLUTION:
$$\dfrac{\overset{12}{\underset{1}{\cancel{36}}}\left(\dfrac{16}{\underset{1}{\cancel{3}}}\right) + \left(\dfrac{11}{\underset{1}{\cancel{4}}}\right)\overset{9}{\underset{1}{\cancel{36}}}}{\overset{4}{\underset{1}{\cancel{36}}}\left(\dfrac{41}{\underset{1}{\cancel{9}}}\right) - \left(\dfrac{4}{\underset{1}{\cancel{3}}}\right)\overset{12}{\underset{1}{\cancel{36}}}} = \dfrac{12(16) + 11(9)}{4(41) - 4(12)} = \dfrac{192 + 99}{164 - 48}$$

$$= \dfrac{291}{116} = 2\frac{59}{116}$$

You probably noticed that we skipped a couple of steps in our solution. As you do more and more of the same mathematical operations, it's okay to do that. In fact, there are no new operations that we need to do in this section. So we'll skip our customary problem set and go directly to the Workbook.

Simplify the following complex fractions. Put your answer in proper form.

1) $\dfrac{\frac{2}{30}}{\frac{4}{9}}$

2) $\dfrac{\frac{2}{7}}{\frac{4}{7}}$

3) $\dfrac{\frac{3}{14} - \frac{1}{7}}{\frac{1}{5} - \frac{1}{10}}$

4) $\dfrac{\frac{3}{5}}{2\frac{7}{10}}$

5) $\dfrac{1\frac{1}{3} + \frac{3}{4}}{\frac{7}{12} + \frac{5}{6}}$

6) $\dfrac{1\frac{1}{5}}{2\frac{2}{5}}$

7) $\dfrac{2\frac{1}{5} - \frac{7}{15}}{\frac{13}{25} - \frac{3}{10}}$

8) $\dfrac{1\frac{1}{6} - \frac{2}{3}}{1\frac{2}{3} + \frac{5}{12}}$

9) $\dfrac{2\frac{4}{5} - \frac{2}{3}}{1\frac{1}{4} + 1\frac{1}{6}}$

10) $\dfrac{1\frac{1}{3} + 2\frac{3}{5}}{2}$

11) $\dfrac{1\frac{2}{5}}{\frac{2}{5}}$

12) $\dfrac{\frac{6}{12} - \frac{1}{2}}{1\frac{1}{10} - \frac{2}{5}}$

13) $\dfrac{\frac{3}{4} + \frac{1}{2}}{\frac{7}{8} - \frac{5}{12}}$

14) $\dfrac{\frac{3}{13} + 2\frac{1}{2}}{\frac{1}{26} + 1\frac{1}{13}}$

15) $\dfrac{\frac{13}{16} - \frac{3}{4}}{1\frac{1}{8} + \frac{1}{3}}$

16) $\dfrac{1\frac{1}{14} + 1\frac{1}{4}}{1\frac{1}{7} + 1\frac{1}{2}}$

17) $\dfrac{\frac{2}{3} + \frac{3}{4}}{\frac{2}{9} - \frac{1}{6}}$

18) $\dfrac{\frac{1}{2} + \frac{1}{3}}{\frac{1}{5} - \frac{1}{15}}$

19) $\dfrac{1\frac{1}{5} + \frac{2}{3}}{1\frac{1}{9} + 1\frac{1}{15}}$

20) $\dfrac{5\frac{1}{5} - 4\frac{1}{2}}{\frac{3}{4} - \frac{3}{20}}$

4.6 Order of Operations Applied to Mixed Numbers

The order of operations we applied to whole numbers and fractions also applies to mixed numbers. In this section we'll give you some examples and problems to practice applying the order of operations to mixed numbers. Remember, you MUST use improper fractions when multiplying or dividing.

Remember, the order of operations is:

Step 1:   Inside groupings first

Step 2:   Exponents

Step 3:   Multiplication/Division (left to right)

Step 4:   Addition/Subtraction (left to right)

Example 1: $6 + \left(2\frac{1}{2}\right)\left(1\frac{3}{5}\right)$

SOLUTION: $6 + \left(\frac{5}{2} \times \frac{8}{5}\right)$

Multiply first, using improper fractions:

$$2\frac{1}{2} = \frac{5}{2}, \quad 1\frac{3}{5} = \frac{8}{5},$$

$$\left(\frac{5}{2} \times \frac{8}{5}\right) = 4$$

Now add the remaining numbers.

$6 + 4$ $6 + 4 = 10$

Example 2: $\left(\frac{1}{5} + \frac{1}{25}\right)\left(25 - 2\frac{1}{2}\right)$

SOLUTION: First do the addition and subtraction inside the grouping. Find the common denominators and covert to the common denominators; then add.

$\left(\frac{5}{25} + \frac{1}{25}\right)\left(\frac{50}{2} - \frac{5}{2}\right)$ $\frac{1}{5} = \frac{5}{25}, \ \frac{5}{25} + \frac{1}{25} = \frac{6}{25}, \ 25 = \frac{50}{2}, \ 2\frac{1}{2} = \frac{5}{2}$

$$\frac{50}{2} - \frac{5}{2} = \frac{45}{2}$$

Now reduce the common factors; then multiply.

$\left(\frac{6}{25} \times \frac{45}{2}\right)$ $\left(\frac{6}{25} \times \frac{45}{2}\right) = \frac{27}{5}$

Convert the improper fraction to a mixed number.

$\frac{27}{5} = 5\frac{2}{5}$ $\frac{27}{5} = 5\frac{2}{5}$

Example 3: $\frac{2}{5} + \frac{1}{4}\left(3\frac{2}{3} + 4\frac{1}{3}\right)^2$

SOLUTION: First do the addition inside the grouping.

$\frac{2}{5} + \frac{1}{4}(8)^2$

$$3\frac{2}{3}$$
$$+4\frac{1}{3}$$
$$\overline{7\frac{3}{3}} = 7 + 1 = 8$$

Next do the exponent.

$(8)^2 = 8(8) = 64$

Now multiply.

$\frac{2}{5} + \frac{1}{\overset{}{\underset{1}{4}}}\left(\frac{\overset{16}{\cancel{64}}}{1}\right)$

$\frac{1}{\overset{}{\underset{1}{4}}}\left(\frac{\overset{16}{\cancel{64}}}{1}\right) = 16$

Finally add what's left.

$\frac{2}{5} + 16 = 16\frac{2}{5}$

$\frac{2}{5} + 16 = 16\frac{2}{5}$

Example 4: $3\frac{3}{4} \div \left(\frac{3}{4}\right)\left(\frac{1}{5}\right)$

SOLUTION: $\overset{1}{\underset{1}{\frac{\overset{\cancel{\overset{1}{\cancel{8}}}}{\cancel{15}}}{4}}} \bullet \overset{1}{\underset{1}{\frac{4}{3}}} \bullet \overset{1}{\underset{1}{\frac{1}{5}}} = 1$

Should we divide or multiply first? Remember, multiplication and division are done left to right. Here we'll divide and then multiply, which leaves us with three fractions multiplied together. Reduce the common factors. Finally, multiply.

Problem Set 4.6:

Perform the following computations.

1) $3 + \left(4\frac{1}{3}\right)\left(2\frac{4}{5}\right)$

2) $8 - \left(3\frac{1}{2}\right)\left(1\frac{5}{6}\right)$

3) $2\frac{2}{7} \div \left(1\frac{3}{4}\right)\left(1\frac{10}{40}\right)$

4) $1\frac{3}{8} \div \left(1\frac{5}{6}\right)\left(1\frac{1}{3}\right)$

5) $\left(\frac{1}{4} + \frac{3}{8}\right)\left(6\frac{1}{9} + 3\frac{2}{9}\right)$

6) $\left(\frac{2}{5} - \frac{1}{3}\right)\left(17\frac{4}{9} - 8\frac{1}{3}\right)$

7) $\frac{3}{8} + \frac{4}{5}\left(6\frac{1}{4} + 3\frac{3}{4}\right)^2$

8) $\frac{9}{10} + \frac{7}{20}\left(8\frac{3}{5} - 2\frac{3}{5}\right)^2$

9) $3\frac{1}{8} + \left(2\frac{5}{8}\right)\left(\frac{1}{2}\right)^2$

10) $4\frac{5}{6} \div \left(\frac{4}{16} - \frac{1}{4}\right)$

11) $\left(5\frac{1}{4} + \frac{7}{8}\right) \div \left(1\frac{4}{5} + 1\frac{1}{5}\right)^2$

12) $\left(6\frac{5}{7} - 1\frac{5}{7}\right)^3 \div \left(2\frac{2}{3} + 2\frac{1}{3}\right)^2$

SOLUTIONS TO PROBLEM SET:

1) $3 + \left(4\frac{1}{3}\right)\left(2\frac{4}{5}\right) = 3 + \left(\frac{13}{3}\right)\left(\frac{14}{5}\right) = 3 + \frac{182}{15} = 3 + 12\frac{2}{15} + 15\frac{2}{15}$

2) $8 - \left(3\frac{1}{2}\right)\left(1\frac{5}{6}\right) = 8 - \left(\frac{7}{2}\right)\left(\frac{11}{6}\right) = 8 - \left(\frac{77}{12}\right) = 8 - 6\frac{5}{12} = 7\frac{12}{12} - 6\frac{5}{12} = 1\frac{7}{12}$

3) $2\frac{2}{7} \div \left(1\frac{3}{4}\right)\left(1\frac{10}{40}\right) = \frac{16}{7} \bullet \overset{1}{\cancel{\frac{4}{7}}} \bullet \overset{}{\underset{1}{\cancel{\frac{5}{4}}}} = \frac{80}{49} = 1\frac{31}{49}$

4) $1\frac{3}{8} \div \left(1\frac{5}{6}\right)\left(1\frac{1}{3}\right) = \overset{\frac{1}{2}}{\underset{\frac{4}{1}}{\cancel{\frac{11}{8}}}} \bullet \overset{\frac{6}{11}}{\cancel{}} \bullet \overset{1}{\underset{1}{\cancel{\frac{4}{3}}}} = 1$

5) $\left(\frac{1}{4} + \frac{3}{8}\right)\left(6\frac{1}{9} + 3\frac{2}{9}\right) = \left(\frac{2}{8} + \frac{3}{8}\right)\left(9\frac{3}{9}\right) = \frac{5}{8} \times 9\frac{1}{3} = \overset{}{\underset{2}{\cancel{\frac{5}{8}}}} \times \overset{7}{\cancel{\frac{28}{3}}} = \frac{35}{6} = 5\frac{5}{6}$

6) $\left(\frac{2}{5} - \frac{1}{3}\right)\left(17\frac{4}{9} - 8\frac{1}{3}\right) = \left(\frac{6}{15} - \frac{5}{15}\right)\left(17\frac{4}{9} - 8\frac{3}{9}\right) = \frac{1}{15}\left(9\frac{1}{9}\right) = \frac{1}{15} \times \frac{82}{9} = \frac{82}{135}$

7) $\frac{3}{8} + \frac{4}{5}\left(6\frac{1}{4} + 3\frac{3}{4}\right)^2 = \frac{3}{8} + \frac{4}{5}(10)^2 = \frac{3}{8} + \frac{4}{5} \times \overset{20}{\cancel{\frac{100}{1}}} = \frac{3}{8} + 80 = 80\frac{3}{8}$

8) $\frac{9}{10} + \frac{7}{20}\left(8\frac{3}{5} - 2\frac{3}{5}\right)^2 = \frac{9}{10} + \frac{7}{20}(6)^2 = \frac{9}{10} + \overset{}{\underset{10}{\cancel{\frac{7}{20}}}} \times \overset{18}{\cancel{\frac{36}{1}}} = \frac{9}{10} + \frac{126}{10}$
$= \frac{135}{10} = 13\frac{5}{10} = 13\frac{1}{2}$

9) $3\frac{1}{8} + \left(2\frac{5}{8}\right)\left(\frac{1}{2}\right)^2 = 3\frac{1}{8} + \left(\frac{21}{8}\right)\left(\frac{1}{4}\right) = 3\frac{1}{8} + \frac{21}{32} = 3\frac{4}{32} + \frac{21}{32} = 3\frac{25}{32}$

10) $4\frac{5}{6} \div \left(\frac{4}{16} - \frac{1}{4}\right) = 4\frac{5}{6} \div 0$ is undefined

11) $\left(5\frac{1}{4} + \frac{7}{8}\right) \div \left(1\frac{4}{5} + 1\frac{1}{5}\right)^2 = \left(5\frac{2}{8} + \frac{7}{8}\right) \div (3)^2 = 5\frac{9}{8} \div \frac{9}{1} = \frac{49}{8} \times \frac{1}{9} = \frac{49}{72}$

12) $\left(6\frac{5}{7} - 1\frac{5}{7}\right)^3 \div \left(2\frac{2}{3} + 2\frac{1}{3}\right)^2 = (5)^3 \div (5)^2 = 125 \div 25 = 5$

Perform the following computations:

1) $4 + \left(7\frac{2}{3}\right)\left(2\frac{1}{4}\right)$

2) $2 + \left(1\frac{1}{3}\right)\left(\frac{9}{16}\right)$

3) $\left(\frac{1}{8} + \frac{2}{5}\right)\left(7\frac{1}{4} + 2\frac{1}{6}\right)$

4) $16 - \left(2\frac{1}{4}\right)\left(2\frac{5}{9}\right)$

5) $\left(\frac{5}{12} + \frac{1}{4}\right)\left(1\frac{4}{5} + 1\frac{5}{8}\right)$

6) $4\frac{2}{5} \div \left(5\frac{1}{15}\right)\left(1\frac{6}{15}\right)$

7) $\frac{8}{9} + \frac{5}{64}\left(9\frac{3}{7} - 1\frac{3}{7}\right)^2$

8) $6 + \left(\frac{1}{3}\right)^2\left(\frac{13}{6}\right)$

9) $39 - \left(4\frac{4}{5}\right)\left(3\frac{3}{4}\right)$

10) $\left[\left(\frac{1}{4}\right)^2 - \left(\frac{1}{2}\right)^4\right] \div \left(6\frac{1}{5} - 2\frac{3}{8}\right)$

11) $\left(\frac{4}{5} - \frac{2}{3}\right)\left(9\frac{4}{9} - 7\frac{1}{3}\right)$

12) $\left(\frac{3}{5} - \frac{1}{3}\right)\left(4\frac{1}{2} - 1\frac{1}{8}\right)$

13) $\frac{4}{5} + \frac{2}{3}\left(4\frac{3}{4} + 3\frac{1}{4}\right)^2$

14) $5 + \frac{3}{4}\left(6\frac{1}{2} - \frac{1}{2}\right)^2$

15) $\left(\frac{7}{15} + \frac{2}{5}\right) \div \left(1\frac{2}{3} - \frac{4}{5}\right)$

16) $\left(\frac{4}{5} + 1\frac{1}{5}\right)\left(2\frac{5}{9} + 2\frac{4}{9}\right)^2$

CHALLENGE PROBLEMS

17) $\frac{1}{16} + \left\{\left[14 - \left(2\frac{1}{2} - \frac{1}{2}\right)^2\right] \div 20\right\}^4$

18) $\left[2\frac{6}{7} \cdot \frac{70}{140} - 1\frac{3}{7}\right] \div \left(2\frac{1}{4} \div 5\frac{1}{8}\right)$

19) $\left(4\frac{1}{3} - 2\frac{3}{9}\right) \div \left[\left(\frac{1}{4}\right)^2 - \left(\frac{1}{2}\right)^4\right]$

20) $\left[\left(2\frac{4}{15} + 3\frac{6}{15}\right) \times 1\frac{4}{17}\right]^3 \div \left(1\frac{1}{5}\right)^2$

Chapter 4 • Review

Write your answers in proper reduced form.

1) $2\frac{1}{2} \div 3\frac{2}{3}$

2) $4\frac{1}{5} \div 0$

3) $0 \div 4\frac{1}{5}$

4) $\left(1\frac{1}{2}\right)^3$

5) $1\frac{2}{3} + 2\frac{1}{8}$

6) Convert $\frac{7}{3}$ into a mixed number.

7) Convert $4\frac{3}{5}$ into an improper fraction.

8) I would like to make 12 shelves that are $4\frac{3}{4}$ feet long. How long a piece of wood would I need?

9) $\dfrac{\frac{1}{3} + \frac{3}{4}}{\frac{5}{6} + \frac{1}{2}}$

10) $\dfrac{1\frac{7}{8} + \frac{7}{12}}{1\frac{5}{6} - \frac{3}{8}}$

11) $5\frac{3}{5} - 2\frac{1}{4}$

12) $\left(1\frac{2}{7}\right)^2$

13) $1\frac{1}{3} \times 2\frac{2}{5}$

14) Convert $4\frac{4}{5}$ into an improper fraction.

15) $3\frac{4}{7}\left(9\frac{4}{5}\right)$

16) $\dfrac{1\frac{1}{8} + 2\frac{2}{3}}{2\frac{3}{4} + 3\frac{1}{6}}$

17) If $\frac{3}{4}$ of 48 people in a room are married, how many people in the room are married?

18) A woman walked $3\frac{1}{2}$ miles in the morning, $4\frac{1}{3}$ miles in the afternoon, and $2\frac{2}{3}$ miles in the evening. How far did she walk that day?

19) Convert $\frac{12}{5}$ into a mixed number.

20) $2\frac{3}{4} \div 3\frac{5}{8}$

21) $\left(1\frac{2}{3} - \frac{1}{6}\right)^2 \div \left(\frac{1}{3}\right)^2$

22) A farmer buys $3\frac{1}{2}$ acres of land in January, $2\frac{3}{4}$ acres in February, and $1\frac{3}{5}$ acres in March. How many acres does he buy altogether?

23) Convert $\frac{12}{5}$ into a mixed number.

24) $4\frac{2}{5} \times 3\frac{3}{4}$

25) $5\frac{7}{8} \div 2\frac{3}{10}$

26) $\left(2\frac{1}{4} - \frac{5}{6}\right) \times \left(\frac{3}{4} - \frac{1}{2}\right)^2$

27) $\dfrac{\frac{4}{5} + \frac{3}{10}}{6 - \frac{8}{15}}$

28) Convert $10\frac{5}{8}$ into an improper fraction.

29) Peri has $12\frac{1}{4}$ yards of cloth. She gives $5\frac{5}{6}$ yards to Jim. How many yards of cloth does she have left?

30) $\frac{2}{3} + \frac{1}{5}\left(2\frac{1}{3} + 1\frac{2}{3}\right)^2$

31) $2\frac{7}{12} + 4\frac{1}{3}$

32) $16 \div 2\frac{2}{3}$

1) If Shawn can run $8\frac{4}{5}$ miles in one hour, how far can he run at that speed in $2\frac{1}{2}$ hours?

2) Convert $\frac{13}{4}$ into a mixed number.

3) Convert $4\frac{3}{8}$ into an improper fraction.

4) $\dfrac{\frac{6}{12} - \frac{1}{2}}{1\frac{1}{10} - \frac{2}{5}}$

5) $\dfrac{\frac{3}{13} + 2\frac{1}{2}}{\frac{1}{26} + 1\frac{1}{13}}$

6) Suppose you walked $1\frac{3}{4}$ miles from your house, stopped for a few minutes to rest, and then continued for another $2\frac{1}{5}$ miles, rested for a few more minutes, and then decided to walk back home. If you walked back $1\frac{1}{3}$ miles, how far would you be from your house?

7) $\left(3\frac{2}{3}\right)^2$

8) $2\frac{5}{8} \div 4\frac{1}{2}$

9) Change $\frac{19}{5}$ into a mixed number.

10) $\left(3\frac{1}{8} - 2\frac{1}{4}\right) + \left(5\frac{2}{3} + 1\frac{1}{3}\right)^2$

11) $13\frac{1}{10} - 1\frac{1}{20} \div 2\frac{1}{10}$

12) Convert $3\frac{2}{9}$ into an improper fraction.

13) $\left(4\frac{7}{8} - 2\frac{1}{4}\right) - \left(2\frac{2}{3} + 1\frac{1}{3}\right)^3$

14) A jar containing $78\frac{1}{2}$ ounces of juice was leaking $2\frac{1}{4}$ ounces an hour. How many ounces of juice were still in the jar after three and a half hours?

15) How many pieces of cloth $1\frac{3}{4}$ inches long can be cut from a piece of cloth that is 70 inches long?

16) $0 \div 5\frac{1}{2}$

17) $\left[2\frac{1}{2} - 1\frac{2}{3}\right] \div \left[\left(\frac{1}{3}\right)^4 - \left(\frac{1}{9}\right)^2\right]$

18) $1\frac{2}{3}\left(1\frac{4}{5}\right)$

19) $\begin{array}{r} 4\frac{1}{3} \\ -\ 2\frac{3}{4} \\ \hline \end{array}$

20) $0 \div 1\frac{3}{8}$

21) $13\frac{1}{3} \ \bullet \ 1\frac{1}{2} \ \div \ \left(1\frac{1}{2}\right)^2$

22) If Sue walked $2\frac{3}{4}$ miles on Monday, $4\frac{5}{6}$ miles on Wednesday, $1\frac{2}{5}$ miles on Friday, and $4\frac{1}{12}$ miles on Sunday, how far did she walk that week?

23) $\left[\left(\frac{1}{4}\right)^3 - \left(\frac{1}{8}\right)^2\right] \div 6\frac{2}{3}$

24) $\left(6 - 2\frac{4}{5}\right) \div \left(3\frac{5}{7} - 2\right)$

25) A cooler contains $71\frac{1}{2}$ ounces of Gatorade. How many $6\frac{1}{2}$ ounce cups can be poured from the cooler?

26) $1\frac{3}{8} \div 0$

Chapter 4 • Test 2

1) Change $12\frac{1}{2}$ to an improper fraction.

2) Change $\frac{42}{11}$ to a mixed number.

3) $1\frac{2}{3}\left(1\frac{4}{5}\right)$

4) $\left(3\frac{1}{2}\right)^3$

5) $1\frac{2}{3} \div 1\frac{4}{5}$

6) $\begin{aligned} 2\tfrac{5}{7} \\ +\,3\tfrac{4}{5} \end{aligned}$

7) $\begin{aligned} 4\tfrac{1}{3} \\ -\,2\tfrac{3}{4} \end{aligned}$

8) $0 = 1\frac{3}{8}$

9) If Sue walked $2\frac{3}{4}$ miles on Monday, $4\frac{5}{6}$ miles on Wednesday, $1\frac{2}{5}$ miles on Friday, and $4\frac{1}{12}$ miles on Sunday, how far did she walk that week?

10) $\dfrac{1\frac{4}{5} + 2\frac{1}{3}}{6\frac{1}{25} - 4\frac{2}{3}}$

11) $13\frac{1}{3} \cdot 1\frac{1}{2} \div \left(1\frac{1}{2}\right)^2$

12) $\left[\left(\frac{1}{4}\right)^3 - \left(\frac{1}{8}\right)^2\right] \div 6\frac{2}{3}$

13) A cooler contains $71\frac{1}{2}$ ounces of Gatorade. How many $6\frac{1}{2}$ ounce cups can be poured from the cooler?

14) $\left(3\frac{1}{2} - 1\frac{1}{2}\right)^2 \div 1\frac{1}{2}$

15) $\left(6 - 2\frac{4}{5}\right) \div \left(3\frac{5}{7} - 2\right)$

16) $\dfrac{1\frac{1}{3} + 1\frac{3}{4}}{2\frac{1}{8} - 1\frac{1}{24}}$

17) $9\frac{3}{5}\left(2\frac{3}{12}\right)\left(2\frac{7}{9}\right)$

18) $9\frac{3}{5} - 2\frac{3}{12} \div 2\frac{7}{9}$

19) $12\frac{1}{3} - 2\frac{4}{5} - 3\frac{1}{15}$

20) $\left(3\frac{1}{2} + 4\frac{2}{3}\right)^2$

21) $\left(2 - 1\frac{1}{5}\right) \div \left(4\frac{2}{5} - \frac{2}{5}\right)$

22) $7\frac{1}{21} + 11\frac{2}{3} - 6\frac{8}{9}$

23) Emilio went on a diet to gain weight. He gained $8\frac{1}{2}$ lbs. in the first week, $6\frac{3}{8}$ lbs. the second week, $5\frac{3}{4}$ lbs. the third week and $2\frac{1}{2}$ lbs. the fourth week. How many pounds did Emilio gain all together.

24) How many $\frac{1}{4}$ lb. bags of coffee can be formed from a 50 pound container?

25) If Steve can run $8\frac{2}{5}$ miles in one hour, how far can he run at that speed in $3\frac{1}{2}$ hours?

Chapter 5. Decimals

If you thought the last couple of chapters were fun, just wait till you sink your teeth into decimals. We're not only going to add, subtract, multiply, and divide decimals, but we're going to do a whole lot more. So what are we waiting for? Let's get started!

5.1 Place Value

In Chapter 1 we talked about the place value of whole numbers. In this chapter we'll talk about the place value of decimals. Decimals have many applications in our world. Our favorite application is money. Let's look at the place values of $24.31. We know from chapter 1 that the 2 is in the tens' place and has a value of 2(10), or 20, and the 4 is in the ones' place, so it has a value of 4(1), or 4. Next comes the decimal point. The place value table below, which you should memorize, shows us the values of the numbers to the right of the decimal point. Notice the first place value to the right of the decimal point is the tenths. There isn't a ones or units place to the right of the decimal point.

TABLE 5.1   DECIMAL PLACE VALUES

Decimal Point	Tenth	Hundredth	Thousandth	Ten Thousandth	Hundred Thousandth
.	$\dfrac{1}{10}$	$\dfrac{1}{100}$	$\dfrac{1}{1,000}$	$\dfrac{1}{10,000}$	$\dfrac{1}{100,000}$

The 3 in $24.31 is in the tenths' place, so it has a value of $3\left(\dfrac{1}{10}\right)$ or $\dfrac{3}{10}$. In money that would be 3 dimes or 30 cents. The one is in the hundredths' place, so it has a value of $1\left(\dfrac{1}{100}\right)$ or $\dfrac{1}{100}$. In money that would be worth 1 cent. Notice that whenever a number is to the right of the decimal point, it has a th at the end of its place value. Let's look at the place value of a few numbers.

Example 1:   Let's list the value of each of the digits in the number 0.12345 and write it in expanded form.

SOLUTION:

the 1 is in the tenths' place, so it has a value of
$$1\left(\frac{1}{10}\right) = \frac{1}{10}.$$

the 2 is in the hundredths' place, so it has a value of
$$2\left(\frac{1}{100}\right) = \frac{2}{100}.$$

the 3 is in the thousandths' place, so it has a value of
$$3\left(\frac{1}{1,000}\right) = \frac{3}{1,000}.$$

the 4 is in the ten thousandths' place, so it has a value of
$$4\left(\frac{1}{10,000}\right) = \frac{4}{10,000}.$$

the 5 is in the hundred thousandths' place, so it has a value of
$$\left(\frac{1}{100,000}\right) = \frac{5}{100,000}.$$

Expanded form: $0.12345 = \dfrac{1}{10} + \dfrac{2}{100} + \dfrac{3}{1,000} + \dfrac{4}{10,000} + \dfrac{5}{100,000}.$

0.12345 is read twelve thousand, three hundred forty-five hundred thousandths, because the last place value occupied in this number is the hundred thousandths' place.

Example 2:   Write 78.256 in expanded form.

SOLUTION: 78.256 is read 78 and 256 thousandths, because the last digit, 6, is in the thousandths' place. Notice the word "and" is used to represent the decimal point.

Expanded form: $78.256 = 70 + 8 + \dfrac{2}{10} + \dfrac{5}{100} + \dfrac{6}{1,000}.$

Example 3:   Write 145.09 in expanded form.

SOLUTION: 145.09 is read one hundred forty-five, and nine hundredths.

Expanded form: $145.09 = 100 + 40 + 5 + \dfrac{9}{100}.$

Now it's your turn to try a few problems.

Problem Set:

State the place value of the 2 in the following numbers; write the number in expanded form; and write the number in words.

1)   36.278

2)   1,567.3682

3)   0.0024

4)   2.01

5)   18.325

1) tenths' place
   36.278 is read thirty-six and two hundred seventy-eight thousandths.

   Expanded form: $36.278 = 30 + 6 + \dfrac{2}{10} + \dfrac{7}{100} + \dfrac{8}{1,000}.$

2) ten thousandths' place
   1,567.3682 is read one thousand, five hundred sixty-seven, and three thousand, six hundred eighty-two ten thousandths.

   Expanded form: $1,567.3682 = 1,000 + 500 + 60 + 7 + \dfrac{3}{10} + \dfrac{6}{100} + \dfrac{8}{1,000} + \dfrac{2}{10,000}.$

3) thousandths' place
   0.0024 is read twenty-four ten thousandths.

   Expanded form: $0.0024 = \dfrac{2}{1,000} + \dfrac{4}{10,000}.$

4) ones' place
   2.01 is read two and one hundredth.

   Expanded form: $2.01 = 2 + \dfrac{1}{100}.$

5) hundredths' place
   18.325 is read eighteen, and three hundred twenty-five thousandths.

   Expanded form: $18.325 = 10 + 8 + \dfrac{3}{10} + \dfrac{2}{100} + \dfrac{5}{1,000}.$

Ordering Decimals

Sometimes it's important for us to be able to look at two decimals and know which one is the higher value. There are times when we need to list decimals in order from, lowest to highest, or highest to lowest. This section will help us do that, but first we have to learn that adding zeros at the end of a decimal doesn't change the value of the decimal. Let us convince you of that by adding zeros to the end of the number 0.1, and comparing the values of the numbers 0.1 (one tenth), 0.10 (ten hundredths), and 0.100 (one hundred thousandths).

$0.1 = \dfrac{1}{10}$ $0.10 = \dfrac{10}{100} = \dfrac{1}{10}$ $0.100 = \dfrac{100}{1000} = \dfrac{1}{10}$

Notice 0.1, and 0.10, and 0.100 all equal $\dfrac{1}{10}$, so they're all the same value. By now you should be convinced that adding zeros at the end of a decimal doesn't change its value.

Example 4: **Which number has the greater value, 0.2 or 0.002?**

SOLUTION: We'll start by adding two zeros to 0.2 to make it 0.200. We chose to add two zeros, not one or three or four, because we want the numbers we're comparing to have the same amount of decimal places. Since 0.002 already has three places, we want 0.2 to have three decimal places; so we'll add two zeros to make it 0.200.

Now let's compare 0.200 (200 thousandths) and 0.002 (2 thousandths). Clearly two hundred thousandths is a greater value than two thousandths, so 0.2 is greater than 0.002.

Example 5: **Which number has the lowest value, 0.34, 0.025, or 1.002?**

SOLUTION: Right away we should notice that 1.002 can't possibly be the lowest value because it has a whole number of 1, which automatically makes it greater than 0.34 and 0.025. We just need to compare 0.34 and 0.025. We'll add a zero to 0.34 to make it 0.340, so 0.340 and 0.025 have the same amount of decimal places. Now it's easy to see that 0.025 (25 thousandths) is a lower value than 0.340 (340 thousandths), so 0.025 is the lowest value.

Example 6: **Write the numbers 1.1, 1.0021, and 1.23 in order from lowest to highest.**

SOLUTION: The first thing we're going to do is add the necessary amount of zeros, so all numbers have the same amount of decimal places. Now it's easy to compare the numbers.

1.1	=	1.1000	(1 and 1,000 ten thousandths)
1.0021	=	1.0021	(1 and 21 ten thousandths)
1.23	=	1.2300	(1 and 2,300 ten thousandths)

From lowest to highest the answer is : 1.0021, then 1.1, and then 1.23.

Example 7: **Write the numbers 6.01, 0.601, and 0.61 from highest to lowest.**

SOLUTION: We know 6.01 has to be the highest value because of the whole number 6, so we really only need to compare the 0.601 and the 0.61.

0.601	=	0.601	(601 thousandths)
0.61	=	0.610	(610 thousandths)

From highest to lowest the answer is 6.01, then 0.61, then 0.601.

Problem Set:

1) Which number has the higher value, 3.018 or 3.108?

2) Which number has the lower value, 0.0194 or 0.00952?

3) Order the following numbers from highest to lowest:
   7.89, 0.789, 0.0789, 0.009.

4) Order the following numbers from lowest to highest:
   0.016, 0.0054, 0.504, 0.0405.

SOLUTIONS TO PROBLEM SET:

1) 3.018 (3 and 18 thousandths)
   3.108 (3 and 108 thousandths)

3.108 has the higher value.

2) 0.0194 = 0.01940 (1,940 hundred thousandths)
   0.00952 = 0.00952 (952 hundred thousandths)

0.00952 has the lower value.

3) 7.89 (because this has the whole number 7, it has to be the highest value).
   0.789 = 0.7890 (7,890 ten thousandths)
   0.0789 = 0.0789 (789 ten thousandths)
   0.009 = 0.0090 (90 ten thousandths)

Highest to lowest: 7.89, 0.789, 0.0789, 0.009

4) 0.016 = 0.0160 (160 ten thousandths)
   0.0054 = 0.0054 (54 ten thousandths)
   0.504 = 0.5040 (5,040 ten thousandths)
   0.0405 = 0.0405 (405 ten thousandths)

Lowest to highest: 0.0054, 0.016, 0.0405, 0.504

State the place value of the 6 in the following problems; write the number in expanded form; and write the number in words.

1) 0.4689

2) 0.678

3) 4.689

4) 0.00678

5) 0.04689

6) 60.78

7) 46.89

8) 0.00678

9) 0.6

10) 0.006

11) 0.0006

12) 0.00006

13) 652.002

14) 16.009

15) Which number has the greater value?   0.003 or 0.03

16) Which number has the greater value?   1.2308 or 1.0238

17) Which number has the lesser value?   0.003 or 0.0303

18) Which number has the greatest value?   5.67 or 0.567 or 0.0567

19) Which number has the greatest value?   0.8905 or 0.0985 or 0.908

20) Which number has the lowest value?   0.001 or 0.04 or 1.0001

21) Which number has the lowest value?   0.3 or 0.33 or 0.0333

22) Order the following numbers from lowest to highest: 5.55, 0.055, 0.0456

23) Order the following numbers from lowest to highest: 2.47, 0.246, 2.0988

24) Order the following numbers from highest to lowest: 0.589, 0.5, 0.056

25) Order the following numbers from highest to lowest: 0.67, 0.087, 0.00921

26) Order the following numbers from highest to lowest: 23.45, 0.2345, 0.02345

27) Order the following numbers from lowest to highest: 0.1, 0.01, 0.001, 1.1

5.2   Rounding Decimals

Rounding decimals is almost the same as rounding whole numbers, which we covered in section 1.2. We'll use the same techniques. In Chapter 1 we would have asked you to round to the nearest tens, in this section we'll ask you to round to the nearest tenths. The ths at the end of the word means we're asking you to round to a place value to the right of the decimal point. If there isn't a ths at the end of the word, we're asking you to round to a place value to the left of the decimal point; which is the whole numbers.

Example 1:   Round 0.6 to the nearest whole number.

SOLUTION: Is 0.6 closer to 1 or 0? It's closer to 1, so we round 0.6 up to 1.

The other way to do this problem is to look one place to the right of the decimal point, where we have a 6, which is five or more, so we round up. From chapter 1 we know that if the number to the right of the place value we're asked to round off to is 5 or greater, we round up by adding 1 to the digit in the place value we're rounding to.

Example 2:   Round 4.04 to the nearest tenth.

SOLUTION: Is 4.04 closer to 4.10 or 4.00? It's closer to 4.00, so we round to 4.0. The other way to do this problem is to look one place to the right of the tenths' place. We see a 4, so we don't add a one to the tenths' place and everything to its right becomes zeros. We don't need to write zeros at the end of a decimal, because it doesn't change the value of the decimal; 0.4 is the same value as 0.40 or 0.400 or 0.4000, and so on.

Example 3:   Round 7.0465 to the nearest hundredth.

SOLUTION: When we look one place to the right of the hundredths' place, we see a 6, so we add a one to the 4 and round to 7.05.

Example 4:   Round 0.0017 to the nearest hundredth.

SOLUTION: When we look one place to the right of the hundredths' place, we see a 1, so we don't add a 1 to the hundredths' place and round to 0.00.

Example 5:   Round 456.7999 to the nearest thousandth.

SOLUTION: When we look one place to the right of the thousandths' place, we see a 9, so we add 1 to the 9 in the thousandths' place and round to 456.800.

$$
\begin{array}{r}
456.7999 \\
+ 1 \\
\hline
456.800
\end{array}
$$

Example 6: Round 657.982 to the nearest hundred.

  SOLUTION: Be careful—this asks for the nearest hundred not hundredth. This is really an example from chapter 1. If we want to round to the nearest hundred, not hundredth, we look one place to the right of the hundreds' place and see a 5, so we round up to 700.

Example 7: Round 6.0471 to the nearest ten thousandth.

  SOLUTION: When we look one place to the right of the ten thousandths' place we don't see a digit, so we assume it's a 0. As we've already seen, adding a 0 to the right of a decimal number doesn't change the number's value. We round to 6.0471.

Example 8: Round 12.895999 to the nearest hundred thousandth.

  SOLUTION: When we look one place to the right of the hundred thousandths' place, we see a 9, so we round up by adding a 1 to the 9 in the hundred thousandths' place. The consecutive nines causes us to carry. The answer is 12.89600.

$$\begin{array}{r} 12.895999 \\ +\qquad 1 \\ \hline 12.89600 \end{array}$$

Example 9: Melanie spent $125.78 shopping at the Gap. How much money did she spend to the nearest:

  a) dollar   b) dime   c) penny

  SOLUTION:
  a) $126   The nearest dollar is the ones' place.
  b) $125.80   The nearest dime is the tenths' place.
  c) $125.78   The nearest penny is the hundredths' place.

Problem Set:

Round each of the following numbers to the nearest tenth, hundredth, and thousandth.

1) 0.96   2) 54.349   3) 5.9951   4) 0.01499

Round each of the following numbers to the nearest ten thousandth and hundred thousandth.

5) 14.582956   6) 0.006492   7) 1.806396   8) 0.003626

SOLUTIONS TO PROBLEM SET:

Number	Tenth	Hundredth	Thousandth
1) 0.96	1.0 or 1	0.96	0.960
2) 54.349	54.3	54.35	54.349
3) 5.9951	6.0	6.00	5.995
4) 0.01499	0.0	0.01	0.015

Number	Ten Thousandth	Hundredth Thousandth
5) 14.582956	14.5830	14.58296
6) 0.006492	0.0065	0.00649
7) 1.806396	1.8064	1.80640
8) 0 .003626	0.0036	0.00363

1) Round 4.68 to the nearest tenth.

2) Round 67.9245 to the nearest thousandth.

3) Round 0.000791 to the nearest hundred thousandth.

4) Round 0.28282 to the nearest hundredth.

5) Round 14.68 to the nearest ten.

6) Round 167.9245 to the nearest hundred.

7) Round each of these numbers to the nearest tenth, hundredth, and thousandth.
   a) 14.491   b) 1.9746   c) 0.48888   d) 0.1209   e) 0.09090909

8) Round each of these numbers to the nearest ten thousandth, and hundred thousandth.
   a) 25.6788   b) 0.09471   c) 9.999999   d) 0.000938   e) 2.22353

9) Round each of these numbers to the nearest tenth.
   a) 0.98   b) 0.889   c) 5.56   d) 0.789   e) 0.099

10) Round each of these numbers to the nearest hundredth.
    a) 2.5678   b) 25.678   c) 256.782   d) 2567.813

11) Round each of these numbers to the nearest thousandth.
    a) 9.9099   b) 99.099   c) 990.990321   d) 9909.90041

12) Round each of these numbers to the nearest ten thousandth and hundred thousandth.
    a) 0.000347   b) 2.209187   c) 23.700999   d) 0.0001234

13) Round each of these numbers to the nearest hundred thousandth.

a) 248.099256
b) 99.0098
c) 12,093.789601
d) 88.95999

14) Jessica spent $124.67 in the grocery store. How much did she spend to the nearest:

a) dollar
b) dime
c) penny

15) Karen spent $209.09 shopping for new clothes. How much did she spend to the nearest:

a) hundred dollars
b) dollar
c) dime
d) cent

5.3 Addition and Subtraction of Decimals

In the first chapter we studied addition and subtraction of whole numbers. Addition and subtraction of decimals is the same as addition and subtraction of whole numbers, except you have to remember one basic rule:

YOU MUST LINE UP THE DECIMAL POINTS TO ADD OR SUBTRACT.

Example 1:   Find the sum of 23.45 and 1.45.

SOLUTION: We will write this problem vertically, being careful to make sure the decimal points are lined up. First we will add the numbers as though they were whole numbers, and then we will bring down the decimal point.

$$
\begin{array}{r}
23.45 \\
+\ 1.45 \\
\hline
24.90
\end{array}
$$
or 24.9   24.90 has the same value as 24.9.

Example 2:   147.908 + 0.0034 + 6.89

SOLUTION: In order to line up the decimal points we'll write 147.908 as 147.9080 and 6.89 as 6.8900.

$$
\begin{array}{r}
147.9080 \\
0.0034 \\
+\quad 6.8900 \\
\hline
154.8014
\end{array}
$$

Line up the decimal points; then add. You can always add zeros at the end of the number if that helps you line up the digits.

Example 3:   3.28 + 15 + 1.5

SOLUTION: In order to line up the decimal points, we'll write 15 as 15.00 and 1.5 as 1.50.

$$
\begin{array}{r}
3.28 \\
15.00 \\
+\quad 1.50 \\
\hline
19.78
\end{array}
$$

Now let's try some subtraction problems. Line up the decimal points and subtract the same way we subtracted in chapter 1. When subtracting, it usually helps to add extra zeros at the end of the numbers, so every number has the same amount of digits to the right of the decimal point.

Example 4:   Find the difference of 24.789 and 2.4785

SOLUTION:
$$
\begin{array}{r}
24.789 \\
-\quad 2.4785
\end{array}
$$

Notice the empty space above the 5. It's okay to fill in the empty space with a zero if it's the last place to the right of the decimal point. As we already know, that doesn't change the number's value. (24.789 = 24.7890). Now we can subtract.

$$
\begin{array}{r}
24.7890 \\
-\quad 2.4785 \\
\hline
22.3105
\end{array}
$$

Remember, you can always check the answer to a subtraction problem using addition.

Check: 22.3105
          + 2.4785
            24.7890

Example 5:   **Find the difference of 56.008 and 26.899.**

SOLUTION:

      56.008   Line up the decimal points and subtract.
   − 26.899   Be careful borrowing.
      29.109

Check: 29.109
          + 26.899
             56.008

Example 6:   **356.09 + 0.102 − 6.0008**

SOLUTION:

      356.090   Line-up the decimal points and add.
   +   0.102   Fill in the empty place with a 0 to subtract.
      356.192

      356.1920
   −   6.0008
      350.1912

Problem Set: Try the following addition and subtraction problems.

1) Find the sum of 109.78 and 0.234.

2) Find the difference of 109.78 and 0.234.

3) 0 .7604 + 4.56 + 0.021.   4) 6.022 - 0.19

SOLUTIONS TO PROBLEM SET:

1)   109.780   2)   109.780
     +   0.234   −   0.234
       110.014   109.546

3)   0.7604   4)   6.022
       4.5600   −   0.190
     + 0.0210   5.832
       5.3414

Perform the following addition and subtraction problems.

1) $2.89 + 135.678$

2) $4.5 - 0.12$

3)   Find the sum of 645.671 and 67.8.

4)   Find the difference of 645.671 and 67.8.

5)   Find the difference of 0.0098 and 0.00098.

6)   Find the sum of 0.0098 and 0.00098.

7) $6.9 + 0.69 + 0.0069$

8) $2.909 + 34.567 + 0.00899$

9) $6.9 - 0.0069$

10) $2.34 - 0.0909$

11) $6 - 0.92$

12) $8 - 0.001$

13) $88 - 0.88$

14) $23 - 0.23$

15) $0.209 - 0.0029$

16) $789.25 - 78.209$

17) $0.209 + 0.0029$

18) $789.25 + 78.209$

19) $0.0231 - 0.00231$

20) $3.02 - 3.0002$

21) $19.0022 + 3.0012$

22) $0.068 + 1.068 + 10.68$

23) $3.3 + 12.45 - 2.009$

24) $0.0087 + 0.103 - 0.099$

25) $4.56 + 0.456 - 2.3451$

26) $2.009 + 0.2009 - 0.002009$

27) $15.1 + 15.01 + 15.001$

28) $678.92 - 0.678 - 0.00678$

5.4   Multiplying Decimals

When multiplying decimals, we are not required to line-up the decimal points. The only difference between multiplying decimals and multiplying whole numbers is figuring out where to place the decimal point in the answer. For instance, when we multiply 0.5 by 0.5, where do we put the decimal in our answer?

Example 1:   **Find the product of 0.5 and 0.5.**

SOLUTION: We know that $5 \times 5 = 25$. So how much is 0.5×0.5? Is it 0.025, 0.25, 2.5, or what? Here's the rule we use: When we multiply numbers with decimals, we add the amount of digits to the right of the decimal point in the two factors, and then, starting from the right, we move the same number of places to find where the decimal point goes in our answer. That probably sounds a lot more complicated than it is. Let's go back to 0.5×0.5. How many digits are after the decimal points in the factors 0.5 and 0.5? There's one digit in each factor, for a total of two. Now we go to our answer of 25 and place the decimal point two places from the right, at 0.25. When you get a few more of these under your belt, you'll be able to do them automatically. We'd like to take this opportunity to explain why this method works. We'll explain this by using the fraction form for 0.5 (five tenths).

$$0.5 \times 0.5 = \frac{5}{10} \times \frac{5}{10} = \frac{25}{100} = 0.25$$

Example 2:   **How much is 0.34(0.63)?**

SOLUTION: There are two places to the right of the decimal point in 0.34, and another two places to the right of the decimal point in 0.63. That gives us a total of four places to the right of the decimal point that have to be marked off in the product.

$$
\begin{array}{r}
0.34 \\
\times\ 0.63 \\
\hline
102 \\
204 \\
\hline
0.2142
\end{array}
$$

If the whole number is zero, you don't have to multiply 6 and 3 by 0.6 and 3 by 0 – only by 34. So we're multiplying 34 by 63, and then placing the decimal point in the product.

Example 3:   **Find the product of 50 and 0.72.**

SOLUTION:

$$
\begin{array}{r}
50 \\
\times\ 0.72 \\
\hline
100 \\
350 \\
\hline
3600
\end{array}
$$

$$36.00 = 36$$

There are a total of two decimal places in the factors. That means we have to mark off two places from the right in the product. The answer is 36.

Example 4: Multiply 0.002 times 0.005.

SOLUTION:

```
  0.002
× 0.005
───────
   010
  000
 000
───────
 00010
.000010 = 0.00001
```

There are a total of six decimal places in the factors. We have to mark off six decimal places from the right in the product. Be careful when you count the six places. You'll find there are only five digits. You have to fill in the missing spot with another 0, so we'll place a 0 to the left. Now place the decimal point in the answer. The product is 0.000010 or 0.00001.

You can easily get the hang of multiplying decimals by working out more problems. So go ahead and do this problem set.

Problem Set:

1) 0.13
   × 0.45

2) 0.0014
   × 6.29

3) 115.81
   × 12.06

4) 10.482
   × 9.64

SOLUTIONS TO PROBLEM SET:

Multiply, then count the number of decimal places in the factors. Count the number of places from right to left in the product, then insert the decimal point in your answer.

1) 0.13
   × 0.45
   ─────
     65
     52
   ─────
    585

0.0585

2) 0.0014
   × 6.29
   ──────
    0126
    0028
    0084
   ──────
   008806

0.008806

3) 115.81
   × 12.06
   ──────
    69486
    00000
    23162
    11581
   ────────
   13966686

1396.6686

4) 10.482
   × 9.64
   ──────
    41928
    62892
    94338
   ────────
   10104648

101.04648

MULTIPLYING DECIMALS BY POWERS OF TEN.

We already did a bit of multiplying by powers of 10 in chapter 2. In chapter 2 we multiplied whole numbers by 10, 100, 1,000, 10,000, and 100,000 by adding zeros to the right of the whole number. When we multiply decimals by powers of ten, we just move the decimal place the appropriate number of places to the right. To find the appropriate number of places we count up the number of zeros in the power of ten.

Example 5: Multiply 0.034 by:

a) 10   b) 100   c) 1,000   d) 10,000   e) 100,000

SOLUTION:

a) There's one zero in 10, so we move the decimal point one place to the right.

0.034

$0.034 \times 10 = 0.34$

To show you the reason why we can say just move the decimal point one place to the right, we'll look at this problem in fraction form.

$0.034 = \dfrac{34}{1,000},$ $\dfrac{34}{1,000} \times \dfrac{10}{1} = \dfrac{34}{100} = 0.34$

b) There are two zeros in 100, so we move the decimal point two places to the right.

0.034

$0.034 \times 100 = 3.4$

In fraction form it would look like this. $\dfrac{34}{1,000} \times \dfrac{100}{1} = \dfrac{34}{10} = 3.4$

c) There are three zeros in 1,000, so we move the decimal point three places to the right.

0.034

$0.034 \times 1,000 = 34$

d) There are four zeros in 10,000, so we add a 0 to 0.034 to make it 0.0340 and move the decimal point four places to the right.

0.0340

$0.034 = 0.0340$ $0.0340 \times 10,000 = 340$

e) There are five zeros in 100,000, so we add two zeros to 0.034 to make it 0.03400 and move the decimal point five places to the right.

$0.034 = 0.03400$ 0.03400 $0.03400 \times 100,000 = 3,400$

Example 6: **Multiply 1.6 by 10.**

SOLUTION: Ten has one zero, so we'll move the decimal point one place to the right.

1.6

$1.6 \times 10 = 16$

Example 7: **Multiply 0.002 by 100.**

SOLUTION: One hundred has two zeros, so we'll move the decimal point two places to the right.

0.002

$0.002 \times 100 = 0.2$

Example 8: Find the product of 1,000 and 5.5.

> SOLUTION: One thousand has three zeros, so we'll move the decimal point three places to the right. Before we move the decimal point in 5.5, we'll write it as 5.500.

$$5.5 = 5.500 = 5.500 \qquad 5.500 \times 1,000 = 5,500$$

Example 9: Multiply 6.89 by:

> a) 10   b) 100   c) 1,000   d) 10,000   e) 100,000
>
> SOLUTION:
>
> a) $6.89 \times 10 = 68.9$ Move the decimal point one place to the right.
>
> b) $6.89 \times 100 = 689$ Move the decimal point two places to the right.
>
> c) $6.89 \times 1,000 = 6,890$ To move the decimal point three places to the right; think of 6.89 as 6.890.
>
> d) $6.89 \times 10,000 = 68,900$ To move the decimal point four places to the right; think of 6.89 as 6.8900.
>
> e) $6.89 \times 100,000 = 689,000$ To move the decimal point five places to the right; think of 6.89 as 6.89000.

Problem Set:

1) Multiply each of these decimals by 10:

    a) 0.2   b) 4.63   c) 20.47

2) Multiply each of these decimals by 100:

    a) 0.001   b) 0.01   c) 0.1

3) Multiply 12.04 by:

    a) 10   b) 100   c) 1,000   d) 10,000

SOLUTIONS TO PROBLEM SET:

1) 10 has one zero so we'll move the decimal point one place to the right.

    a) $0.2 \times 10 = 2$ b) $4.63 \times 10 = 46.3$ c) $20.47 \times 10 = 204.7$

2) 100 has two zeros so we'll move the decimal point two places to the right.

    a) $0.001 \times 100 = 0.1$ b) $0.01 \times 100 = 1$ c) $0.1 \times 100 = 10$

3) Count up the number of decimal places in each power of ten and move the decimal point that number of places to the right.

a) 12.04 x 10 = 120.4

b) 12.04 x 100 = 1,204

c) 12.04 x 1,000 = 12,040

d) 12.04 x 10,000 = 120,400

Exponents

In chapters 2 and 3 we looked at exponents involving whole numbers and fractions. Now it's time for us to apply the same procedure to exponents involving decimals. Remember, when multiplying decimals, all we have to do to figure out how many decimal places we should mark off in our answer is count up the total number of decimal places in the factors. Let's look at a few examples.

Example 10: $(0.2)^2$

SOLUTION: Something squared means multiply the number times itself. If we multiply 2 times 2 we get 4.

0.2(0.2)

0.04

There's one decimal place in 0.2 and another one in the second 0.2, so we have to count two places from the right to the left. If we move the decimal point two places to the left of 4, we get our answer of 0.04.

Example 11: $(0.02)^2$

SOLUTION: Again, we are squaring a number, so we multiply the number times itself, and this time we mark off four decimal places, because the factors

(0.02)(0.02)

0.0004

0.02 have two decimal places each for a total of four decimal places. Count four places from the right to the left.

Example 12: $(0.2)^3$

SOLUTION: This is a cube, so write the 0.2 three times and multiply. Each factor

(0.2)(0.2)(0.2)

0.008

of 0.2 has one decimal place, and we have three 0.2s for a total of three decimal places that need to be marked off in our answer.

Example 13: $(2.6)^3$

SOLUTION: We have to multiply 2.6 times 2.6 times 2.6, so we multiply 2.6 times

(2.6)(2.6)(2.6)

6.76 (2.6)

17.576

2.6 and get 6.76. Then we multiply 6.76 times 2.6 and get our answer of 17.576.

Example 14: $(9.0134)^0$

SOLUTION: Anything, (except 0) even decimals, to the zero power is 1.

$(9.0134)^0 = 1$

Problem Set: Simplify the following problems.

1) $(0.5)^2$ 2) $(0.05)^2$ 3) $(0.1)^4$ 4) $(4.1)^3$

SOLUTIONS TO PROBLEM SET:

1) $(0.5)(0.5) = 0.25$

$$
\begin{array}{r}
0.5 \\
\times 0.5 \\
\hline
0.25
\end{array}
$$

2) $(0.05)(0.05) = 0.0025$

$$
\begin{array}{r}
0.05 \\
\times 0.05 \\
\hline
0.0025
\end{array}
$$

3) $(0.1)(0.1)(0.1)(0.1)$
   $\quad (0.01)(0.1)(0.1)$
   $\qquad (0.001)(0.1)$
   $\qquad\quad 0.0001$

4) $(4.1)(4.1)(4.1)$
   $\quad (16.81)(4.1)$
   $\qquad 68.921$

Simplify the following problems.

1)
$$\begin{array}{r} 0.35 \\ \times\, 0.9 \\ \hline \end{array}$$

2)
$$\begin{array}{r} 1.62 \\ \times 0.8 \\ \hline \end{array}$$

3)
$$\begin{array}{r} 6.33 \\ \times 0.72 \\ \hline \end{array}$$

4)
$$\begin{array}{r} 0.69 \\ \times 0.4 \\ \hline \end{array}$$

5)
$$\begin{array}{r} 4.53 \\ \times 0.7 \\ \hline \end{array}$$

6)
$$\begin{array}{r} 12.96 \\ \times 0.5 \\ \hline \end{array}$$

7)
$$\begin{array}{r} 4.29 \\ \times 0.75 \\ \hline \end{array}$$

8)
$$\begin{array}{r} 10.88 \\ \times 0.57 \\ \hline \end{array}$$

9)
$$\begin{array}{r} 10.6 \\ \times 1.45 \\ \hline \end{array}$$

10)
$$\begin{array}{r} 37.45 \\ \times 1.46 \\ \hline \end{array}$$

11)
$$\begin{array}{r} 156.29 \\ \times 5.82 \\ \hline \end{array}$$

12)
$$\begin{array}{r} 116.78 \\ \times 5.96 \\ \hline \end{array}$$

13) Find the product of 37.02 and 0.55.

14) Find the product of 129.53 and 14.21.

15) (1.1)(2.1)(3.01)

16) (45.6)(0.001)(2.5)

17)   408.13
        ×71.04

18)   731.86
        ×35.41

19)   300.875
        ×29.34

20)   417.936
        ×291.53

21)   Multiply 0.456 by:

a) 10   b) 100   c) 1,000   d) 10,000

22)   Multiply 4.56 by:

a) 10   b) 100   c) 1,000   d) 10,000

23)   Find the product of 3.45 and:

a) 100   b) 10   c) 1   d) 0.1   e) 0.01   f) 0.001

24)   Find the product of 2.7 and:

a) 100   b) 10   c) 1   d) 0.1   e) 0.01   f) 0.001

25) $(0.3)^2$

26) $(0.45)^2$

27) $(0.03)^2$

28) $(0.045)^2$

29) $(1.1)^2$

30) $(0.22)^3$

31) $(2.04)^2$

32) $(0.22)^2$

33) $(0.1)^2$

34) $(0.3)^2$

35) $(0.1)^3$

36) $(0.3)^3$

37) $(0.1)^4$

38) $(0.3)^4$

39) $(0.01)^2$

40) $(0.03)^2$

41) $(0.01)^3$

42) $(0.03)^3$

43) $(0.002)^0$

44) $(5.6)^0$

CHALLENGE PROBLEMS

45) $(0.1)^{10}$

46) $(0.2)^6$

47) $(0.01)^6$

48) $(0.02)^5$

5.5   Dividing Decimals

Once again, the only difference between dividing decimals and dividing whole numbers is figuring out where to place the decimal point. But instead of applying an arithmetic rule as we did when we multiplied decimals, we'll just get rid of the decimal point in the divisor (the number we divide by) and do regular division. When we divide, the divisor must always be a whole number. To make it a whole number, just move the decimal point as many places to the right as is necessary. Be sure to move the decimal point in the dividend the same number of places. As we work out the following examples, remember the divisor must be a whole number, and you must move the decimal point the same number of places to the right in the divided as the divisor.

$$dividend \div divisor = quotient.$$

$$divisor \overline{)dividend}^{\textstyle quotient}$$

Example 1:   **How much is 1.888 divided by 0.02?**

SOLUTION: Let's start by writing this problem in a division box.

$0.02\overline{)1.888}$ The divisor 0.02 has two digits to the right of the decimal point. We have to move the decimal point two places to the right to make 0.02 the whole number 2. Therefore we also have to move the decimal point in the dividend 1.888 two places to the right, which makes it 188.8. 1.888 ÷ 0.02 is the same as 188.8 ÷ 2. Bring up the decimal point. Now divide to find the quotient, 94.4.

$$
\begin{array}{r}
94.4 \\
2\overline{)188.8} \\
-18 \\
\hline
08 \\
-\ 8 \\
\hline
08 \\
-\ 8 \\
\hline
0
\end{array}
$$

Before we move on to the next example, we'd like to explain why we can say that 1.888 ÷ .02 is the same as 188.8 ÷ 2. Remember, another way to show division is to use a fraction bar. 1.888 ÷ 0.02 can be written as $\dfrac{188}{0.02}$. Earlier in this example we moved the decimal point in 0.02 two places to the right to make it the whole number 2. We also moved the decimal point in 1.888 two places to the right, which made it 188.8. If we multiply the numerator (top) and denominator (bottom) of $\dfrac{188}{0.02}$ by 100,

$\left(\dfrac{1.888 \times 100}{0.02 \times 100} = \dfrac{188.8}{2} \right)$ the divisor becomes a whole number and we get the fraction $\dfrac{188.8}{2}$, which is

the same as $2\overline{)188.8}$.

Remember, if we multiply the numerator (top) and the denominator (bottom) of a fraction by the same amount (in this case 100), it's the same as multiplying by one, because $\frac{100}{100} = 1$. Multiplying by one doesn't change the value of the problem.

Example 2: **Find the quotient of 0.159 and 0.002.**

SOLUTION: The first thing we have to do is move the decimal point three places to the right in the divisor 0.002 to make it the whole number 2. We now have to move the decimal point three places to the right in the dividend 0.159, which makes it 159. (0.159 ÷ 0.002 = 159 ÷ 2)

Bring up the decimal point and divide.

$$
\begin{array}{r}
79.5 \\
0.002\overline{)0.159{.}0} \\
-14 \\
\hline
19 \\
-18 \\
\hline
10 \\
-10 \\
\hline
0
\end{array}
$$

Notice we had to add a 0 to the dividend. We can do that because section 5.1 showed us that 159 has the same value as 159.0.

Example 3: **Find the quotient of 0.4 and 0.25.**

SOLUTION: To convert the divisor 0.25 to a whole number, we move the decimal point two places to the right. In order for us to move the decimal point in 0.4 two places to the right, we write 0.4 as 0.40. Move the decimal point two places to the right, and we have a dividend of 40. Bring up the decimal point and divide.

$$
\begin{array}{r}
1.6 \\
25\overline{)\ 40.0} \\
-25 \\
\hline
150 \\
-150 \\
\hline
0
\end{array}
$$

Another way to prove our answer is correct is to convert the decimals to fractions and then perform division.

$$
\frac{.4}{.25} = \frac{\frac{4}{10}}{\frac{25}{100}} = \frac{\overset{2}{\cancel{4}}}{\underset{5}{\cancel{10}}} \cdot \frac{\overset{4}{\cancel{100}}}{\underset{1}{\cancel{25}}} = \frac{2}{5} \cdot \frac{4}{1} = \frac{8}{5} = 1.6
$$

Example 4: **Round the quotient of $1 \div 3$ to the nearest hundredth.**

SOLUTION: Be careful when setting up the division box on this one. Don't assume that the larger number goes inside the box and the smaller one goes outside the box. That's a very common error. The 1 is divided *by* the 3. It should look like this $3\overline{)1}$ (read 1 divided by 3), not like this $1\overline{)3}$ (read 3 divided by 1). 1 and 3 are already whole numbers, so we don't have to worry about moving the decimal point. Bring up the decimal point. Our answer will be a decimal.

```
        0.333
    3)1.000
     −9
     ────
      10
    −  9
     ────
       10
      −9
     ────
        1
```

To round to the hundredths' place, we divide to the thousandths' place. In order to do this we have to write 1 as 1.000.

You should have noticed a pattern of a continuous loop in this example. We keep dividing 3 into 10 with no end in sight. This is called a repeating decimal. In the next chapter you'll learn about repeating decimals, terminating decimals, and continuous decimals.

Since practice makes perfect in math, let's get in some more practice dividing with decimals. See if you can do all these problems correctly.

Problem Set:

1) $0.5\overline{)10}$

2) $0.42\overline{)1.3734}$

3) Find the quotient of 3 and 0.33 rounded to the nearest thousandth.

4) Find the quotient of 5 and 7 rounded to the nearest tenth.

SOLUTIONS TO PROBLEM SET:

1)
```
          20
    0.5) 10.0
     −10
    ─────
        00
    −   0
    ─────
         0
```

2)
```
           3.27
    0.42) 1.3734
     −126
    ──────
       113
     −  84
    ──────
        294
      − 294
    ──────
          0
```

3)
$$
\begin{array}{r}
9.0909 \\
0.33{\overline{\smash{)}}}3.00000 \\
-297 \\
\hline
300 \\
-297 \\
\hline
3
\end{array}
$$

4)
$$
\begin{array}{r}
0.71 \\
7{\overline{\smash{)}}}5.00 \\
-49 \\
\hline
10 \\
-7 \\
\hline
3
\end{array}
$$

Rounds to 9.091

Rounds to 0.7

Dividing Decimals by Powers of Ten

In section 5.4 we showed you how to multiply by powers of 10. In this section we'll show you how to divide by powers of 10. Dividing by powers of 10 is the same as multiplying by powers of 10 except, this time we move the decimal point to the left.

Example 5: What's $2.6 \div 10$?

SOLUTION: We'll do this problem two ways, using a division box and using a shortcut.

$$
\begin{array}{r}
0.26 \\
10{\overline{\smash{)}}}2.60 \\
-20 \\
\hline
60 \\
-60 \\
\hline
0
\end{array}
$$

Or the faster way to do this problem is to say the divisor 10 has one zero, so move the decimal point in 2.6 one place to the left to find the quotient 0.26.

Example 6: **Find the quotient of 23.47 and 100.**

SOLUTION: There's two zeros in 100, so move the decimal point two places to the left to find the quotient 0.2347.   .23.47

$23.47 \div 100 = 0.2347$.

Example 7: **Divide 678.90013 by:**

a) 10   b) 100   c) 1,000   d) 10,000

SOLUTION:

a) Ten has one zero, so move the decimal point one place to the left.

$678.90013 \div 10 = 6\,7.890013$

b) One hundred has two zeros, so move the decimal point two places to the left.

$678.90013 \div 100 = 6.7890013$

c) One thousand has three zeros, so move the decimal point three places to the left.

$$678.90013 \div 1,000 = 0.67890013$$

d) Ten thousand has four zeros, so move the decimal point four places to the left. Don't forget to add the extra zero in front of the 6.

$$678.90013 \div 10,000 = 0.067890013$$

Problem Set:

1) Divide 2.567 by:

   a) 10   b) 100   c) 1,000   d) 10,000   e) 100,000

2) $23,008.1 \div 1,000$ 3) $56.78 \div 100$

4) Divide 1,430 by:

   a) 10   b) 100   c) 1,000   d) 10,000   e) 100,000

SOLUTIONS TO PROBLEM SET:

1) a) 0.2567   b) 0.02567   c) 0.002567   d) 0.0002567   e) 0.00002567

2) 23.0081   3) 0.5678

4) a) 143   b) 14.3   c) 1.43   d) 0.143   e) 0.0143

Perform the following divisions.

1) $0.6\overline{)1.5}$

2) $0.4\overline{)1}$

3) $0.07\overline{)1.47}$

4) $0.35\overline{)1.05}$

5) $0.24\overline{)10.8}$

6) $0.012\overline{)9.6}$

7) $0.3\overline{)1.86}$

8) $0.8\overline{)2}$

9) $0.05\overline{)6}$

10) $0.07\overline{)2.94}$

11) $0.24\overline{).144}$

12) $0.45\overline{)27}$

13) $1.5\overline{)9}$

14) $0.15\overline{)9}$

Round your answers to the nearest thousandth.

15) $0.83\overline{)5.9}$

16) $0.37\overline{)0.841}$

17) $0.37\overline{)0.841}$

18) $1.65\overline{)0.9}$

19) $0.026\overline{)1}$

20) $0.713\overline{)0.1}$

21) $1.963\overline{)5.307}$

22) $0.13\overline{)37.3}$

23) $0.153\overline{)19}$ 24) $0.469\overline{)1}$

25) $0.709\overline{)0.3}$ 26) $1.143\overline{)8.076}$

27) Find the quotient of 1 and 8.

28) Find the quotient of 6 and 12.

29) Find the quotient of 8 and 27. Round your answer to the nearest tenth.

30) Find the quotient of 1 and 6. Round your answer to the nearest thousandth.

31) Divide 1,829.034 by:

  a) 10   b) 100   c) 1,000   d) 10,000

32) Divide 89,278.88 by:

  a) 10   b) 100   c) 1,000   d) 10,000

CHALLENGE PROBLEMS

33) Find the quotient of the sum of 0.2 and 0.003 and the difference of 0.33 and 0.001. Round your answer to the nearest hundredth.

34) $2.2 \div 0.2 \div 0.01$

35) A business firm began the month with a checking account balance of $145,678.91, and wrote checks for $4,678.01, $567.89, $23,456.78, and $6,782.20. They made deposits of $6,345.12, $3,452.12, and $6,783.23. At the end of the month they gave each of their 212 employees a Christmas bonus of $500. How much was their balance at the end of the month?

5.6   Square Roots of Decimals

In the last section we worked with problems involving squares of decimals. In the second chapter we showed you how to undo squares by using its inverse operation, square roots. Let's apply this procedure to decimals. Don't worry – it's easy.

Example 1: $\sqrt{0.04}$

SOLUTION: This example is read the square root of 0.04. What it's really asking us is, what number times itself is 0.04? Let's logically reason this out.

$\sqrt{4}$ is 2. Should the $\sqrt{0.04}$ be 2, 0.2, 0.02, 0.002 or 0.0002? The easy way to figure this out is to remember that when we multiply decimals, their product has the total number of decimal places as its factors. If the product, 0.04, of two identical factors, has two decimal places, the factors must have one decimal place each. Now we know $\sqrt{0.04} = 0.2$. Another way to find the number of decimal places in the answer is to divide the number of decimal places in the radicand, (the number under the $\sqrt{}$) by 2, because we're working with *square* roots. 0.04 has two decimal places, so 2 decimal places ÷ 2 decimal places, leaves us with one decimal place in the answer. If you want to check your answer, just square 0.2 to see if you get 0.04.

Check: $(0.2)^2 = (0.2)(0.2) = .04$.

Example 2: $\sqrt{0.0001}$

SOLUTION: We have four decimal places under the radical, so our answer should have 4 ÷ 2, which is 2 decimal places.

$$\sqrt{1} = 1 \qquad \text{Now mark off two decimal places.}$$

$\sqrt{0.0001} = 0.01$

Check: $(0.01)^2 = 0.0001$

Example 3: $\sqrt{0.16}$

SOLUTION: There are two decimal places under the radical, so there should be 2 ÷ 2 is one decimal place in the answer.

$$\sqrt{16} = 4 \qquad \text{Now mark off one decimal place.}$$

$\sqrt{0.16} = 0.4$

Check: $(0.4)^2 = 0.16$

Example 4: $\sqrt{0.000025}$

> SOLUTION: The answer should have $6 \div 2$, which is three places past the decimal point, because there are six places under the radical.

$$\sqrt{25} = 5 \qquad \text{Now mark off three decimal places.}$$

$$\sqrt{0.000025} = 0.005$$

Check: $(0.005)^2 = 0.000025$

Example 5: $\sqrt{0.0009}$

> SOLUTION: The answer should have two decimal places because 0.0009 has four.

$$\sqrt{9} = 3 \qquad \text{Now mark off two decimal places.}$$

$$\sqrt{0.0009} = 0.03$$

Check: $(0.03)^2 = 0.0009$

Problem Set: Simplify the following problems.

1) $\sqrt{0.01}$ 2) $\sqrt{0.0001}$ 3) $\sqrt{0.000001}$

4) $\sqrt{0.25}$ 5) $\sqrt{0.0025}$ 6) $\sqrt{0.81}$

SOLUTIONS TO PROBLEM SET:

1) 0.1   2) 0.01   3) 0.001   4) 0.5   5) 0.05   6) 0.9

Workbook

Simplify the following problems.

1) $\sqrt{0.36}$

2) $\sqrt{0.0036}$

3) $\sqrt{0.000049}$

4) $\sqrt{0.01}$

5) $\sqrt{0.0081}$

6) $\sqrt{0.09}$

7) $\sqrt{0.04}$

8) $\sqrt{0.0004}$

9) $\sqrt{0.64}$

10) $\sqrt{0.0064}$

11) $\sqrt{0.0004}$

12) $\sqrt{0.0016}$

13) $\sqrt{0.25}$

14) $\sqrt{0.0144}$

15) $\sqrt{0.0001}$

16) $\sqrt{0.000001}$

17) $\sqrt{0.49}$

18) $\sqrt{0.0049}$

19) $\sqrt{0.81}$

20) $\sqrt{0.0025}$

5.7   Order of Operations Applied to Decimals

In earlier chapters we applied the order of operations to whole numbers, fractions, and mixed numbers. In this chapter we'll apply the order of operations to decimals. For your benefit, we'll list the steps for the order of operations.

Order of Operations:

Step 1: Inside groupings first : { }, [ ] , ( ), and | |.

Step 2: Exponents

Step 3: Multiplication / Division (left to right)

Step 4: Addition / Subtraction (left to right)

Example 1: $2.1 + 4 \div 0.2$

SOLUTION:

$2.1 + 4 \div 0.2$ Remember to divide $(4 \div 0.2 = 20)$ before you add.

$2.1 + 20$ Line up the decimal points when adding.

$$\begin{array}{r} 2.1 \\ +20.0 \\ \hline 22.1 \end{array}$$

Example 2: $7 - (2.4)^2$

SOLUTION:

$7 - (2.4)^2$ Exponents first, $(2.4)^2 = 5.76$; then line up the decimal
$7 - 5.76$ points and subtract.
$\quad 1.24$

$$\begin{array}{r} 7.00 \\ -5.76 \\ \hline 1.24 \end{array}$$

Example 3: $(6.8 - 0.02) \div (0.1 + 0.1)$

SOLUTION: Inside the groupings first; then divide.

$(6.8 - 0.02) \div (0.1 + 0.1)$

$\quad 6.78 \div 0.2$

$\quad\quad 33.9$

$$\begin{array}{r} 6.80 \\ -.02 \\ \hline 6.78 \end{array} \qquad \begin{array}{r} 0.1 \\ +0.1 \\ \hline 0.2 \end{array}$$

Example 4: $(1.1)^2 + (0.01)^3 + (0.2)^4$

SOLUTION:

$(1.1)^2 + (0.01)^3 + (0.2)^4$ Exponents first; then add. $(1.1)^2 = (1.1)(1.1) = 1.21$.

$1.21 + 0.000001 + 0.0016$ $(0.01)^3 = (0.01)(0.01)(0.01) = (0.0001)(0.01) = 0.000001$.

$\quad\quad 1.211601$ $(0.2)^4 = (0.2)(0.2)(0.2)(0.2) = (0.04)(0.2)(0.2) =$
$(0.008)(0.2) = 0.0016$

Problem Set: Follow the order of operations to simplify the following problems.

1) $7.01 + 4.2(0.06)$

2) $100.5 - (0.5)^2$

3) $(0.8 - 0.002) \div (0.1 + 0.2)$

4) $(0.3 - 0.27) \div (0.12)$

SOLUTIONS TO PROBLEM SET:

1) $7.01 + 4.2(0.06)$ Multiply first, $(4.2)(0.06) = 0.252$; then add.
    $7.01 + 0.252$ Line up the decimal points.
      7.262

$$\begin{array}{r} 7.010 \\ +0.252 \\ \hline 7.262 \end{array}$$

2) $100.5 - (0.5)^2$ Exponents first, $(0.5)^2 = 0.25$; then subtract.
    $100.5 - 0.25$ Line up the decimal points.
      100.25

$$\begin{array}{r} 100.50 \\ -0.25 \\ \hline 100.25 \end{array}$$

3) $(0.8 - 0.002) \div (0.1 + 0.2)$ Inside groupings first; Then divide.
      $0.789 \div 0.3$
        2.66

$$\begin{array}{r} 0.800 \\ -0.002 \\ \hline 0.798 \end{array} \qquad \begin{array}{r} 0.1 \\ +0.2 \\ \hline 0.3 \end{array}$$

4) $(0.3 - 0.27) \div (0.12)$ Inside groupings first.
    $0.03 \div 0.12$ Then divide.
      0.25

$$\begin{array}{r} 0.30 \\ -0.27 \\ \hline 0.03 \end{array}$$

Follow the order of operations to simplify the following problems.

1) $4.56 + 0.3 \div 0.1$

2) $12.42 - 0.25 \div 0.5$

3) $(0.2)^2 \div (0.1)^3$

4) $(0.6 \div 0.2)^2$

5) $(4.1) + (2.3)^2$

6) $(4.62 - 0.003) \div (0.1 - 0.1)$

7) $0.5[0.04 - (0.2)^2]$

8) $0.15 \div [0.02 \div (0.2)^2]$

9) $6.03 - 0.55 \div 0.5$

10) $7.2 \div [(0.7)^2 - 0.49)]$

11) $(0.6)^2 + (1.3)^2$

12) $(2.3 - 0.23) \div (0.2 + 0.1)$

13) $[(0.3)^2 - 0.09]$

14) $(4 \div 80) - (0.2 \div 5)$

15) $2.0625 - [0.16 \div (0.4)^2]$

16) $0.25[0.02 \, (0.2)^3]$

17) $(0.3 - 0.001) + (1.6 - 0.8)^2$

18) $0.16 \div (0.2)^3$

19) $2.1 + 0.3(0.6 - 0.06)$

20) $(0.4)^3 + (0.2)^2 \div (0.1)^3$

CHALLENGE PROBLEMS

21) $(1.5 \div 0.3)^2 + (0.63 \div 0.3)^3$

22) $10.875 - [(0.2)(2.2)]^2$

23) $0.34[0.81 \div (0.3)^4]^2$

24) $(0.65)^2 + (0.5)^3 \div (0.5 + 0.5)$

5.8 Applications

Decimals have many applications to everyday life, especially when money is involved. Money problems are our favorite type of application. When money's involved there isn't anything less than a penny; therefore it's assumed we'll round all money answers to the nearest penny, which is to the nearest hundredth. In the following problems you'll need to figure out whether to add, subtract, multiply, or divide, or apply some combination of these operations.

Example 1: Phil walked 2.4 miles on Monday, 3.1 miles on Wednesday, 3.4 miles on Thursday, and 4.0 miles on Saturday. How many miles did he walk that week?

> SOLUTION: This is an addition problem because we want to know how many miles altogether.

$$
\begin{array}{r}
2.4 \\
3.1 \\
3.4 \\
+4.0 \\
\hline
12.9 \text{ miles}
\end{array}
$$

Example 2: Bunky spent $45.67 shopping in Kings grocery store. If he gave the cashier a $50 bill, how much change should he receive?

> SOLUTION:
$$
\begin{array}{r}
\$50.00 \\
-\$45.67 \\
\hline
\$4.33
\end{array}
$$

Example 3: Jerome bought three six-packs of Sprite for $1.60 a six pack. He paid with a $20 bill. How much change did he get back?

> SOLUTION: Multiply first ($1.60 x 3 = $4.80); then subtract.

$$
\begin{array}{r}
\$20.00 \\
-\$4.80 \\
\hline
\$15.20 \text{ change.}
\end{array}
$$

Example 4: In her gymnastics floor routine, Martha received scores of 9.4, 9.6, 9.7, 9.2, 9.8, 9.3, 9.5, and 9.6. Find her average score, rounded to the nearest tenth.

> SOLUTION: To find her average, add up all the scores to find their sum (76.1), and divide by the number of scores (8). Divide to two decimal places to round to one place.

$$
\begin{array}{r}
9.5 \\
8{\overline{\smash{\big)}\,76.1}} \\
-72 \\
\hline
41 \\
-40 \\
\hline
10 \\
-\ 8 \\
\hline
2
\end{array}
$$

9.51 rounded to the nearest tenth is 9.5.

Martha's average score rounded to the nearest tenth is 9.5.

Example 5: At the beginning of the month Hans Brinker had a checking account balance of $3,072.96. During the month he made deposits of $467.89, $2,345.61, and $56.09. He wrote checks for $2,789, $23.50, and $899.55. What was his balance at the end of the month?

SOLUTION: We begin with $3,072.96, add $2,869.59 in deposits for a total of $5,942.55. Then we add the three checks he wrote.

$$\begin{array}{r} 3,072.96 \\ +2,869.59 \\ \hline 5,942.55 \end{array} \qquad \begin{array}{r} 467.89 \\ 2,345.61 \\ +56.09 \\ \hline 2,869.59 \end{array}$$

$$\begin{array}{r} \$5,942.55 \\ -\$2,869.59 \\ \hline \$3,072.96 \end{array}$$

Subtract the sum of the checks from the $5,942.55 balance to find $3,072.96 is the balance at the end of the month.

Example 6: Mr. and Mrs. Suarez obtain a $90,000 home mortgage. If they need to make a monthly payment of $7.98 for every $1,000 that they borrowed, how much is their monthly payment?

SOLUTION: The first question we have to ask ourselves is, how many thousands are in 90,000? The answer is 90. Their monthly payment is $7.98 for every $1,000. The mortgage of $90,000 is 90 thousands; so we'll multiply $7.98 by 90 to find our answer of $718.20.

The following problem is one we know you all can identify with.

Example 7: Stacey Ponticello's test grades in her math course are 95, 85, 54, 80, and 76. What's her average in the course?

SOLUTION: We'll begin by finding the sum of the grades and we'll divide by 5 because there are 5 grades.

$$95 + 85 + 54 + 80 + 76 = 390$$

$$\begin{array}{r} 78 \\ 5\overline{)390} \\ -35 \\ \hline 40 \\ -40 \\ \hline 0 \end{array}$$

Stacey's average grade is 78.

Now that we've shown you a few examples, you should be ready to do the following problem set. Good luck.

Problem Set:

1) Four people in a family held jobs. In one month they earned these amounts: $5,121.98, $3,950.23, $4,698.85, and $2,629.58. How much did the family earn that month?

2) Ms. Schultz bought 4.75 yards of material to make an outfit. If she paid $6.98 a yard, how much did the material cost her?

3) Mike left home with $212.94. He spent $57.91 on shoes, $5.42 on a tie, and $34.79 on shirts. How much money did he have left?

4) At the Miss America Pageant, Miss Florida received these scores in the swimsuit competition : 4.8, 4.7, 4.4, 4.9, 4.6, 4.8, and 4.5. What was her average score, rounded to the nearest thousandth?

5) If heavy-duty cable cost $5.99 a foot, how much would 8.75 feet of cable cost?

6) You can get 4.974 francs for a dollar. How many francs could you get for $12.25? Round off your answer to three decimal places.

7) Teresa Papagna's grades in her chemistry course are; 78, 92, 88, 70, 95, and 87. What's her average grade?

SOLUTIONS TO PROBLEM SET:

1)
$$
\begin{array}{r}
\$\ 5,121.98 \\
3,950.23 \\
4,698.85 \\
+\ 2,629.58 \\
\hline
\$16,400.64
\end{array}
$$

The family earned $16,400.64

2)
$$
\begin{array}{r}
\$6.98 \\
\times 4.75 \\
\hline
3490 \\
4886 \\
2792 \\
\hline
\$33.1550 = \$33.16
\end{array}
$$

The material cost $33.16

3)
$$
\begin{array}{r}
\$\ 57.91 \\
5.42 \\
+\ 34.79 \\
\hline
\$212.94
\end{array}
\qquad
\begin{array}{r}
\$212.94 \\
-\ 98.12 \\
\hline
\$114.82
\end{array}
$$

He had $114.82 left.

4)
$$
\begin{array}{r}
4.8 \\
4.7 \\
4.4 \\
4.9 \\
4.6 \\
4.8 \\
+4.5 \\
\hline
32.7
\end{array}
\qquad 32.7 \div 7 = 4.6714
$$

Her average score is 4.671.

5) $5.99
    ×8.75
    ‾‾‾‾‾‾
    2995
    4293
    4792
    ‾‾‾‾‾‾
    $52.4125

8.75 feet of cable would cost $52.41.

6)   4.974
    ×   12.25
    ‾‾‾‾‾‾‾‾
    24870
    9948
    9948
    4994
    ‾‾‾‾‾‾‾‾
    60.9315

You could get 60.932 francs.

7)   78 $510 \div 6 = 85$
      92
      88   Teresa's average grade is 85.
      70
      95
    + 87
    ‾‾‾‾
     510

1) On a cross-country trip, the Luciano family bought gas eight times. Find out how much gas they bought all together if these were their purchases in gallons: 13.2, 12.6, 14.9, 15.1, 13.4, 15.0, 14.5, and 14.

2) Kuldip Singh earned $96.34 on Monday, $126.75 Tuesday, $114.39 Wednesday, $138.05 Thursday and $98.73 on Friday. How much did he earn during this week?

3) At the beginning of the month Sylvia Cortez had $3,116.98 in her checking account. During the month she made deposits totaling $4,908.22 and she wrote checks totaling $3,792.11. How much money did she have in her account at the end of this month?

4) A company has 254 workers on its payroll. If their average weekly pay is $473.29, how much is the company's weekly payroll?

5) The O'Connor family has five depositors at the same savings and loan association. How many shares does the family own if they own these individual shares 31.372, 34.890, 20.187, 13.296, and 15.991?

6) Pedro had the following expenses on a vacation trip: gas, $39.75; meals, $94.18; lodging, $149.22; and tolls, $6.90. What was the total cost of his vacation?

7) Marla ran six days last week. Her distances were 2.8 miles, 4.2 miles, 3.5 miles, 2.9 miles, 4.4 miles, and 5.6 miles. How far did she run during that week?

8) Stan bought 12.5 square yards of carpeting at $12.99 a square yard. How much did he spend?

9) Rainfall recorded in centimeters (cm.) during the first four months of 2008 in Baltimore was: 5.83, in January 3.79, in February, 5.65 in March, and 8.44 in April. How much rain fell during those four months?

10) On a home mortgage you have to make a monthly payment of $8.46 for every $1,000 that you borrow. What is the monthly payment on this loan if $75,000 is borrowed?

11) The electric company charges 4.67 cents per kilowatt hour. If you use 516 kilowatt hours, how much is your bill?

12) You can get 167.396 yen for a dollar. How many yen could you get for $7.75?

13) At a bingo hall, the average person spends $23.52. If 584 people attend, how much money does the bingo hall take in?

14) Snowfall recorded during the winter season in Syracuse, New York, is as follows: December, 18.34 inches; January, 19.02 inches; February, 24.79 inches; and March, 10.48 inches. How much snow fell during the winter season?

15) If your car averages 22.07 miles per gallon, how many miles would you expect to drive on 13.243 gallons of gas? Round your answer to the nearest thousandth.

16) A lottery prize of $1.47 million dollars was divided equally by 100 people. How much money did each person get?

17) If silver cost $6.238 an ounce, how much would 10.35 ounces cost?

18) Claire began the day with 12.334 ounces of gold on hand. If she used 5.564 ounces to make jewelry, how much gold did she have left at the end of the day?

19) If artificial turf cost $4.75 a square yard, how much would it cost to cover a field that is 99.25 square yards?

20) On a cross-country trip, the Flynn family bought gas eight times. Their purchases were : 14.5, 13.6, 14.8, 15.3, 12.9, 14.4, 13.7, and 12.6 gallons. Find the amount of their average purchase rounded to the nearest gallon.

21) Karl Malone worked as a waiter and earned these tips: $54.22 on Thursday, $74.50 on Friday, $96.60 on Saturday, and $88.31 on Sunday. How much did he earn over these four days?

22) At the beginning of the month Marshall Anker had $5,129.37 in his checking account. During the month he made the, following deposits; $125.67, $983.76, and $1,000. He wrote checks for the following amounts; $2,345.20, $567.01, and $3,900.90. How much money did he have in his account at the end of the month?

23) John weighed 64.632 kilograms (km.) on his last birthday. If he weighs 70.194 km. now, how much weight did he gain since his last birthday?

24) The scale on a map is 1 inch = 3.5 miles. How many miles does 6.25 inches represent? Round your answer to the nearest hundredth.

25) On a home mortgage you have to make a monthly payment of $9.29 for every $1,000 that you borrow. What is the monthly payment on a loan of $125,000?

26) The electric company charges 5.12 cents per kilowatt hour. If you use 649 kilowatt hours, how much is your bill?

27) You can get 1.4176 marks for a dollar. How many marks could you get for $94.75? Round your answer to three decimal places.

28) A box contains 425.13 grams of cereal. If one portion is 38.3 grams, how many portions are in the box? Round to one decimal place.

29) Tommy Larkin's test grades in his pre-calculus course are; 76, 82, 90, 45, and 96. Tommy's sister, Jeannie, is taking the same course. Her grades are; 45, 66, 88, 90, and 100. Who has the higher average? How much higher is it? Round the averages to the nearest whole number.

Chapter 5 Review

1)   1,407.26
       4,219.85
         734.18
   +10,063.91

2)   519,640.39
   −389,761.40

3) Write 119.95 in words.

4) $0.28\overline{)4079.012}$

5) Round each of these numbers to the nearest tenth:
    a) 5.28   b) 0.528   c) 0.0528

6) Round each of these numbers to the nearest hundredth:
    a) 7.995   b) 0.7995   c) 0.07995

7) State the place value of the digit 5 in each of the following numbers; write the number in words and in expanded form.
    a) 2.345   b) 5.902   c) 0.592   d) 450.02

8) Round each of these numbers to the nearest ten:
    a) 23.456   b) 456.789   c) 199.0999

9) Find the difference of 341.6 and 0.711.

10) Find the product of 3,042.33 and 17.7.

11) Find the quotient of 3,042.33 and 17.7. Round your answer to the nearest tenth.

12) Write three hundred sixteen thousandths as a decimal.

13) Multiply each of the following by 10, 100, 1,000, and 10,000.
    a) 0.3   b) 0.0003   c) 3.33   d) 0.303

14) Divide each of the following by 10, 100, 1,000, and 10,000.
    a) 0.3   b) 300   c) 3.33   d) 0.303

15) In a diving competition Juan received scores of 5.6, 5.5, 5.8, 5.5, 5.7, 5.5, 5.4, and 5.7. Find his average score rounded to two decimal places.

16) Susan has a very large dog. For breakfast the dog ate 2.9 pounds of dog food; for lunch, 3.2 pounds; for dinner, 4 pounds, and 2.3 pounds for a bedtime snack. How many pounds of food did her dog eat that day?

17) If optical fiber wiring cost $1.99 a foot, how much would 10.25 feet of wire cost?

18) A company has 309 workers on its payroll. If their average monthly pay is $4,922.34, how much is the company's monthly payroll?

19) Write 3,046.1005 in expanded form.

20) State the place value of the 3 in each of the following numbers.
    a) 3.45   b) 0.345   c) 0.0034   d) 30.45

21)   15.197   22) $24.98 + 54.5 - 12.0005$
    -14.199

23) Find the product of 15.197 and 7.48 rounded to the nearest tenth.

24) Find the quotient of 2 and 5.

25) Find the quotient of 5 and 2.

26) Find the quotient of 1.96 and 5.41 rounded to the nearest thousandth.

Chapter 5 • Test 1

1) State the place value of the digit 5 in the following numbers and write the number in expanded form.

   a)   3.56

   b)   0.0045

   c)   65.92

   d)   0.905

2) Round the following numbers to the nearest tenth, hundredth, and thousandth.

   a)   0.0398

   b)   23.756

   c)   0.00999

   d)   0.33333

3) Find the product of 1.25 and 0.001.

4) Multiply 1.239 by

   a)   0.001

   b)   0.01

   c)   0.1

   d)   10

   e)   100

5) 23.0015×0.0024

6) $23.0015 + 0.0024$

7) $23.004 - 0.002$

8) $23.004 \div 0.0024$

9) What number is 4.5 more than the difference of 89 and 6.7?

10) If Rawl left his house with $50 and bought $10.25 worth of candy and a shirt for $32, how much money did he have left?

11) A store has 206 employees on its payroll. If their average monthly pay is $5,234.45, how much is the store's monthly payroll?

12) In an ice skating competition Christie received scores of 8.8, 8.9, 7.8, 9, 7.7, and 8.7. What was her average score to the nearest tenth?

13) Write 345.801 in words.

14) Write six hundred and thirty-four thousandths as a number.

15) Multiply each of the following by 10, 100, 1,000, and 10,000.

a) 34.7 _____ _____ _____ _____

b) 0.347 _____ _____ _____ _____

c) 0.00347 _____ _____ _____ _____

16) Divide each of the following by 10, 100, 1,000, and 10,000.

a) 34700 _____   _____   _____   _____

b) 0.347 _____   _____   _____   _____

c) 34.7 _____   _____   _____   _____

17) $(0.07)^2$

18) $\sqrt{0.0025}$

19) $20.4 + 0.8 \div (0.2)^3$

20) $(0.2 - 0.0002) \div (0.3 - 0.001)$ Round your answer to the nearest hundredth.

Chapter 5 • Test 2

1) State the place value of the digit 4 in the following numbers and write the number in expanded form.

    a)   0.456   b)   0.00456

    c)   45.6   d)   0.0456

2) Round the following numbers to the nearest tenth, hundredth, and thousandth.

    a)   0.082   b)   0.00912

    c) 0.00998   d)   34.09532

3) Find the product of 0.0023 and 23.45

4) Multiply 4.561 by

    a) 0.001   b)   0.01

    c) 0.1   d)   10

    e) 100

5) 3.5 - 0.0203

6) 3.5 x 0.0203

7) 3.5 + 0.0203

8) 3.5 ÷ 0.0203

9) What number is 6.7 less than the product of 2.3 and 4.5?

10) Mackenzie's checking account balance at the beginning of the month was $506.34. She spent $56.34 on a Christmas gift for Billy, $20.46 on a gift for Britany, $35.02 on a gift for Reanna, and $65.85 on a gift for Kianna, how much money did she have left in her account?

11) The employees of a company are paid the following salaries: $45,678.23, $23,567.54, $10,567.89, and $35,672.01. What's the average salary of the employees?

12) The number of miles driven by Jose in one week are: 6.5, 17.9, 45.7, 55.6, 2.2, and 23.3. What's the average number of miles he drove per day to the nearest tenth?

13) Write 4,500.056 in words.

14) Write seventy-eight and six hundred three ten thousandths as a number.

15) Multiply each of the following by 10, 100, 1,000, and 10,000.

   a) 4.05   _____   _____   _____   _____

   b) 0.4402   _____   _____   _____   _____

   c) 0.00034   _____   _____   _____   _____

16) Divide each of the following by 10, 100, 1,000 and 10,000.

    a) 3.45   _____ _____ _____ _____

    b) 278.009   _____ _____ _____ _____

    c) 5   _____ _____ _____ _____

17) $(0.1)^4$ 18) $\sqrt{0.0009}$

19) $12.09 + 0.12 \div 0.004$

20) Round your answer to the nearest tenths. $(0.3 - 0.003) \div (1 \div 8)$

Chapter 6.   Conversions, Ratios and Proportions

In this chapter we'll convert fractions to decimals and decimals to fractions. We'll also work on many word problems involving rates, ratios, and proportions, which will allow us to see some real life applications of mathematics. We hope you enjoy this chapter as much as we enjoyed writing it.

6.1   Fraction and Decimal Conversions

Sometimes it's necessary to convert a fraction to a decimal or a decimal to a fraction. You'll see as we work our way through this section that these conversions are not difficult.

Converting Decimals to Fractions

In Section 5.1 we studied the place value of decimals. We'll use our knowledge of the place value of decimals to help us convert decimals to fractions.

To convert a decimal to a fraction, drop the decimal point and put the number over the place value of the last digit to its right, then reduce.

Example 1:   **Convert 0.5 to a fraction.**

SOLUTION: 0.5 is read five tenths. The 5 is in the tenths' place, so we'll write 5 over 10 and reduce.

$0.5 = \frac{5}{10} = \frac{1}{2}$

Example 2:   **Convert 0.05 to a fraction.**

SOLUTION: 0.05 is read five hundredths. The 5 is in the hundredths' place, so we'll write 5 over 100 and reduce.

$0.05 = \frac{5}{100} = \frac{1}{20}$

Example 3:   **Convert 0.34 to a fraction.**

SOLUTION: 0.34 is read thirty-four hundredths. The 4 is in the hundredths' place, so we'll write 34 over 100 and reduce.

$0.34 = \frac{34}{100} = \frac{17}{50}$

Example 4:   **Convert 0.237 to a fraction.**

SOLUTION: 0.237 is read two hundred thirty-seven thousandths. The 7 is in the thousandths' place, so we'll write 237 over 1,000, which doesn't reduce.

$0.237 = \frac{237}{1,000}$

Example 5:   **Convert 6.75 to a fraction.**

> SOLUTION: 6.75 is read six and seventy-five hundredths. The whole number 6 remains a whole number. The 5 is in the hundredths' place so we'll write 75 over 100 and reduce.
>
> $$\frac{75}{100} = \frac{3}{4} \qquad 6.75 = 6\frac{3}{4}$$

Problem Set:

Convert each of these decimals to fractions in lowest terms:

1) 0.6   2) 0.06   3) 0.006   4) 6.0606   5) 1.125   6) 12.123

1) $0.6 = \frac{6}{10} = \frac{3}{5}$

2) $0.06 = \frac{6}{100} = \frac{3}{50}$

3) $0.006 = \frac{6}{1,000} = \frac{3}{500}$

4) $6.0606 = 6\frac{606}{10,000} = 6\frac{303}{5,000}$

5) $1.125 = 1\frac{125}{1,000} = 1\frac{25}{200} = 1\frac{5}{40} = 1\frac{1}{8}$

6) $12.123 = 12\frac{123}{1,000}$

Converting Fractions to Decimals

Example 6:   **Convert $\frac{1}{10}$ to a decimal.**

> SOLUTION: We'll work this problem out using two techniques. The first technique works only if the denominator of the fraction is a power of ten, or can be converted to a power of ten. The first digit past the decimal point is the tenths' place. Therefore $\frac{1}{10} = 0.1$.
>
> The other method works on any fraction whether or not its denominator is or can be converted to a power of ten. The rule for the other method is: *To convert a fraction to a decimal, divide the numerator by the denominator.*
>
> $$\overset{decimal}{denominator\,)numerator} \qquad \frac{1}{10} = 10\overline{)1} = 0.1$$

Example 7:   **Convert $\frac{7}{100}$ to a decimal.**

> SOLUTION: The denominator is a power of ten, so we'll use the first method.
>
> $$\frac{7}{100} = 0.07.$$

Example 8: Convert $\frac{4}{25}$ to a decimal.

SOLUTION: The denominator is not a power of ten, but twenty-fifths can easily be converted to hundredths, which is a power of ten.

$$\frac{4}{25} = \frac{16}{100} = 0.16$$

Exampe 9: Convert $\frac{6}{2,000}$ to a decimal.

SOLUTION: The denominator is not a power of ten, but can easily be converted to thousandths if we reduce the fraction.

$$\frac{6}{2,000} = \frac{3}{1,000} = 0.003$$

Example 10: Convert $\frac{1}{8}$ to a decimal.

SOLUTION: The denominator is not a power of ten and can not be converted to a power of ten; so we have to use the second technique. We'll divide the numerator 1 by the denominator 8.

$$8\overline{)1.000}^{\,0.125} \qquad \frac{1}{8} = 0.125$$

Now that you know one eighth is 0.125, can you guess what three eighths is? It's three times the value of one eighth. Three eighths is $3(0.125) = 0.375$. Can you guess what seven eighths is? It's seven times the value of one-eighth. Seven eighths is $7(0.125) = 0.875$. We hope you see a pattern here. This pattern works for all fractions—not just eighths.

When dividing the numerator by the denominator results in a decimal that ends, it's called a *terminating decimal*. The fractions in examples six through ten resulted in terminating decimals. There are three types of decimals—terminating, repeating and continuous. Next we'll look at a few examples of repeating decimals. If dividing the numerator by the denominator results in a quotient that has a repeated pattern that never ends, it's called a *repeating decimal*. To show a decimal is a repeating decimal, we'll put a line over the part that repeats.

Example 11: Convert $\frac{1}{3}$ to a decimal.

SOLUTION: The denominator is not a power of ten and can't be converted to a power of ten, so we'll begin by dividing the numerator by the denominator. This time the division doesn't end; instead there's a pattern of a repeated 3. To show this is a repeating decimal, we'll put a line over the part that repeats.

$\frac{1}{3} = 1\overline{)3} = 0.\overline{3}$ This is a good one to memorize.

Sometimes people make the mistake of writing $\frac{1}{3}$ as 0.3 not $0.\overline{3}$.

$0.3 = \frac{3}{10} \qquad 0.\overline{3} = \frac{1}{3} \qquad \frac{3}{10} \neq \frac{1}{3} \qquad 0.3 \neq 0.\overline{3}$

0.3 is not the same as $0.\overline{3}$: be careful not to make this very common mistake!

Example 12: Convert $\frac{2}{3}$ to a decimal.

> SOLUTION: We'll begin by dividing two by three, and soon find a repeating 6.
>
> $\frac{2}{3} = 3\overline{)2} = 0.\overline{6}$ This is a another good one to memorize.
>
> 0.6 is not the same as $0.\overline{6}$.
>
> $0.6 = \frac{6}{10} = \frac{3}{5}$ $0.\overline{6} = \frac{2}{3}$ $\frac{3}{5} \neq \frac{2}{3}$

Example 13: Convert $\frac{21}{22}$ to a decimal.

> SOLUTION: This time the 5 and the 4 repeat, and not the 9, so we'll draw a line over the 5 and 4, and not the 9.
>
> $\frac{21}{22} = 22\overline{)21} = 0.9\overline{54}$ Put the line over only the part that repeats.

So far we've seen examples of terminating and repeating decimals. Now it's time to look at our last type of decimal, continuous decimals. If we divide the numerator by the denominator, and it doesn't terminate or produce a repeated pattern, it's called a ***continuous decimal***. When we ask you to convert a fraction to a decimal, that's continuous; we don't expect you to keep dividing forever. We'll ask you to round a continuous decimal to a specified place value.

Example 14: Convert $\frac{5}{7}$ to a decimal. Round your answer to the nearest hundred thousandth.

> SOLUTION: When we divide the numerator by the denominator, it doesn't terminate or repeat. This is not a terminating or a repeating decimal. When a decimal continues with no pattern, it's called a continuous decimal. The … at the end of the decimal means it continues without out a pattern.
>
> $\frac{5}{7} = 0.714286\ldots$
>
> To round this off to the nearest hundred thousandth, we look at the millionths' place where there's a 6, so we round up to 0.71429.

The following table lists the most commonly used fraction and decimal conversions. You probably have some of these memorized already; it would save you some time if you memorize all of them.

TABLE 6.1   FRACTION AND DECIMAL CONVERSIONS

Fraction	$\frac{1}{2}$	$\frac{1}{3}$	$\frac{1}{4}$	$\frac{1}{5}$	$\frac{1}{8}$	$\frac{1}{10}$	$\frac{2}{3}$
Decimal	0.5	$0.\overline{3}$	0.25	0.2	0.125	0.1	$0.\overline{6}$

Problem Set:

Convert the following fractions to decimals. Round all continuous decimals two decimal places. State whether the fraction is a terminating, repeating or a continuous decimal.

1) $\frac{2}{5}$ 2) $\frac{17}{19}$ 3) $\frac{1}{9}$ 4) $\frac{32}{35}$ 5) $\frac{12}{50}$ 6) $\frac{5}{15}$

SOLUTIONS TO PROBLEM SET:

1) $\frac{2}{5} = \frac{4}{10} = 0.4$ terminating decimal   2) $\frac{17}{19} = 0.894...$ continuous decimal

3) $\frac{1}{9} = 0.\overline{1}$ repeating decimal $= 0.11$ 4) $\frac{32}{35} = 0.914...$ continuous decimal $= 0.91$

5) $\frac{12}{50} = \frac{24}{100} = 0.24$ terminating decimal   6) $\frac{5}{15} = \frac{1}{3} = 0.\overline{3}$ repeating decimal

In the next few examples, we'll perform arithmetic operations on problems that have fractions and decimals in them. We'll either convert all the fractions to decimals, or all the decimals to fractions, before we perform the arithmetic operations. If you remember the following rule, you'll be fine.

Arithmetic Operations With Fractions

Arithmetic operations can be performed using only terminating decimals.

If a problem contains a repeating decimal or a continuous decimal, you **must** convert to all fractions before simplifying the problem.

Example 15: $\frac{2}{5} + 0.1$

SOLUTION: The first thing we have to check is whether or not the fraction is a terminating decimal. $\frac{2}{5} = 0.4$, so it is a terminating decimal. We have the option of simplifying this problem, using all fractions or all decimals. Just for the fun of it, we'll simplify this problem using all decimals first, then all fractions. You'll see that we'll get the same answer, except one will be in fraction form and the other will be in decimal form.

Using all decimals: $\frac{2}{5} + 0.1 = 0.4 + 0.1 = 0.5$

Using all fractions: First we'll write 0.1 as $\frac{1}{10}$. To add $\frac{2}{5} + \frac{1}{10}$, we have to convert the $\frac{2}{5}$ to the common denominator, tenths: $\frac{2}{5} = \frac{4}{10}$.

$\frac{2}{5} + \frac{1}{10} = \frac{4}{10} + \frac{1}{10} = \frac{5}{10} = \frac{1}{2}$. In fraction form the answer is $\frac{1}{2}$.

In decimal form the answer is 0.5; $\frac{1}{2} = 0.5$

Example 16: $\frac{1}{6} + \frac{1}{4}\left(0.1\right)$

SOLUTION: First we have to decide if either $\frac{1}{6}$ or $\frac{1}{4}$ is a non-terminating decimal. We cannot perform arithmetic operations with non-terminating decimals.

$\frac{1}{6} = 0.1\overline{6}$, which is a repeating decimal. We have no choice but to use all fractions. $0.1 = \frac{1}{10}$.

$$\frac{1}{6} + \frac{1}{4}\left(0.1\right) = \frac{1}{6} + \frac{1}{4}\left(\frac{1}{10}\right) = \frac{1}{6} + \frac{1}{40} = \frac{20}{120} + \frac{3}{120} = \frac{23}{120}$$

The next example can be simplified using all fractions or all decimals. We'll show you both ways. See if one way is easier than the other.

Example 17: $\frac{3}{8}\left(8\right)^2 + 100\left(0.5\right)^2$

SOLUTION: First we have to check to see if $\frac{3}{8}$ is a terminating decimal. We can perform arithmetic operations only with terminating decimals.

$\frac{3}{8} = 0.375$, which is a terminating decimal. We'll simplify this problem using decimals first.

$$\frac{3}{8}\left(8\right)^2 + 100\left(0.5\right)^2 = 0.375\left(64\right) + 100\left(0.25\right) = 24 + 25 = 49$$

Now we'll simplify the same problem using fractions. $0.5 = \frac{1}{2}$.

$$\frac{3}{8}\left(8\right)^2 + 100\left(\frac{1}{2}\right)^2 = \frac{3}{8}\left(\frac{64}{1}\right) + \frac{100}{1}\left(\frac{1}{4}\right) = 3(8) + 25(1) = 24 + 25 = 49$$

Problem Set:

Simplify the following problems. If possible, simplify the problem twice, once using all fractions, and once using all decimals.

1) $\frac{1}{3} + 0.4$

2) $1.2 - \left(\frac{1}{5}\right)^2$

3) $\frac{3}{8}\left(0.6\right) + \frac{1}{8}\left(0.7\right)$

4) $\left(\frac{2}{3}\right)^3\left(\frac{1}{2}\right)^2\left(13.5\right)$

SOLUTIONS TO PROBLEM SET:

1) $\frac{1}{3}$ is not a terminating decimal; therefore this problem can be simplified only by converting 0.4 to the fraction $\frac{4}{10}$, which reduces to $\frac{2}{5}$.

$$\frac{1}{3} + 0.4 = \frac{1}{3} + \frac{2}{5} = \frac{5}{15} + \frac{6}{15} = \frac{11}{15}$$

2) $\frac{1}{5}$ is the terminating decimal 0.2; therefore this problem can be simplified two ways. We'll begin with all decimals and follow the order of operations.

$$1.2 - \left(\frac{1}{5}\right)^2 = 1.2 - \left(0.2\right)^2 = 1.2 - 0.04 = 1.16$$

Now we'll simplify this problem using all fractions. $1.2 = 1\frac{2}{10} = 1\frac{1}{5}$.

$$1.2 - \left(\tfrac{1}{5}\right)^2 = 1\tfrac{1}{5} - \tfrac{1}{25} = 1\tfrac{5}{25} - \tfrac{1}{25} = 1\tfrac{4}{25}$$

The answers, 1.16 and $1\frac{4}{25}$, are equal.

3) $\frac{3}{8}$ is the terminating decimal 0.375 and $\frac{1}{8}$ is the terminating decimal 0.125; therefore this problem can be simplified using all decimals or using all fractions.

First we'll simplify this problem using all decimals.

$$\tfrac{3}{8}\left(0.6\right) + \tfrac{1}{8}\left(0.7\right) = 0.375\left(0.6\right) + 0.125\left(0.7\right) = 0.225 + 0.0875 = 0.3125$$

Now we'll simplify this problem using all fractions. $0.6 = \frac{6}{10} = \frac{3}{5}$ $0.7 = \frac{7}{10}$

$$\tfrac{3}{8}\left(\tfrac{3}{5}\right) + \tfrac{1}{8}\left(\tfrac{7}{10}\right) = \tfrac{9}{40} + \tfrac{7}{80} = \tfrac{18}{80} + \tfrac{7}{80} = \tfrac{25}{80} = \tfrac{5}{16}$$

$$0.3125 = \tfrac{3,125}{10,000} = \tfrac{125}{400} = \tfrac{5}{16}$$

The answers, 0.3125 and $\frac{5}{16}$, are equal.

4) $\frac{2}{3}$ is a repeating decimal; therefore it can be simplified only using all fractions.

$$13.5 = 13\tfrac{5}{10} = 13\tfrac{1}{2} = \tfrac{27}{2}$$

$$\left(\tfrac{2}{3}\right)^3 \left(\tfrac{1}{2}\right)^2 \left(13.5\right) = \tfrac{8}{27}\left(\tfrac{1}{4}\right)\left(\tfrac{27}{2}\right) = 1$$

Workbook

Convert each of these fractions into decimals. Round all continuous decimals to the nearest hundredth. State whether the fraction is a terminating, repeating, or continuous decimal.

1) $\frac{4}{5}$

2) $\frac{8}{50}$

3) $\frac{66}{300}$

4) $\frac{9}{15}$

5) $\frac{0}{2}$

6) $\frac{954}{1,000}$

7) $\frac{77}{10,000}$

8) $\frac{59}{100}$

9) $\frac{59}{10}$

10) $\frac{48}{200}$

11) $\frac{1}{3}$

12) $\frac{2}{3}$

13) $\frac{25}{4}$

14) $\frac{17}{13}$

15) $\frac{11}{9}$

16) $\frac{20}{50}$

17) $\frac{8}{15}$

18) $\frac{25}{35}$

19) $\frac{105}{15}$

20) $\frac{9}{8}$

Convert each of the following decimals to fractions in reduced form.

21) 0.8

22) 0.08

23) 0.008

24) 8.8

25) 13.125

26) 0.33

27) 3.33

28) 0.00045

29) 0.0875

30) 0.7

31) 0.77

32) 0.009

33) 1.67

34) 0.0808

35) 66.6

36) $2.\overline{3}$

37) $6.\overline{6}$

38) $6.\overline{3}$

Perform the following operations. Write your answers as a fraction, and as a decimal if possible.

39) $0.8 + \frac{1}{5}$

40) $1.3 - \frac{1}{8}$

41) $40.5 + \frac{2}{3}$

42) $40.5 + \frac{1}{15}$

43) $10.5 + 1\frac{1}{6}$

44) $\frac{1}{11} + \left(0.2\right)^2$

45) $\left(0.3\right)^2 + \left(\frac{4}{5}\right)^2$

46) $0.1 \div \frac{1}{10}$

47) $0.1 \div \frac{1}{9}$

48) $\frac{1}{3}\left(0.3\right) + \frac{2}{3}$

CHALLENGE PROBLEMS:

49) $\frac{2}{5} + \left(0.1\right)^2 \div \frac{1}{2}$

50) $\left(\frac{1}{3}\right)^2 + \left(0.3\right)^2$

51) $12\frac{4}{5} \div 5\frac{3}{25} + 1.5$

52) $\frac{3}{5}\left(8\frac{1}{3}\right) + \frac{4}{9}\left(20\frac{1}{4}\right)$

6.2   Rate and Ratio

How much gas does your car use? If you drove 100 miles and used two gallons of gas, how many miles did you get per gallon? You got 50 miles per gallon (100 ÷ 2). You've just done your first rate problem. Before we study rates, let's look at ratios.

A ratio compares the relationship between two quantities of the same type.

A ratio is actually another way of showing a fraction. For example, the fraction $\frac{3}{4}$ can be written as 3 to 4, 3 /4, or 3:4. Any of these forms, including the fraction form, represents the ratio of 3 to 4.

Example 1:   **Write the ratio of 5 to 6 using each of the standard forms listed above.**

SOLUTION: $\frac{5}{6}$, 5 to 6, 5/6 and 5:6.

Example 2:   **Write $\frac{8}{10}$ as a ratio.**

SOLUTION: First we'll reduce the fraction $\frac{8}{10}$ to $\frac{4}{5}$. Now we'll write it as $\frac{4}{5}$, 4 to 5, 4 /5 and 4:5.

Example 2:   **Write the ratio of $\frac{3}{4}$ to $2\frac{1}{4}$.**

SOLUTION: First we'll write the ratio as the fraction $\dfrac{\frac{3}{4}}{2\frac{1}{4}}$.
Now we'll simplify.

$$\frac{\frac{3}{4}}{2\frac{1}{4}} = \frac{\frac{3}{4}}{\frac{9}{4}} = \frac{\overset{1}{\cancel{3}}}{\underset{1}{\cancel{4}}} \times \frac{\overset{1}{\cancel{4}}}{\underset{3}{\cancel{9}}} = \frac{1}{3}$$

The ratio is $\frac{1}{3}$, 1 to 3, 1/3 or 1:3.

Example 4:   **Write the ratio of 0.5 to $\frac{2}{3}$.**

SOLUTION: We'll begin by writing the ratio in fraction form: $\dfrac{0.5}{\frac{2}{3}}$.
We can't simplify this fraction until the numerator and denominator are both fractions or both decimals. $\frac{2}{3}$ is a repeating decimal. We can't perform arithmetic operations with repeating decimals, so we'll convert the 0.5 to $\frac{1}{2}$.

$$\frac{0.5}{\frac{2}{3}} = \frac{\frac{1}{2}}{\frac{2}{3}} = \frac{1}{2} \cdot \frac{3}{2} = \frac{3}{4}$$

The ratio is 3 to 4, 3 /4, or 3:4.

Problem Set:

Write each ratio as a fraction in simplest form. Then write each fraction using another ratio form.

1) 14:16

2) 1.5 to $2\frac{1}{4}$

3) $\frac{1}{3}$ to $\frac{2}{3}$

4) $\dfrac{\frac{1}{4}}{\frac{2}{5}}$

5) $\frac{3}{8}$

6) $\frac{0.1}{7}$

SOLUTIONS TO PROBLEM SET:

1) $\frac{14}{16} = \frac{7}{8}$ or 7:8 or 7 to 8 or 7/8

2) $\dfrac{1.5}{2\frac{1}{4}} = \dfrac{\frac{3}{2}}{\frac{9}{4}} = \frac{3}{2} \cdot \frac{4}{9} = \frac{2}{3}$

3) $\dfrac{\frac{1}{3}}{\frac{2}{3}} = \frac{1}{2}$ or 1:2 or 1 to 2 or 1/2

4) $\frac{1}{4}\left(\frac{5}{2}\right) = \frac{5}{8}$ or 5:8 or 5 to 8 or 5/8

5) 3 to 8, 3 /8 or 3:8

6) $\dfrac{\frac{1}{10}}{7} = \frac{1}{10}\left(\frac{1}{7}\right) = \frac{1}{70}$ or 1 to 70, 1:70, 1/70

Example 5: In a school there are five hundred students and twenty-five instructors. What is the ratio of students to faculty? What is the ratio of faculty to students?

SOLUTION: We'll write this as a fraction and reduce. The first question asks for student to faculty ratio. The number of students goes in the numerator and the number of faculty goes in the denominator.

$$\frac{500}{25} = \frac{20}{1}$$

We would say that the student-faculty ratio is 20 to 1 or 20:1. The faculty to student ratio is 1 to 20 or 1:20. For one faculty member there are 20 students.

Notice the order of the question reverses the ratio. The faculty to student ratio is the reverse of the student to faculty ratio.

When a ratio is used to compare two different kinds of quantities, the ratio is called a *rate*.

Miles and gallons are two different kinds of quantities. At the beginning of the chapter, you calculated your miles per gallon, which is a rate. Sometimes rates are also called ratios. Let's look at some examples of rates and ratios.

Example 6: A plane travels 2,250 miles in 5 hours. How many miles per hour does it fly?

SOLUTION: This is a rate of miles per hour. $\dfrac{2{,}250\ miles}{5\ hours} = 450$ miles per hour.

Example 7: A man drinks 22 ounces of soda in 10 minutes. How many ounces of soda does he drink per minute?

> SOLUTION: This is a rate of ounces per minute. $\frac{22}{10} = 2.2$ ounces per minute.

Example 8: A basketball team had 20 wins and 30 losses. What is the rate of wins to losses? What is the ratio of losses to games?

> SOLUTION: The ratio of wins to losses is $\frac{20}{30} = \frac{2}{3}$; for every 2 wins there are 3 losses. The ratio of losses to games is $\frac{30}{50} = \frac{3}{5}$; there will be 3 losses every 5 games.

Example 9: A senior citizens residence has 25 men and 75 women. What is the ratio of (a) men to women? (b) women to men?

> SOLUTION: (a) $\frac{25}{75} = \frac{1}{3}$ 1 man to 3 women
>
> (b) $\frac{75}{25} = \frac{3}{1}$ 3 women to 1 man

Problem Set:

1) A 20-ounce nut mixture contains 12 ounces of peanuts. What is the ratio of (a) peanuts to other nuts? and (b) peanuts to all nuts (including peanuts)?

2) A car travels 280 miles on 7 gallons of gas. How many miles per gallon does is get?

3) A class has 12 girls and 18 boys. (a) What is the ratio of boys to girls? (b) What is the ratio of girls to children?

4) A plane travels 1,250 miles in 5 hours. At what rate is the plane traveling?

SOLUTIONS TO PROBLEM SET:

1) (a) $\frac{12}{8} = \frac{3}{2}$ 3 ounces of peanuts to 2 ounces of other nuts.

   (b) $\frac{12}{20} = \frac{3}{5}$ 3 ounces of peanuts to 5 ounces of mixture.

2) $\frac{280}{7} = \frac{40}{1}$ 40 miles per gallon.

3) (a) $\frac{18}{12} = \frac{3}{2}$ 3 boys to 2 girls.

   (b) $\frac{12}{30} = \frac{2}{5}$ 2 girls to 5 children.

4) $\frac{1,250}{5} = \frac{250}{1}$ 250 miles per hour.

1) A baseball team had 50 wins and 20 losses. What is the ratio of (a) its wins to losses? b) its losses to wins? c) its wins to games played?

2) Fifty businesses are members of the Smithtown Business Association. Ten are losing money and the rest are profitable. Find the ratio of (a) profitable businesses to unprofitable businesses. (b) unprofitable businesses to profitable businesses.

3) A department has 60 employees, 12 of whom are welders, and the rest, riveters. What is the ratio of (a) welders to riveters? (b) riveters to welders? and (c) welders to all employees?

4) A retail store has 48 employees, 6 of whom are managers, and the rest, sales clerks. What is the ratio of (a) managers to sale clerks? (b) sales clerks to managers? (c) managers to all employees?

5) A train travels 450 miles in 9 hours. How many miles per hour does is travel?

6) A car travels 432 miles on 9 gallons of gas. How many miles per gallon does it get?

7) There are 120 boy scouts and 6 scout masters in F Troop. What is the ratio of scouts to scout masters?

8) A woman eats 42 crackers over a six-hour period. How many crackers does she eat per hour?

9) Mr. Williams worked 80 days and was out sick 16 days. What is the ratio of his (a) days worked to sick days? (b) sick days to days worked?

10) A school has 120 lower classmen and 180 upper classmen. What is the ratio of (a) lower classmen to upper classmen? (b) upper classmen to lower classmen? (c) lower classmen to classmen?

11) There are 100 cars at an auto dealership. If 70 are new and the rest used, what is the ratio of (a) new cars to used cars? (b) used cars to new cars? (c) new cars to cars?

12) A box with 40 cookies has 20 chocolate cookies; the rest are vanilla cookies. What is the ratio of (a) chocolate cookies to vanilla cookies? (b) chocolate cookies to all cookies (including chocolate)?

13) Company B has 120 soldiers, of whom 110 are enlisted personnel, and the rest officers. What is the ratio of (a) officers to enlisted personnel? (b) enlisted personnel to officers? (c) officers to all soldiers?

14) For every 6 games of tennis I play, I drink 36 ounces of water. What's the rate of ounces of water to games of tennis played?

15) If it took Mr. Maccaluso 2 hours to grade 22 lab reports, at what rate did he grade the reports?

16) If it took Melanie 3 hours to walk 9 miles; what's her rate in miles per hour?

17) If it took us 320 days to write a book containing 10 chapters; at what rate did we write the book?

18) In a group of 220 employees, 80 received merit bonuses for outstanding work. (a) what's the ratio of employees who received bonuses to employees who didn't receive bonuses? (b) what's the ratio of employees who received bonuses to all employees?

6.3   Length, Distance, and Weight

Some of the word problems you will encounter in life will involve length, distance, and weights. This section will help you review a few of the basics you may have forgotten.

We're sure you know the answers, but we need to ask you these questions:

> (1)   How many inches are in a foot?
> (2)   How many feet are in a yard?
> (3)   How many feet are in a mile?

> (1)   There are *12 inches in a foot*.
> (2)   There are *3 feet in a yard*.
> (3)   There are *5,280 feet in a mile*.

We know 1 mile is larger than 1 yard, which is larger than 1 foot, which is larger than 1 inch. When we want to convert from a larger measurement to a smaller, such as yards to inches we multiply. When we want to convert from a smaller measurement to a larger, such as inches to feet, we divide.

Example 1:   **Convert 4 feet into inches.**

> SOLUTION: Since 1 foot = 12 inches, we multiply 4 feet by 12 inches to find that 4 feet = 4(12) = 48 inches.

Example 2:   **How many feet are there in two miles?**

> SOLUTION: Since 1 mile = 5,280 feet, we multiply 2 miles by 5,280 feet to find that 2 miles = 2(5,280) = 10,560 feet.

Example 3:   **How many yards is 16 feet?**

> SOLUTION: We're converting from a smaller measurement, feet to a larger measurement—yards—so we'll divide. We know that 1 yard = 3 feet. To find how many yards there are in 16 feet, we need to divide 16 feet by 3 to convert feet into yards:
> $\frac{16}{3} = 5\frac{1}{3}$ yards

Example 4:   **A woman is 5 feet 3 inches tall. How tall is she in inches?**

> SOLUTION: To convert from feet to inches we multiply, because feet is a larger measurement than inches. 1 foot = 12 inches.
>
> 5 feet 3 inches = 5(12) + 3 = 60 + 3 = 63 inches.

Example 5:   A football player ran just one foot short of 10 yards. How many feet did she run?

> SOLUTION: To convert from yards to feet we multiply. 1 yard = 3 feet. One foot short of 10 yards is one foot less than 10 yards; therefore, we also have to subtract.
>
> 10 yards − 1 foot = 10(3) − 1 = 30 − 1 = 29 feet.

Example 6:   The distance from Farmer Fred's front door to his mailbox is 20 feet further than a quarter mile. What is the distance from his front door to his mailbox in feet?

> SOLUTION: To convert from miles to feet we have to multiply. 1 mile = 5,280 feet. To show 20 feet further than a quarter mile, we have to add 20 feet to the number of feet that represents a quarter mile. To find the number of feet that represents a quarter mile, we multiply $\frac{1}{4}(5,280) = 1,320$ feet.
>
> 20 feet further than a quarter mile =
> $20 + \frac{1}{4}(5,280) = 20 + 1,320 = 1,340$ feet.

Now let's take a look at some examples involving weights. We'll look at three measures of weight—ounces, pounds, and tons.

We're sure you know the answers, but we need to ask you these questions:

(1)   How many ounces are in a pound?
(2)   How many pounds are in a ton?

(1)   There are *16 ounces in a pound*.
(2)   There are *2,000 pounds in a ton*.

Example 7:   A baby weighs 8 pounds 14 ounces. What is her weight in ounces?

> SOLUTION: 1 pound = 16 ounces.
> 8 pounds 14 ounces = 8(16) + 14 = 128 + 14 = 142 ounces.

Example 8:   If a miner delivered 336 ounces of gold to a refinery, how many pounds did that gold weigh?

> SOLUTION: 1 pound = 16 ounces. To convert from a smaller measurement, ounces, to a larger one, pounds, we divide.
>
> $\frac{336}{16} = 21$ pounds

Example 9: Express 100,000 pounds in terms of tons.

SOLUTION: 2,000 pounds = 1 ton. To convert from a smaller measurement, pounds, to a larger one, tons, we divide.

$$\frac{100,000}{2,000} = 50 \text{ tons}$$

Problem Set:

1) How many inches is 3 yards 6 inches?

2) If someone is 6 feet 6 inches tall, how many inches tall are they?

3) How many feet is 54 inches?

4) During a three hour period, a turtle covered a distance that was 20 feet short of an eighth of a mile. How many feet did the turtle travel?

5) Convert 96 ounces to pounds.

6) If a car weighs three quarters of a ton, how much does it weigh in pounds?

7) A steak weighs 4 pounds 3 ounces. How much is its weight in ounces?

SOLUTIONS TO PROBLEM SET:

1) 3 feet = 1 yard. 1 foot = 12 inches. 3 yards = 3(3) feet = 9 feet
   9 feet = 9(12)inches = 108 inches.
   3 yards 6 inches = 108 + 6 = 114 inches.

2) 6 feet 6 inches = 6(12) + 6 = 72 + 6 = 78 inches tall.

3) To convert from inches to feet we divide. 12 inches = 1 foot.
   54 inches = $\frac{54}{12}$ = $4\frac{1}{2}$ feet, or 4.5 feet.

4) To convert miles to feet we'll multiply. 5,280 feet = 1 mile.
   $\frac{1}{8}$(5,280) − 20 = 660 feet − 20 feet = 640 feet.

5) 16 ounces = 1 pound.
   96 ounces = $\frac{96}{16}$ = 6 pounds.

6) 1 ton = 2,000 pounds.
   $\frac{3}{4}$(2,000) = 1,500 pounds.

7) 16 ounces = 1 pound.
   4 pounds 3 ounces = 4(16) + 3 = 64 + 3 = 67 ounces.

1) How many inches tall is a person who is 5 feet and 9 inches tall?

2) How many ounces is a person who weighs 145 pounds?

3) How many feet is 48 inches?

4) How many yards is 4 feet?

5) How many yards is 45 inches?

6) If a runner ran 6.5 miles, how many feet did he run?

7) If a basketball player ran 10 feet short of a mile, how many feet did she run?

8) Convert 108 inches into yards.

9) During a half-hour Tim walked a distance of 2 miles and 100 feet. How many feet did he walk?

10) A child weighs 25 lbs. and 3 ounces, how many ounces does he weigh?

11) If Charlie's cat, C.B., weighs 25 pounds, how many tons is that? Write your answer as a decimal.

12) Convert 4.5 tons into pounds.

13) How many yards is 5 feet and 2 inches?

14) How many pounds is 50 ounces?

15) The distance from Ginny's home to school is 5 feet more than three quarters of a mile. How many feet is this distance?

16) A man weighs 10 pounds less than a tenth of a ton. How many pounds does he weigh?

17) A horse weighs 12 pounds more than a half a ton. How many pounds does he weigh?

6.4 Unit Pricing

More and more, grocery stores and supermarkets are stating how much something costs per ounce, per pound, or per some other unit of measurement. This is called the price per unit or the unit price. The *unit price* is the price paid for one unit of an item. Before they started doing this it was difficult to know whether or not a large container of a product or a small container was the better buy. Sometimes a consumer assumes the larger the package of a particular item, the better the buy. This isn't always the case. For example, suppose a 64 ounce bottle of detergent costs $4.48, while a 28 ounce bottle costs $1.40. Would you know that the 28 ounce bottle is the better buy? The larger bottle costs 7 cents per ounce while the smaller bottle costs 5 cents per ounce. By the time you've finished this section, you'll be a smarter shopper. Be careful not to assume that the same product is measured in the same units. Sometimes when we're shopping, we'll look at the unit price of two different brands of the same product, and assume the unit of measurement is the same. Be careful the same product could be measured in liters, ounces, quarts, or some other unit. Read the label carefully.

$$Unit\ Price = (price) \div (number\ of\ units)$$

Example 1: A 12-ounce can of soda costs $.60. How much is the unit price?

SOLUTION: To find the unit price, we have to divide the price by the number of ounces.

unit price = $0.60 ÷ 12 = 0.05 The unit price is $0.05 per ounce.

Example 2: A 6-yard piece of material costs $15. How much does it cost per yard?

SOLUTION: To find the price per yard which is the unit price, divide.
unit price = 15 ÷ 6 = $2.50
The unit price is $2.50 per yard.

Example 3: A $7\frac{1}{2}$ foot length of knotty pine costs $22.50. How much does it cost per foot?

SOLUTION: unit price = $22.50 ÷ 7.5 = $3.00
The unit price is $3.00 per foot.

Example 4: What's the unit price (a) per pound, (b) per ounce of a $19.95 steak that weighs 2 pounds and 4 ounces?

SOLUTION: (a) 1 pound = 16 ounces. 4 ounces is $\frac{4}{16}$ or $\frac{1}{4}$ or 0.25 pounds.
unit price = $19.95 ÷ 2.25 pounds = $8.87 per pound.

(b) 2 pounds = 2(16) = 32 ounces.
2 pounds and 4 ounces = 32 + 4 = 36 ounces.
unit price = $19.95 ÷ 36 ounces = $.55 per ounce.

Example 5: An 11 ounce can of hot chocolate mix costs $2.30. A box of hot chocolate mix containing eight 1.5 ounce envelopes costs $3.00. Find the unit price of each to answer the question, which one is the better buy?

SOLUTION: The unit price of the can = $2.30 ÷ 11, or approximately $0.21 per ounce. To find the unit price of the box we multiply 8 packets times 1.5 ounces, which gives us 12 ounces of mix.

The unit price of the box = $3.00 ÷ 12, or approximately $0.25 per ounce. The can is the better buy.

Example 6: A beam of wood 7 feet 6 inches long costs $30. Find the unit price (a)per inch, and (b)per foot.

SOLUTION: (a) 7 feet 6 inches = 7(12) + 6 = 84 + 6 = 90 inches.
The unit price per inch = $30 ÷ 90 = $0.33 per inch.

(b) 6 inches is half a foot; 7 feet 6 inches is 7.5 feet.
The unit price per foot = $30 ÷ 7.5 = $4 per foot.

Problem Set:

1) A 48-ounce jar of pickles costs $1.89. What's the unit price?

2) The price of an 8.5 foot length of pipe is $34. What's the price per foot?

3) The price of a $3\frac{1}{4}$ pound fish is $16. What's the price per (a) ounce, (b) pound?

4) A 16-ounce jar of peanut butter costs $1.80. A 20-ounce jar of peanut butter costs $2.60. What's the unit price of each jar? Which is the better buy?

SOLUTIONS TO PROBLEM SET:

1) The unit price = $1.89 ÷ 48, approximately $.04 per ounce.

2) The unit price = $34 ÷ 8.5, approximately $4.00 per foot.

3) (a) 3 pounds = 3(16) = 48 ounces. $\frac{1}{4}$ pound = $\frac{1}{4}$(16) = 4 ounces. $3\frac{1}{4}$ pounds = 48 + 4 = 52 ounces. The unit price = $16 ÷ 52, approximately $0.31 per ounce.

(b) The unit price = $16 ÷ 3.25, approximately $4.92 per pound.

4) The unit price of the 16-ounce jar = $1.80 ÷ 16, is approximately $.11 per ounce. The unit price of the 20-ounce jar = $2.60 ÷ 20, approximately $.13 per ounce. The 16-ounce jar is the better buy.

Find the unit price for each of the following.

1) A 16-ounce can of beans costs $1.99.

2) A 12-foot beam of wood cost $15.

3) The price of a piece of fabric 9.2 yards long is $29.50.

4) A goose that weighs 8 pounds and 2 ounces costs $16.39. Find the price per ounce.

5) A tinsmith buys 8 pounds 10 ounces of tin for $16.39. Find the price per ounce.

6) A 15-foot beam of wood cost $18.

7) The price of a piece of fabric 8.3 yards long is $31.25.

8) A 12 foot 9 inch piece of ribbon cost $3.35. Find the unit price per (a) foot, (b) inch?

9) A can of green peas weighing 1 pound and 4 ounces sells for $1.69. Find the unit price per (a) ounce, (b) pound.

10) A jar of ketchup that weighs 1 pound and 12 ounces costs $2.29.

11) A piece of cloth three yards, 1 foot long costs $10. What's the price per foot?

12) A turkey that weighs 9 pounds 12 ounces, costs $14.59. Find the unit price per (a) pound (b) ounce.

13) A silversmith buys 8 pounds, and 7 ounces of silver for $596. How much is the unit price per ounce?

14) A 10 foot 3 inch length of pipe costs $12.17. Find the unit price per (a) foot, (b) inch?

15) A jar of meatballs weighing 2 pounds and 4 ounces sells for $5.99. Find the unit price per (a) pound (b) ounce.

16) A 64-ounce bottle of soda costs $1.99. An 18-ounce bottle of soda costs $1.45. What's the unit price of each item? Which one is the better buy?

17) A 12-ounce can of corn costs $1.39. An 8-ounce can of corn costs $0.99. What's the unit price of each can? Which one is the better buy?

18) A box of 100 index cards cost $0.88. A box of 80 index cards cost $0.65. What's the unit price of each box? Which one is the better buy?

19) It cost $6.99 to develop a roll of film with 24 exposures. It cost $9.99 to develop a roll of film with 36 exposures. What's the unit price of each roll? Which is the better buy?

20) Ten pairs of socks cost $12.50. Eight pairs of socks cost $10.25. What's the unit price of each pack of socks? Which is the better buy?

6.5 Solving Equations by Division

Later in this chapter we'll have to find the answer to a problem by dividing an equation. Suppose we asked you what number times 3 is 12. You would say the number is 4. If we wanted to ask you the same question in mathematical symbols instead of words, it would look like this: $3x = 12$. Any of the twenty-six letters of the alphabet can be used to represent an unknown value. **A letter that represents an unknown value is called a** *variable*. In the equation $3x = 12$, it's assumed the 3x stands for 3 times x. The variable is x and the 3 is called the *coefficient*. The number in front of a variable is called its coefficient. We want to find the value for x, which is assumed to mean 1x. In order for us to find the value for 1x, not 3x, we need to change the 3x to an x. That's very easy for us to do, since 3x means 3 times x. We'll undo the multiplication by dividing both sides of the equation by 3. Remember, anything (except 0) divided by itself is 1.

When we reduce the common factor of 3 in the numerator and denominator, we have

$\frac{3x}{3} = 1x$ or x To solve for the variable, divide both sides of the equation by the coefficient of the variable. The coefficient is 3.

$3x = 12$

$\frac{3x}{3} = \frac{12}{3}$ We have to be fair about this; whatever we do to one side of an equation, we must do the same thing to the other side of the equation. This keeps the equation balanced.

$x = 4$

Example 1: Solve $2y = 24$

SOLUTION: $\frac{2y}{2} = \frac{24}{2}$ Divide both sides of the equation by 2.

$y = 12$

Example 2: Solve $0.3a = 1.8$

SOLUTION: $\frac{0.3a}{0.3} = \frac{1.8}{0.3}$ Divide both sides of the equation by 0.3.

$a = 6$

Example 3: Solve $\frac{2}{3}x = 12$

SOLUTION: We can solve this fractional equation in two ways.

First Method: $\frac{\frac{2}{3}x}{\frac{2}{3}} = \frac{12}{\frac{2}{3}}$ Divide both sides by $\frac{2}{3}$.

$\frac{2}{3} \cdot \frac{3}{2}x = \frac{\overset{6}{\cancel{12}}}{1} \cdot \frac{3}{\cancel{2}}$ Division by $\frac{2}{3}$ becomes multiplication by $\frac{3}{2}$.

$x = 18$ Simplify.

Second Method: When a problem has a coefficient that's a fraction, you can save one step by multiplying both sides of the equation by the reciprocal of the fractional coefficient. Remember division by $\frac{2}{3}$ becomes multiplication by $\frac{3}{2}$.

$$\overset{1}{\underset{1}{\cancel{\frac{2}{3}}}} \overset{1}{\underset{1}{\cancel{\frac{3}{2}}}} x = \overset{6}{\underset{1}{\cancel{\frac{12}{1}}}} \cancel{\frac{3}{2}} \quad \text{Now reduce the common factors.}$$

$$x = 18$$

Example 4: Solve $0.4z = \frac{2}{3}$.

SOLUTION: This problem has a fraction and a decimal in it. In Section 6.1 we learned that when a problem has a fraction and a decimal in it, you must convert to all fractions or all decimals. We already know that $\frac{2}{3}$ is a repeating decimal, and that we can't perform arithmetic operations with repeating decimals. We have no choice but to convert the decimal 0.4 to the fraction $\frac{4}{10}$, which reduces to $\frac{2}{5}$.

$$\frac{2}{5}z = \frac{2}{3} \quad\quad \text{Multiply both sides by } \frac{5}{2}, \text{ which is the same as dividing by } \frac{2}{5}.$$

$$\overset{1}{\underset{1}{\cancel{\frac{5}{2}}}} \overset{1}{\underset{1}{\cancel{\frac{2}{5}}}} z = \frac{2}{3} \cdot \overset{1}{\cancel{\frac{5}{2}}}$$

$$z = \frac{5}{3} = 1\frac{2}{3} \quad\quad \text{Reduce the common factors.}$$

Problem Set:

Solve the following equations.

1) $5x = 125$ 2) $0.05y = 1.25$ 3) $\frac{1}{5}z = 125$ 4) $\frac{3}{4}d = \frac{75}{16}$

SOLUTIONS TO PROBLEM SET:

1) $\frac{5}{5}x = \frac{125}{5}$

     $x = 25$

2) $\frac{0.05}{0.05}y = \frac{1.25}{0.05}$

     $y = 25$

3) $\overset{1}{\cancel{\frac{5}{1}}} \cdot \underset{1}{\cancel{\frac{1}{5}}} z = \frac{125}{1} \cdot \frac{5}{1}$

     $z = 625$

4) $\overset{1}{\underset{1}{\cancel{\frac{4}{3}}}} \cdot \overset{1}{\underset{1}{\cancel{\frac{3}{4}}}} d = \overset{25}{\underset{4}{\cancel{\frac{75}{16}}}} \cdot \overset{1}{\underset{1}{\cancel{\frac{4}{3}}}}$

     $d = \frac{25}{4} = 6\frac{1}{4}$

Workbook

Solve the following equations.

1) $14a = 28$

2) $39 = 3r$

3) $6b = 36$

4) $72 = 6c$

5) $7x = 70$

6) $96 = 4w$

7) $66 = 11y$

8) $11 = 66t$

9) $0.1c = 7.7$

10) $0.03 = 24x$

11) $2.2y = 7.04$

12) $0.5z = 0$

13) $20.4a = 102$

14) $0.9 = 8.1b$

15) $0.3x = 0.009$

16) $1.1x = 0.11$

17) $\frac{4}{5}x = 20$

18) $1\frac{2}{3}y = 25$

19) $3\frac{1}{2}t = 12\frac{1}{4}$

20) $\frac{1}{8}y = 0.125$

21) $1.2s = 1\frac{11}{25}$

22) $\frac{2}{3}q = 2.1$

23) $3\frac{1}{5} = 2.2x$

24) $1\frac{1}{2}c = 0.7$

25) $0.3 = 0.9s$

26) $5.5f = 0.055$

27) $6\frac{1}{4} = 25y$

28) $1\frac{1}{6} = 1\frac{1}{6}w$

29) $0.1p = 0.001$

30) $7.7x = 847$

6.6 Proportions

The Means and the Extremes

Remember the ratios we worked with earlier in this chapter? When two ratios are equal, such as 2:1 and 4:2, we say they are *proportional*. Here's another example of two ratios that are equal: $\frac{2}{3} = \frac{4}{6}$. Two equivalent fractions joined by an equal sign is called a *proportion*. Here are three ways to write the same proportion.

1)	5 is to 10 as 25 is to 50	Written in words
2)	5:10 = 25:50	Written with a:
3)	$\frac{5}{10} = \frac{25}{50}$	Written in fraction form

We'll use fraction form because it will allow us to easily solve word problems involving proportions.

How would we read this proportion: $\frac{a}{b} = \frac{c}{d}$?

It's read, "a is to b as c is to d." The letter "a" is the first term, "b" is the second term, "c" is the third term, and "d" is the fourth term:

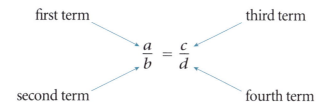

The second and third terms are called the *means*, while the first and fourth terms are the *extremes* of the proportion. Let's put these terms to work. For this proportion, see if you can identity the means and the extremes:

$\frac{1}{5} = \frac{2}{10}$ 5 and 2 are the means and 1 and 10 are the extremes. Remember, the second and third terms are the means and the first and fourth terms are the extremes.

Identify the means and the extremes of this proportion: $\frac{3}{4} = \frac{6}{8}$

4 and 6 are the means and 3 and 8 are the extremes. In a true proportion, the product of the means is equal to the product of the extremes. We'll use this rule later in this section to solve proportions. Let's look at the last proportion.

$\frac{3}{4} = \frac{6}{8}$ The product of the means $6 \times 4 = 24$, equals the product of the extremes $3 \times 8 = 24$. This type of multiplication is *called cross multiplication*.

Example 1:   **Cross multiply $\frac{2}{3} = \frac{5}{7}$ to see if it is a true proportion.**

SOLUTION: $2(7) = 5(3)$

$14 = 15$ This is a false statement, so $\frac{2}{3} = \frac{5}{7}$ is not a true proportion.

Example 2:   **Solve $\frac{4}{5} = \frac{x}{25}$.**

SOLUTION: Now it's time for us to solve a proportion. The following proportion has a missing part.

$\frac{4}{5} = \frac{x}{25}$

We are given the values for three of the four parts of the proportion. The fourth part we labeled x. To find the value for x we will cross multiply and solve for x.

$4(25) = 5x.$ Remember, $5x$ means 5 times x or x times 5. Multiplication is commutative, so it doesn't matter if we say 5 times x or x times 5.

We'll always write the number first, then the variable: $5x$.

$100 = 5x$ Multiply 4 times 25.

$20 = x$ Divide both sides of the equation by 5.

To check if $x = 20$ is the correct answer, return to the original problem $\frac{4}{5} = \frac{x}{25}$, but this time substitute the 20 for the x: $\frac{4}{5} = \frac{20}{25}$ Cross multiply to see if this is a true proportion.

$4(25) = 20(5)$

$100 = 100$ x = 20 checks

Example 3:   **Solve $\frac{7}{10} = \frac{5}{y}$ and check your answer.**

SOLUTION: We'll begin by cross multiplying.

$7y = 5(10)$

$7y = 50$ Divide both sides by 7.

$y = \frac{50}{7} = 7\frac{1}{7}$

Check: $\frac{7}{10} = \frac{5}{7\frac{1}{7}}$ Substitute $7\frac{1}{7}$ for y.

$\frac{7}{10} = \frac{5}{\frac{50}{7}}$ You must change to improper fraction form to multiply.

$\frac{7}{1}\left(\frac{50}{7}\right) = 10(5)$ Cross multiply and reduce common factors.

$50 = 50$

Example 4: Solve $\frac{0.2}{z} = \frac{\frac{2}{3}}{4}$

SOLUTION: We know the right side of the quation is $\frac{2}{3} \div 4$, not $2 \div \frac{3}{4}$, because the fraction bar on both sides of the equal sign lines up under the 3. We have fractions and decimals in one problem, so we have to change to all fractions or all decimals. In this case we'll use all fractions because $\frac{2}{3}$ is a repeating decimal and we know we can't perform arithmetic operations with repeating decimals. $0.2 = \frac{2}{10} = \frac{1}{5}$

$\quad \dfrac{\frac{1}{5}}{z} = \dfrac{\frac{2}{3}}{4}$ Begin the proportion using $\frac{1}{5}$ instead of 0.2.

$\quad \frac{1}{5}\left(\frac{4}{1}\right) = \frac{2}{3}z$ Cross multiply.

$\quad \frac{4}{5} = \frac{2}{3}z$ Multiply.

$\quad \frac{3}{\cancel{2}}\left(\frac{\cancel{4}}{5}\right) = \frac{\cancel{3}}{\cancel{2}}\left(\frac{\cancel{2}}{\cancel{3}}\right)z$ Multiply by the reciprocal of the coefficient of z.

$\quad \frac{6}{5} = 1\frac{1}{5} = z$ Reduce the common factors and multiply.

Example 5: Solve $\frac{10}{40} = \frac{x}{3}$.

SOLUTION: Before we cross multiply, we'll reduce the fraction. Reducing first makes the problem easier.

$\quad \frac{1}{4} = \frac{x}{3}$ Cross multiply.

$\quad 1(3) = 4x$ Divide by 4.

$\quad \frac{3}{4} = x$

Problem Set:

Solve the following proportions.

1) $\frac{6}{7} = \frac{a}{14}$

2) $\frac{x}{15} = \frac{4}{5}$

3) $\dfrac{1\frac{1}{2}}{y} = \dfrac{2\frac{2}{3}}{4}$

4) $\dfrac{0.7}{z} = \dfrac{\frac{1}{4}}{0.5}$

1) $6(14) = 7a$
   $84 = 7a$
   $\frac{84}{7} = 12 = a$

2) $5x = 4(15)$
   $5x = 60$
   $x = 12$

3) $1\frac{1}{2}\left(\frac{4}{1}\right) = 2\frac{2}{3}y$

   $\frac{3}{2}\left(\frac{\overset{2}{\cancel{4}}}{1}\right) = \frac{8}{3}y$

   $6 = \frac{8}{3}y$

   $\frac{3}{\cancel{8}}\left(\frac{\cancel{6}}{1}\right) = \frac{\cancel{3}}{\cancel{8}}\left(\frac{8}{3}\right)y$

   $\frac{9}{4} = 2\frac{1}{4} = y$

4) $\frac{\frac{7}{10}}{z} = \frac{\frac{1}{4}}{\frac{5}{10}}$

   $\frac{7}{\cancel{10}}\left(\frac{\cancel{5}}{10}\right) = \frac{1}{4}z$

   $\frac{7}{20} = \frac{1}{4}z$

   $\frac{4}{1}\left(\frac{7}{20}\right) = \frac{4}{1}\left(\frac{1}{4}\right)z$

   $\frac{7}{5} = 1\frac{2}{5} = z$

Applications

Now that you know how to solve a proportion it's time to apply what you just learned to word problems. The following word problems will involve two ratios that are related. You'll be given the value of three of the four parts of a proportion and be asked to find the fourth value. When setting up the equation make sure the labels of both numerators are the same and the labels of both denominators are the same.

Example 6: Barry Bonds hits 10 home runs in the first 15 games of the season. If he keeps up that pace, how many home runs will he hit during a 162 game season?

SOLUTION: Barry Bonds hit 10 home runs in 15 games.

We'll show that this way: $\dfrac{10\ home\ runs}{15\ games}$

To show an unknown number of home runs out of 162 games,

we'll use $\dfrac{x\ home\ runs}{162\ games}$

Notice we put the number of home runs in the numerator of both fractions and the number of games in the denominator of both fractions. Make sure the labels are the same in the numerators and denominators of the proportion you use to represent the problem. Now we'll join the two fractions by an equal sign to form a proportion.

$\dfrac{10\ home\ runs}{15\ games} = \dfrac{x\ home\ runs}{162\ games}$ Reduce the $\frac{10}{15}$ to $\frac{2}{3}$.

$\dfrac{2}{3} = \dfrac{x}{162}$ Cross multiply.

$2(162) = 3x$ Multiply.

$324 = 3x$ Divide by 3.

$108\ home\ runs = x$

Example 7: Wei Wong drives 220 miles in 4 hours. If he continues at the same rate, how far will he travel in 7 hours?

SOLUTION: Let m be the miles traveled in 7 hours.

$$\frac{220 \text{ miles}}{4 \text{ hours}} = \frac{m \text{ miles}}{7 \text{ hours}} \qquad \text{Reduce the } \tfrac{220}{4} \text{ to } \tfrac{55}{1}.$$

$$\frac{55}{1} = \frac{m}{7} \qquad \text{Cross multiply.}$$

$$55(7) = m$$

$$m = 385 \text{ miles}$$

Example 8: A map is drawn on a scale of $\frac{1}{2}$ inch = 3 miles. If a distance shown on the map is 2 inches, how many miles is that distance?

SOLUTION: Let d = distance in miles.

$$\frac{\frac{1}{2} \text{ inch}}{3 \text{ miles}} = \frac{2 \text{ inches}}{d \text{ miles}} \qquad \text{Cross multiply.}$$

$$\tfrac{1}{2} d = 2(3) \qquad \text{Multiply.}$$

$$\tfrac{1}{2} d = 6 \qquad \text{Multiply by the reciprocal of the coefficient.}$$

$$\frac{\cancel{2}^{1}}{1}\left(\frac{1}{\cancel{2}_{1}}\right) d = \frac{2}{1}\left(\frac{6}{1}\right) \qquad \text{Reduce the common factor.}$$

$$d = 12 \text{ miles}$$

Example 9: Ms. Aziz is paid $20 for 2.4 hours of work. How much would she earn if she worked 8 hours?

SOLUTION: Let x = pay for 8 hours.

$$\frac{\$20}{2.4 \text{ hours}} = \frac{\$ x}{8 \text{ hours}} \qquad \text{Cross multiply.}$$

$$20(8) = 2.4x \qquad \text{Multiply.}$$

$$160 = 2.4x \qquad \text{Divide both sides by the coefficient 2.4.}$$

$$\frac{160}{2.4} = x = \$66.67$$

Example 10: A manufacturer knows from experience that for every 100 widgets that are produced at her factory, 3 will be defective. If she manufacturers 350 widgets, how many can she expect to be defective?

SOLUTION: Let w = defective widgets out of 350 manufactured.

$$\frac{100\ produced}{3\ defective} = \frac{350\ produced}{w\ defective}$$ Cross multiply.

$100w = 350(3)$ Multiply.

$100w = 1{,}050$ Divide both sides by 100.

$\dfrac{100w}{100} = \dfrac{1050}{100}$ Reduce the common factors.

$w = 10.5$ defective   10.5 widgets out of 350 can be expected to be defective.

Problem Set:

1) Marty Liquori ran 11.5 miles in an hour. If he keeps up the same pace, how far will he run in an hour and a half?

2) Mrs. Chillingotn drove 165 miles in 4 hours. If she continues driving at the same speed, how far will she have driven in 6 hours?

3) If 4 pounds of chopped meat cost $6.50, how much does 10 pounds cost?

4) A map is drawn on a scale where a quarter inch equals two miles.
   (a) If the distance on the map is 3 inches, how many miles is that distance?
   (b) How many inches would represent a distance of 120 miles?

5) Henry O'Toole is paid $50 for 4.6 hours of work. How much would he earn if he worked 10 hours?

SOLUTIONS TO PROBLEM SET:

1) $\frac{11.5}{1} = \frac{x}{1.5}$ Miles are in both numerators and hours in both denominators.

     $11.5(1.5) = 1x$ Cross multiply.

     $17.25 = x$ The answer is: he will run 17.25 miles in an hour and a half.

2) $\frac{165}{4} = \frac{x}{6}$ Miles are in both numerators and hours in both denominators.

     $165(6) = 4x$ Cross multiply.

     $990 = 4x$ Multiply.

     $247.5 = x$ Divide by 4 on both sides of the equal sign.

     The answer is: she will have driven 247.5 miles in 6 hours.

3) $\frac{4}{6.50} = \frac{10}{x}$ Pounds are in both numerators and dollars in both denominators.

    $4x = 6.5(10)$ Cross multiply.

    $4x = 65$ Multiply.

    $x = \frac{65}{4}$ Divide by 4 on both sides of the equal sign.

    $x = \$16.25$ The answer is: 10 pounds cost $16.25.

4) (a) $\frac{0.25}{2} = \frac{3}{x}$ Inches are in both numerators and miles in both denominators.

    $0.25x = 2(3)$ Cross multiply.

    $0.25x = 6$ Multiply.

    $x = \frac{6}{0.25}$ Divide both sides of the equation by 0.25.

    $x = 24$ The answer is 24 miles.

  (b) $\frac{0.25}{2} = \frac{x}{120}$ Inches are in both numerators and miles in both denominators.

    $0.25(120) = 2x$ Cross multiply.

    $30 = 2x$ Multiply.

    $15 = x$ Divide both sides of the equation by 2.
                         The answer is 15 inches.

5) $\frac{50}{4.6} = \frac{x}{10}$ Dollars is in both numerators and hours in both denominators.

    $50(10) = 4.6x$ Cross multiply.

    $500 = 4.6x$ Multiply.

    $\frac{500}{4.6} = x$ Divide both sides of the equation by 4.6.

    $x = \$108.70$ The answer is; he earned $108.70.

1) For each of theses expressions, state the means and the extremes.

   a) $2:3 = 6:9$ b) $4:1 = 16:4$ c) $\frac{5}{7} = \frac{10}{14}$ d) $\frac{w}{x} = \frac{y}{z}$

For problems 2–20 find the unknown value in each proportion.

2) $\frac{2}{5} = \frac{n}{20}$

3) $\frac{3}{m} = \frac{2}{6}$

4) $\frac{5}{7} = \frac{5}{p}$

5) $\frac{8}{r} = \frac{2}{5}$

6) $\frac{s}{10} = \frac{3}{2}$

7) $\frac{1}{9} = \frac{2}{t}$

8) $\frac{5}{7} = \frac{x}{49}$

9) $\frac{a}{3} = \frac{7}{21}$

10) $\frac{16}{n} = \frac{64}{3}$

11) $\frac{2\frac{1}{2}}{y} = \frac{5}{4}$

12) $\frac{0.1}{\frac{1}{2}} = \frac{0.5}{z}$

13) $\frac{2}{0.3} = \frac{n}{0.2}$

14) $\frac{z}{0.4} = \frac{2}{5}$

15) $\frac{2.5}{x} = \frac{75}{0.3}$

16) $\frac{\frac{1}{3}}{5} = \frac{2}{c}$

17) $\frac{\frac{1}{6}}{\frac{1}{12}} = \frac{x}{0.1}$

18) $\frac{\frac{1}{2}}{1\frac{2}{3}} = \frac{r}{\frac{2}{3}}$

19) $\frac{\frac{11}{12}}{t} = \frac{3\frac{5}{12}}{41}$

20) $\frac{0.01}{\frac{1}{10}} = \frac{x}{0.2}$

21) John Landy ran three miles in 14 minutes. If he keeps up this pace, how far will he go in 40 minutes?

22) If 3 pounds of chicken cost $5 how much would 8 pounds cost?

23) A map is drawn on a scale of $\frac{1}{2}$ inch = 10 miles.
   a) If a distance shown on the map is inches, how many miles is that distance?
   b) How many inches would represent a distance of 15 miles?

24) Yassir Hussein is paid $45 for 5.3 hours of work. How much would he earn if he worked 10 hours?

25) A furniture store advertised chairs on sale at 4 for $69. At this rate, what would it cost for 6 chairs?

26) A 24-ounce jar of mixed nuts contains 15 ounces of peanuts. How many ounces of peanuts would be in a 32-ounce jar?

27) If 3 out of every 4 dentists surveyed recommend sugarless gum to their patients who chew gum, how many dentists would you expect to recommend sugarless gum if 480 dentists are surveyed?

28) If a new Toyota Camry gets 32 miles per gallon of gas, how many gallons of gas are necessary to drive 352 miles?

29) If 5 dresses can be sewn from $6\frac{1}{2}$ yards of material, how much material is needed to sew 12 dresses?

30) A recipe calls for $\frac{3}{4}$ of a cup of sugar to make 36 cookies, how much sugar is needed to make 48 cookies?

31) On the average, 15 out of every 100 students who take the basic skills exam fail it. If 165 people failed the exam, how many people took the exam?

32) Walt Frazier made 8 foul shots out of 10 attempts. If he continues to shoot at the same rate, how many shots would he make out of 90 attempts?

33) For every $5 invested in a stock, Nahil Patel makes a return of $2.50. How much would her return be on $26 invested?

34) For every 20 tables Melanie waits on, she makes $40 in tips, how much would she make in tips if she waits on 50 tables?

35) For every 5 newspapers Nancy delivers, she makes $2.75 in tips. How much in tips would she make if she delivered 120 newspapers?

Chapter 6 • Review

Write the following decimals as fractions or mixed numbers in reduced form.

1) 0.2

2) 0.02

3) 0.002

4) 0.0002

5) 2.2

6) 0.091

7) 34.0001

8) 100.6

Write the following fractions as decimals. Label them as terminating, repeating, or continuous decimals. Round all continuous decimals to the nearest ten thousandths.

9) $\frac{1}{10}$

10) $\frac{1}{100}$

11) $\frac{1}{1,000}$

12) $\frac{1}{10,000}$

13) $\frac{1}{3}$

14) $\frac{11}{13}$

15) $\frac{2}{3}$

16) $4\frac{1}{9}$

17) $\frac{2}{5}$

18) $\frac{2}{500}$

19) $6\frac{8}{42}$

20) $3\frac{4}{90}$

Simplify the following problems. Whenever possible, simplify using all decimals and then all fractions.

21) $2\frac{1}{2} + 0.1$

22) $\left(0.2\right)^3 + \frac{1}{3}$

23) $2\frac{1}{10} \div 0.7$

24) $0.\overline{3}\left(1\frac{1}{5}\right)$

25) $1\frac{7}{9}\left(1.5\right)^4$

26) $\frac{9}{16} \div \left(\frac{3}{4}\right)^2$

27) A man drove 200 miles on 12 gallons of gas. What's his rate of miles per gallon?

28) A group of 10 women and 12 men went to the U.S. Open. What's the ratio of (a) men to women, (b) women to men?

29) Sam drove 450 miles in 6 hours. What's his rate in miles per hour?

30) If a man lost 65 pounds in 5 months, what's his average rate of loss per month?

31) In a shipment of 200 television sets, 20 were defective. What's the ratio of (a) defective sets to non defective sets, (b) non-defective sets to defective sets, and (c) defective sets to sets in the shipment?

Solve the following equations.

32) $3x = 3.3$

33) $0.1 = 10x$

34) $1\frac{1}{2}y = 12$

35) $\frac{3}{4}r = 27$

36) $0.25z = 75$

37) $\frac{1}{3} = 0.6s$

38) Convert 144 inches into yards.

39) How many inches are there in 11.5 feet?

40) During a 30 minute walk, Kiichi Miyazawa walked 1 mile and 700 feet. How many feet did he walk?

41) Convert 192 ounces into pounds.

42) If a truck weighs 3.5 tons, how much does it weigh in pounds?

43) How many feet are there in 78 inches?

For questions 44–50, find the unit price.

44) A 64-ounce bottle of soda costs $1.89.

45) The price of a three and three-quarter pound fish is $15.

46) A piece of cloth six yards and two feet long costs $20. Find the unit price in (a) yards, and (b) feet.

47) A 10 pound 11 ounce turkey costs $15.89. Find the unit price in (a) pounds,(b) ounces.

48) A goldsmith buys 2 pounds 3 ounces of gold for $12,040. How much is the unit price per (a) pound, (b) ounce?

49) A store advertised a sale of 3 eight-ounce cans of corn for $1.55. Another store advertised a sale of 3 ten-ounce cans of corn for $1.95. What's the unit price per ounce for each deal? Which one is the better buy?

50) A 12-ounce box of Cheerios is on sale for $3.50. A 16-ounce box of Cheerios costs $4.25. What's the unit price of each box? Which one is the better buy?

Solve the following proportions.

51) $\frac{4}{x} = \frac{1}{7}$

52) $\frac{3}{4} = \frac{x}{12}$

53) $\frac{8}{5} = \frac{24}{t}$

54) Victor Newman ran four miles in 19 miles. If he keeps up this pace, how far will he run in 1 hour?

55) If 5 pounds of steak cost $19, how much would 12 pounds cost?

56) A map is drawn on a scale of a half-inch equals one mile. (a) If a distance shown on the map is three and a half inches, how many miles is that distance? (b) How many inches would represent a distance of 12 miles?

57) Brad is paid $35 for 3.8 hours of work. How much would he earn if he worked 8 hours?

58) A grocery store advertised 2 cans of chili for $1.39 or 5 cans for $5.25. Which one is the better deal?

59) A 16-ounce jar of mixed nuts contains 10 ounces of peanuts. How many ounces of peanuts would be in a 28-ounce jar?

Chapter 6 • Test 1

Round all continuous decimals to the nearest hundredth. Reduce all fractions.

1) Convert $5\frac{2}{11}$ to a decimal.

2) Convert 2.35 to a fraction.

3) Convert $4.\overline{3}$ to a fraction.

4) Convert $\frac{3}{26}$ to a decimal.

5) Simplify $\frac{2}{5} + 0.2$

6) Simplify $\frac{5}{8} + 0.00025$

7) Simplify $\frac{1}{12} + \left(0.2\right)^2$

8) Solve for x: $\frac{2}{3}x + 16$

9) Solve for y: $\frac{6}{y} = \frac{3}{\frac{1}{2}}$

10) Solve for z: $\frac{3}{4} = \frac{z}{0.4}$

11) A train travels 350 miles in 7 hours. How many miles per hour does it travel?

12) There are 840 students and 70 instructors at Rutgers prep. What is the ratio of students to instructors?

13) A school has 150 lower classmen and 180 upper classmen. What is the ratio of a) lower classmen to upper classmen? b) upper classmen to lower classmen? c) lower classmen to classmen?

14) Convert 180 inches to yards.

15) Convert one and a half miles to feet.

16) A 48-ounce bottle of soda costs $1.59. Find the unit price.

17) A 12-foot beam of wood costs $15. Find the price per foot.

18) A carton containing one dozen eggs costs $2.64. A carton containing a dozen and a half eggs costs $3.42. What's the price per egg for each carton? Which one is the better buy?

19) A map is drawn on a scale of $1\frac{1}{2}$ inch = 1 mile. a) If a distance shown on the map is 6 inches, how many miles is that distance? b) How many inches would represent a distance of 15 miles?

20) Jim Ryan ran two miles in 9 minutes. If he keeps up the pace, how far will he go in 40 minutes?

Round all continuous decimals to the nearest thousandth. Reduce all fractions.

1) Convert $6\frac{5}{6}$ to a decimal.

2) Convert 3.488 to a fraction.

3) Convert $15.\overline{6}$ to a fraction.

4) Convert $\frac{3}{26}$ to a decimal.

5) Simplify $\frac{7}{8} + 0.4$.

6) Simplify $\frac{2}{3} - 0.003$.

7) Simplify $\left(0.5\right)^2 + \frac{3}{10}$.

8) Solve $\frac{2}{5}x = 40$.

9) Solve for y: $\frac{8}{y} = \frac{\frac{1}{2}}{12}$.

10) Solve for z: $\frac{2}{3} = \frac{z}{0.3}$.

11) A car travels 468 miles on 12 gallons of gas. How many miles per gallon does it get?

12) Kathy Passanisi worked 70 days and was out sick 5 days. What is the ratio of a) days worked to sick days? b) sick days to days worked?

13) There are 100 cars at an auto dealership. If 80 are new and the rest used, what is the ratio of a) new cars to used cars? b) used cars to new cars? c) new cars to cars?

14) How many inches are in one and a half feet?

15) Convert 208 ounces into pounds.

16) Express 62,000 pounds into terms of tons.

17) A 24-ounce can of green peas is $1.49. What's the unit price?

18) Haile Gebrselassie is paid $50 for 6.7 hours work. How much would he earn if he worked 8 hours?

19) A 20-ounce jar of mixed nuts contains 12.5 ounces of peanuts. How many ounces of peanuts would be in a 28-ounce jar?

20) A 12-ounce bottle of ketchup costs $1.32. A 16-ounce bottle of ketchup costs $2.08. What's the unit price of each bottle? Which one is the better buy?

Chapter 7.   Percents

We encounter percentages virtually every day of our lives. There are sales when prices are reduced by 20%, 50%, or even 80%. We earn 3 or 4 percent interest on our savings accounts. And we're supposed to leave the waitress a tip of 18% of our checks.

Suppose you got 12 out of 20 questions correct on a test. Would you know your grade should be 60%? Suppose you just received an eight percent pay increase. Would you believe that many college graduates could not figure out how much more money you would be making? Before you've finished this chapter, you'll be able to do problems like these in less than a minute.

7.1   What are Percents?

A percent is a portion of a total. Percents are very useful when comparing two quantities. For example, suppose we wanted to compare the field goal percentage of two basketball players. If the first player made 10 out of 25 shots and the second player made 9 out of 20 shots, would you be able to determine which one is the better shot? To compare the success of these basketball players we would write this information in a fraction form, using the common denominator of 100. $\frac{10}{25} = \frac{40}{100}$ and $\frac{9}{20} = \frac{45}{100}$. Now that the given information is in a fraction form with a common denominator, it's easy to see that the second player is the better shot. Sometimes it's not easy to write fractions with a common denominator. When this occurs, it's much easier to compare two ratios in percent form. For example, if a basketball player makes 5 out of 7 shots and another player makes 11 out of 19 shots, it's easier to compare them if we say the first player made about 71% of her shots and the second player made about 58% of his shots. By the end of this chapter, you'll have no trouble converting to percent form and comparing ratios using percents.

Converting Percents to Fractions

When a percent is converted to fraction form, it's assumed that percents are based on one hundred.

So 47 percent is 47 parts of 100 or $\frac{47}{100}$. And 93 percent is 93 parts of 100, or $\frac{93}{100}$.

To Convert Percents to Fractions—drop the % symbol and place the number over one hundred; then reduce to lowest terms.

Example 1:   Convert 10% to a fraction.

SOLUTION: $10\% = \frac{10}{100} = \frac{1}{10}$ Put 10 over 100 and reduce.

Example 2:   Convert 20% to a fraction.

SOLUTION: $20\% = \frac{20}{100} = \frac{1}{5}$ Put 20 over 100 and reduce.

Example 3:   Convert 25% to a fraction.

SOLUTION: $25\% = \frac{25}{100} = \frac{1}{4}$ Put 25 over 100 and reduce.

Example 4: Convert 50% to a fraction.

SOLUTION: $50\% = \dfrac{50}{100} = \dfrac{1}{2}$ Put 50 over 100 and reduce.

Example 5: Convert 75% to a fraction.

SOLUTION: $75\% = \dfrac{75}{100} = \dfrac{3}{4}$ Put 75 over 100 and reduce.

Example 6: Convert 100% to a fraction.

SOLUTION: $100\% = \dfrac{100}{100} = 1$ Put 100 over 100 and reduce.

Example 7: Convert 150% to a fraction.

SOLUTION: $150\% = \dfrac{150}{100} = 1\dfrac{50}{100} = 1\dfrac{1}{2}$ Put 150 over 100 and reduce.

Example 8: Convert 200% to a fraction.

SOLUTION: $200\% = \dfrac{200}{100} = 2$ Put 200 over 100 and reduce.

Examples 1 through 8 are commonly used percents. It's a good idea for you to memorize them. Notice in examples 6, 7, and 8, the percents are at least 100% and the fractions are at least 1.

Let's try a few more difficult problems. We'll do them the same way we did examples one through eight. Drop the % symbol, write the number over 100 and reduce.

Example 9: Convert $\dfrac{1}{2}\%$ to a fraction.

SOLUTION: $\dfrac{1}{2}\% = \dfrac{\frac{1}{2}}{100} = \dfrac{1}{2} \cdot \dfrac{1}{100} = \dfrac{1}{200}$ Put $\dfrac{1}{2}$ over 100 and reduce.

Example 10: Convert $60\dfrac{2}{3}\%$ to a fraction.

SOLUTION: $\dfrac{60\frac{2}{3}}{100} = \dfrac{\frac{182}{3}}{100} = \dfrac{\overset{91}{\cancel{182}}}{3} \cdot \dfrac{1}{\underset{50}{\cancel{100}}} = \dfrac{91}{150}$ Put $60\dfrac{2}{3}$ over 100 and reduce.

Example 11: Convert 5.5% to a fraction.

SOLUTION: This problem can be done in two ways. The first method doesn't use fractions.

First Method:

$\dfrac{5.5}{100}$ Put the 5.5 over 100.

0.055   We move the decimal two places to the left because we're dividing by 100.

$$\frac{55}{1000}$$ Put the 55 over 1,000.

$$\frac{11}{200}$$ Reduce the fraction to lowest terms.

The second method converts the 5.5 to the mixed number $5\frac{5}{10}=5\frac{1}{2}$.

Second Method:

$$\frac{5\frac{1}{2}}{100}$$ Write $5\frac{1}{2}$ as the improper fraction $\frac{11}{2}$.

$$\frac{11}{2}\cdot\frac{1}{100}$$ Division by 100 becomes multiplication by $\frac{1}{100}$.

$$\frac{11}{200}$$

Problem Set: Convert the following percents to fractions:

1) 3%

2) 30%

3) 300%

4) $\frac{2}{3}$%

5) $3\frac{1}{6}$%

6) $8\frac{7}{8}$%

7) 6.6%

8) 0.66%

SOLUTIONS TO PROBLEM SET:

1) $\dfrac{3}{100}$

2) $\dfrac{30}{100}=\dfrac{3}{10}$

3) $\dfrac{300}{100}=3$

4) $\dfrac{\frac{2}{3}}{100}=\dfrac{\overset{1}{\cancel{2}}}{3}\cdot\dfrac{1}{\underset{50}{\cancel{100}}}=\dfrac{1}{150}$

5) $\dfrac{3\frac{1}{6}}{100}=\dfrac{\frac{19}{6}}{100}=\dfrac{19}{6}\cdot\dfrac{1}{100}=\dfrac{19}{600}$

6) $\dfrac{8\frac{7}{8}}{100}=\dfrac{\frac{71}{8}}{100}=\dfrac{71}{8}\cdot\dfrac{1}{100}=\dfrac{71}{800}$

7) $\dfrac{6.6}{100}=0.066=\dfrac{66}{1000}=\dfrac{33}{500}$

8) $\dfrac{0.66}{100}=0.0066=\dfrac{66}{10000}=\dfrac{33}{5000}$

OR $\dfrac{6\frac{3}{5}}{100}=\dfrac{\frac{33}{5}}{100}=\dfrac{33}{5}\cdot\dfrac{1}{100}=\dfrac{33}{500}$

OR $\dfrac{\frac{66}{100}}{100}=\dfrac{\overset{33}{\cancel{66}}}{100}\cdot\dfrac{1}{\underset{50}{\cancel{100}}}=\dfrac{33}{5000}$

Converting Percents to Decimals

Sometimes it's necessary to write a percent in decimal form. Suppose we wanted to write 60% as a decimal. We know percents are based on one hundred, which is the same as dividing by one hundred.

In the previous chapter we learned that if we divide by one hundred, we move the decimal point two places to the left. If we move the decimal two places to the left in 60%, we get its decimal form 0.60, which is the same as 0.6. Remember, extra zeros at the end of a decimal are not always necessary.

To Convert a Percent to a Decimal—drop the percent symbol, (%), and move the decimal point two places to the left.

Example 12: Convert 10% to a decimal.
    SOLUTION: $10\% = 0.10 = 0.1$ Move the decimal point two places to the left.

Example 13: Convert 20% to a decimal.
    SOLUTION: $20\% = 0.20 = 0.2$ Move the decimal point two places to the left.

Example 14: Convert 25% to a decimal.
    SOLUTION: $25\% = 0.25$ Move the decimal point two places to the left.

Example 15: Convert 50% to a decimal.
    SOLUTION: $50\% = 0.50 = 0.5$ Move the decimal point two places to the left.

Example 16: Convert 75% to a decimal.
    SOLUTION: $75\% = 0.75$ Move the decimal point two places to the left.

Example 17: Convert 100% to a decimal.
    SOLUTION: $100\% = 1.00 = 1$ Move the decimal point two places to the left.

Example 18: Convert 150% to a decimal.
    SOLUTION: $150\% = 1.50 = 1.5$ Move the decimal point two places to the left.

Example 19: Convert 200% to a decimal.
    SOLUTION: $200\% = 2.00 = 2$ Move the decimal point two places to the left.

Examples 12 through 19 are the same percents as examples 1 through 8. As we've already stated, these are commonly used percents. It's a good idea to memorize them in fraction and decimal form.

Problem Set: Convert the following percents to decimals:

1) 7%   2) 70%   3) 700%   4) 0.7%

5) 5.9%   6) 35.8%   7) 245.7%   8) 0.16%

SOLUTIONS TO PROBLEM SET:

1) 0.07   2) 0.70 = 0.7   3) 7   4) 0.007

5) 0.059   6) 0.358   7) 2.457   8) 0.0016

Converting Decimals to Percents

We just learned that to convert a percent to a decimal, we move the decimal point two places to the left. Now it's time for us to learn how to convert decimals to percents.

To Convert a Decimal to a Percent – move the decimal point two places to the right and add on the percent symbol %.

Suppose we wanted to convert 0.45 to a percent. Just move the decimal two places to the right. 0.45 is 45%. Convert 0.8 to a percent. When you move the decimal two places to the right, don't forget to add a 0. 0.8 = 0.80 = 80%.

Example 20: **Convert 0.1 to a percent.**

SOLUTION: 0.1 = 0.10 = 10% Add a zero, and then move the decimal point two places to the right.

Example 21: **Convert 0.2 to a percent.**

SOLUTION: 0.2 = 0.20 = 20% Add a zero, and then move the decimal point two places to the right.

Example 22: **Convert 0.25 to a percent.**

SOLUTION: 0.25 = 25% Move the decimal point two places to the right.

Example 23: **Convert 0.5 to a percent.**

SOLUTION: 0.5 = 0.50 = 50% Add a zero, and then move the decimal point two places to the right.

Example 24: **Convert 0.75 to a percent.**

SOLUTION: 0.75 = 75% Move the decimal point two places to the right.

Example 25: **Convert 1 to a percent.**

SOLUTION: 1 = 1.00 = 100% Add two zeros, and then move the decimal point two places to the right.

Example 26: **Convert 1.5 to a percent.**

SOLUTION: 1.5 = 1.50 = 150% Add a zero, and then move the decimal point two places to the right.

Example 27: **Convert 2 to a percent.**

SOLUTION: 2 = 2.00 = 200% Add two zeros, and then move the decimal point two places to the right.

Problem Set: Convert the following decimals to percents:

1) 0.7   2) 7.7   3) 0.77   4) 0.777

5) 0.125   6) 0.33   7) 0.67   8) 0.0001

SOLUTIONS TO PROBLEM SET:

1) 0.7 = 0.70 = 70%   2) 7.7 = 7.70 = 770%   3) 0.77 = 77%

4) 0.777 = 77.7%   5) 0.125 = 12.5%   6) 0.33 = 33%

7) 0.67 = 67%   8) 0.0001 = 0.01%

Converting Fractions to Percents

In chapter 6 you learned how to convert a fraction to a decimal, and a decimal to a fraction, so we have only one more type of conversion to perform. It's time to learn how to convert a fraction into a percent. We're going to show you two ways this can be done. The first method can be used only on a fraction whose denominator can be converted to one hundred. One hundred is the magic number because we know percents are based on 100. The next few examples will show you how easy it is to convert this type of fraction into a percent.

To Convert a Fraction to a Percent—convert the fraction to a denominator of 100, and put a % symbol after the numerator.

Example 28: Convert $\dfrac{2}{5}$ to a percent.

SOLUTION: $\dfrac{2}{5} = \dfrac{40}{100} = 40\%$ Convert the fraction to hundredths.

Example 29: Convert $\dfrac{3}{10}$ to a percent.

SOLUTION: $\dfrac{3}{10} = \dfrac{30}{100} = 30\%$ Convert the fraction to hundredths.

Example 30: Convert $2\dfrac{4}{25}$ to a percent.

SOLUTION: $2\dfrac{4}{25} = \dfrac{54}{25} = \dfrac{216}{100} = 216\%$ Write a mixed number as an improper fraction with a denominator of 100.

Example 31: Convert $\dfrac{200}{400}$ to a percent.

SOLUTION: $\dfrac{200}{400} = \dfrac{50}{100} = 50\%$ Sometimes we have to reduce the fraction to convert to a denominator of 100. In this example we divided by 4.

Example 32: Convert $1\dfrac{50}{500}$ to a percent.

SOLUTION: $1\dfrac{50}{500} = 1\dfrac{10}{100} = \dfrac{110}{100} = 110\%$ Divide the numerator and denominator by 5 to reduce to a denominator of 100.

Problem Set: Convert the following fractions to percents:

1) $\dfrac{11}{20}$ 2) $\dfrac{10}{25}$ 3) $\dfrac{6}{200}$ 4) $\dfrac{3}{10}$

5) $2\dfrac{30}{50}$ 6) $1\dfrac{2}{5}$ 7) $\dfrac{60}{1000}$ 8) $5\dfrac{1}{2}$

1) $\dfrac{11}{20} = \dfrac{55}{100} = 55\%$

2) $\dfrac{10}{25} = \dfrac{40}{100} = 40\%$

3) $\dfrac{6}{200} = \dfrac{3}{100} = 3\%$

4) $\dfrac{3}{10} = \dfrac{30}{100} = 30\%$

5) $2\dfrac{30}{50} = 2\dfrac{60}{100} = \dfrac{260}{100} = 260\%$

6) $1\dfrac{2}{5} = \dfrac{7}{5} = \dfrac{140}{100} = 140\%$

7) $\dfrac{60}{1000} = \dfrac{6}{100} = 6\%$

8) $5\dfrac{1}{2} = \dfrac{11}{2} = \dfrac{550}{100} = 550\%$

To Convert a Fraction to a Percent —convert the fraction to a decimal by dividing the numerator by the denominator, then move the decimal point two places to the right.

Suppose we wanted to convert $\dfrac{1}{8}$ to a percent. Eighths can't be converted to hundredths, so we'll have to use a different method from the one we just used on the last problem set. The other method is to convert the fraction into a decimal and then move the decimal point two places to the right. In Chapter 6 you learned that to convert a fraction to a decimal, all you have to do is divide the numerator (top) by the denominator (bottom). Remember, the numerator goes inside the division box and the denominator goes outside the division box.

Example 33: Convert $\dfrac{1}{8}$ to a percent.

SOLUTION: $\dfrac{1}{8} = 0.125$ Divide 1 by 8 to convert $\dfrac{1}{8}$ to a decimal.

$0.125 = 12.5\%$ Move the decimal point two places to the right.

Example 34: Convert $\dfrac{3}{200}$ to a percent.

SOLUTION: $\dfrac{3}{200} = 0.015$ Divide 3 by 200.

$0.015 = 1.5\%$ Move the decimal point two places to the right.

Example 35: Convert $\dfrac{1}{3}$ to a percent.

SOLUTION: $\dfrac{1}{3} = 0.\overline{3}$ Divide 1 by 3. This gives us a repeating decimal.

$\overline{3} = 33.\overline{3}\% = 33\dfrac{1}{3}\%$ Move the decimal two places to the right.

Remember, any numbers to the right of the decimal point are threes, not zeros. The $33.\overline{3}$ is written as $33\dfrac{1}{3}$ because $0.\overline{3} = \dfrac{1}{3}$.

Example 36: Convert $\dfrac{2}{9}$ to a percent.

SOLUTION: $\dfrac{2}{9} = 0.\overline{2}$ Divide 2 by 9.

$0.\overline{2} = 22.\overline{2} = 22\dfrac{2}{9}\%$ Move the decimal point two places to the right.

Sometimes we're asked to write a non-terminating, non-repeating fraction as a percent. When this occurs we'll be instructed to round to a specified decimal place.

Example 37: Convert $\dfrac{2}{7}$ to the nearest tenth of a percent.

SOLUTION: $\dfrac{2}{7}$ is not a terminating decimal, so we'll divide until we get to the ten-thousandths' place. We chose to stop at the ten-thousandths' place because we're going to round to the nearest tenth of a percent. Since we have to move the decimal point two places to the right, it's not necessary to divide out any further than the ten-thousandths' place.

$\dfrac{2}{7} = 0.2857$ Divide to the ten-thousandths' place.

$0.2857 = 0.286$ Round to three decimal places.

$0.286 = 28.6\%$ Move the decimal two places to the right.

Problem Set: Convert the following fractions to percents. Round non-terminating, non-repeating decimals to the nearest tenth of a percent:

1) $\dfrac{5}{8}$ 2) $\dfrac{17}{40}$ 3) $\dfrac{10}{27}$ 4) $\dfrac{3}{15}$

5) $\dfrac{13}{22}$ 6) $\dfrac{2}{3}$ 7) $\dfrac{19}{30}$ 8) $\dfrac{123}{600}$

SOLUTIONS TO PROBLEM SET:

1) $\dfrac{5}{8} = 0.625 = 62.5\% = 62\dfrac{1}{2}\%$

2) $\dfrac{17}{40} = 0.425 = 42.5\% = 42\dfrac{1}{2}\%$

3) $\dfrac{10}{27} = 0.\overline{370} = 37\%$

4) $\dfrac{3}{15} = \dfrac{1}{5} = 0.2 = 0.20 = 20\%$

5) $\dfrac{13}{22} = 0.5\overline{90} = 59.1\%$

6) $\dfrac{2}{3} = 0.\overline{6} = 66.\overline{6}\% = 66\dfrac{2}{3}\%$

7) $\dfrac{19}{30} = 0.6\overline{3} = 63.\overline{3}\% = 63\dfrac{1}{3}\%$

8) $\dfrac{123}{600} = 0.205 = 20.5\% = 20\dfrac{1}{2}\%$

Table 7.1 lists the most commonly used percents with their fraction and decimal forms.

FRACTION	DECIMAL	PERCENT
$\dfrac{1}{2}$	0.5	50%
$\dfrac{1}{3}$	$0.\overline{3}$	$33.\overline{3}\%$ or $33\dfrac{1}{3}\%$
$\dfrac{1}{4}$	0.25	25%
$\dfrac{1}{5}$	0.2	20%
$\dfrac{1}{8}$	0.125	12.5% or $12\dfrac{1}{2}\%$
$\dfrac{1}{10}$	0.1	10%
$\dfrac{2}{3}$	$0.\overline{6}$	$66.\overline{6}$ or $66\dfrac{2}{3}\%$
1	1.00	100%

TABLE 7.1

Write the following percents as fractions.

1) 10%   2) 32%   3) 45%   4) 20%

5) 2%   6) .2%   7) 200%   8) 250%

9) 73.5%   10) 1.9%   11) .09%   12) .008%

13) $\frac{1}{4}\%$ 14) $\frac{2}{9}\%$ 15) $3\frac{1}{6}\%$ 16) $101\frac{2}{3}\%$

17) $3\frac{1}{8}\%$ 18) $12\frac{1}{10}\%$ 19) $1\frac{6}{9}\%$ 20) $33\frac{1}{3}\%$

Write the following fractions as percents. Do not round your answers.

21) $\frac{1}{10}$ 22) $\frac{8}{25}$ 23) $\frac{9}{20}$ 24) $\frac{1}{5}$

25) $\frac{1}{50}$ 26) $\frac{1}{500}$ 27) 2   28) $2\frac{1}{2}$

29) $\frac{147}{200}$ 30) $\frac{19}{1000}$ 31) $\frac{9}{10000}$ 32) $\frac{1}{125}$

33) $\frac{1}{400}$ 34) $\frac{1}{450}$ 35) $\frac{19}{600}$ 36) $1\frac{1}{60}$

37) $\frac{1}{32}$ 38) $\frac{121}{1000}$ 39) $\frac{1}{60}$ 40) $\frac{1}{3}$

Write the following percents as decimals.

41) 20%   42) 35%   43) 25%   44) 3.5%

45) 2%   46) 0.2%   47) 0.002%   48) 200%

49) 300.59%   50) 100%   51) $5\frac{1}{5}\%$ 52) 3.8%

53) 38%   54) 380%   55) $10\frac{7}{8}\%$ 56) $202\frac{1}{4}\%$

Write the following decimals as percents.

57) 0.35   58) 0.035   59) 3.5   60) 35

61) 0.0035   62) .001   63) .01   64) 0.1

65) 1   66) 10   67) 0.5   68) 0.7

69) 0.75   70) 0.33   71) 0.3   72) $0.\overline{3}$

73) $0.\overline{6}$ 74) $1.\overline{16}$ 75) $22.\overline{2}$ 76) $0.0\overline{3}$

7.2 Basic Percent Problems

In this section we'll solve basic percent problems two ways. The first method uses proportions, and the second uses the technique of translating sentences in English to mathematical symbols.

Example 1: **What number is 50% of 200?**

SOLUTION: You've probably figured out that 50% of 200 is 100. You know 50% is $\frac{1}{2}$ and half of 200 is 100. Most of the problems in this section will not be as easy, so we'll show you two procedures you can chose from to find the answer to this type of problem.

The first method uses proportions. For us to use this method, we have to write 50% as a fraction.

$$50\% = \frac{50}{100} = \frac{1}{2}.$$

Next we have to decide whether the 200 is a portion or a total. Because we're asked to find 50% of 200, the 200 is the total and the 50% is the portion of the total, 200. The total 200 goes in the denominator and the unknown portion we'll call x will go in the numerator. The proportion should look like the following equation. In Chapter 6 we showed you that there is more than one way to solve this proportion.

$$\frac{1}{2} = \frac{x}{200} \qquad \text{Cross multiply.}$$

$2x = 200$ Divide both sides by 2.

$x = 100$ 100 is 50% of 200.

We'll work out a few more examples using the proportion method; then we'll go back and redo the same problems using the second method, the translation method.

Example 2: **100 is 50% of what number?**

SOLUTION: This problem is very similar to the last, except this time we're looking for the total, not the portion. 100 is the portion, so it belongs in the numerator. The 50% has to be written as a fraction. The total we'll call x belongs in the denominator.

$$\frac{50}{100} = \frac{100}{x} \qquad \text{We'll reduce the fraction } \frac{50}{100} = \frac{1}{2}.$$

$$\frac{1}{2} = \frac{100}{x} \qquad \text{Cross multiply.}$$

$x = 200$ 100 is 50% of 200.

Example 3: **What percent of 200 is 100?**

SOLUTION: This problem is very similar to the last two, except this time we're given the portion 100, and the total 200, and we're asked to find the percent. The portion 100 belongs in the numerator and the total 200 belongs in the denominator. We'll let x represent the unknown percent, which we'll put over 100.

$\dfrac{100}{200} = \dfrac{x}{100}$ Reduce the fraction $\dfrac{100}{200} = \dfrac{1}{2}$

$\dfrac{1}{2} = \dfrac{x}{100}$ Cross multiply.

$100 = 2x$ Divide by 2 on both sides of the equation.

$x = 50\,\%$ Add the percent symbol %. 50% of 200 is 100.

Examples 1, 2, and 3 are the same problem in a different order. In example 1 we're asked to find the portion. In example 2 we're asked to find the total, and in example 3 we're asked to find the percent. The method we used to solve these examples, we call the proportion method. Now we'll solve these examples again using a method we call the translation method. When we use the translation method, we translate the words into mathematical symbols and solve the equation for the variable. When we refer to an unknown value, we'll use x or any letter to represent the unknown value. "Of" means multiply and "is" means equals.

Example 4: **What number is 50% of 200?**

SOLUTION: Substitute x to represent "what number," replace the word is by = . We can't perform arithmetic operations with a number in percent form, so we'll change the 50% to a fraction or a decimal. "Of" means multiply.

What number is 50% of 200?   50% = 0.50 = 0.5

$x = 0.5\,(200)$ Multiply.

$x = 100$ 100 is 50% of 200.

Example 5: **100 is 50% of what number?**

SOLUTION: $100 = 0.5\,(x)$ 0.5(x) is generally written 0.5x, which means 0.5 times x.

$\dfrac{100}{0.5} = \dfrac{.05x}{0.5}$ Divide both sides by 0.5.

$200 = x$ 100 is 50% of 200.

Example 6: **What percent of 200 is 100?**

SOLUTION: We'll replace "what percent" by x; "of" means times; "is" means equals.

$x\,(200) = 100$ x(200) is generally written 200x, which means 200 times x.

$200\,x = 100$ Divide both sides by 200.

$$\frac{200x}{200} = \frac{100}{200} = \frac{1}{2} \qquad \text{Convert the } \frac{1}{2} \text{ to } 0.5 = 0.50 = 50\%$$

$$x = \frac{1}{2} = 50\% \qquad 50\% \text{ of } 200 \text{ is } 100.$$

We've done three problems using two methods— the proportion method and the translation method. You may like one method more than the other, but it's best to practice both methods. The more ways you know to solve a problem, the better off you are. Let's look at a few more examples. We'll solve them using both methods.

Example 7:   **155 is 25% of what number?**

SOLUTION: Remember to convert 25% to the fraction $\frac{1}{4}$ or the decimal 0.25 before starting the problem. We'll start with the translation method.

155 is 25% of what number?   "Is" means equals; "of" means times. We'll

$155 = 0.25\, x$ use 0.25 for 25%. and x for "what number".

$$\frac{155}{0.25} = \frac{0.25x}{0.25} \qquad\qquad \text{Divide both sides by 0.25.}$$

$620 = x$ 155 is 25% of 620.

Now we'll use the proportion method. To use this method, put the portion over the total.

SOLUTION: Remember, percents are based on 100, so we'll write 25% as $\frac{25}{100} = \frac{1}{4}$.

Percents are written as portion over total. 155 is the portion, so it goes in the numerator. The unknown amount is the total, which we'll call x. The x goes in the denominator.

$$\frac{1}{4} = \frac{155}{x} \qquad \text{Cross multiply.}$$

$4(155) = x$

$x = 620$ 155 is 25% of 620.

Now that we've done a few problems, it's your turn to try the following example using both methods.

Example 8:   **What percent of 360 is 90?**

SOLUTION: Using the translation method, we'll use x for "what percent," times for "of" and = for "is."

What percent of 360 is 90?

$x\,(360) = 90$ Write x(360) as 360x.

$360x = 90$ Divide both sides by 360.

$$\frac{360x}{360} = \frac{90}{360} = \frac{1}{4} = 0.25 = 25\% \qquad \text{Write } \frac{1}{4} \text{ as 25\%.}$$

$x = 25\%$ 25% of 360 is 90.

SOLUTION: Using the proportion method, we'll write x over 100 to represent "what percent." The 90 is a portion of 360, so it goes in the numerator over 360.

What percent of 360 is 90?

$$\frac{x}{100} = \frac{90}{360}$$ Cross multiply,

$360x = 90(100)$ Multiply.

$360x = 9000$ Divide both sides by 360.

$$\frac{360x}{360} = \frac{9000}{360}$$

$x = 25$ 25% of 360 is 90.

Example 9: 1 is $\frac{1}{3}$% of what number?

SOLUTION: We'll solve this by the translation method first, and then by the proportion method. The only thing new about this problem is the fractional percent. $\frac{1}{3}$ is a repeating decimal. We can't perform arithmetic operations with repeating decimals, so we'll have to write $\frac{1}{3}$% in fraction form. Let's review how to write $\frac{1}{3}$% as a fraction. Percents are based on 100, so we'll write $\frac{1}{3}$% as $\frac{\frac{1}{3}}{100}$. Multiply the numerator by the reciprocal of the denominator.

$$\frac{1}{3}\% = \frac{\frac{1}{3}}{100} = \frac{\frac{1}{3}}{\frac{100}{1}} = \frac{1}{3}\left(\frac{1}{100}\right) = \frac{1}{300}.$$

Now we're ready to finish this problem.

1 is $\frac{1}{3}$% of what number?

$$1 = \frac{1}{300}x$$

$$\frac{1}{\frac{1}{300}} = \frac{\frac{1}{300}x}{\frac{1}{300}}$$ Divide both sides of the equation by $\frac{1}{300}$.

$$1(300) = x \qquad \text{Multiply 1 by the reciprocal of } \frac{1}{300}, \text{ which is 300.}$$

$$x = 300 \qquad \text{1 is } \frac{1}{3}\% \text{ of 300.}$$

Now we'll do this problem by the proportion method.

1 is $\frac{1}{3}$% of what number?   1 is the portion, so it belongs in the numerator.

The unknown x is the total, so it belongs in the denominator.

$$\frac{\frac{1}{3}}{100} = \frac{1}{x} \qquad \text{Cross multiply.}$$

$$\frac{1}{3}x = 1(100) \qquad \text{Multiply.}$$

$$\frac{1}{3}x = 100 \qquad \text{Divide both sides of the equation by } \frac{1}{3}.$$

$$\frac{\frac{1}{3}x}{\frac{1}{3}} = \frac{100}{\frac{1}{3}} \qquad \text{Multiply by the reciprocal of } \frac{1}{3}, \text{ which is 3.}$$

$$x = 100(3)$$

$$x = 300 \qquad \text{1 is } \frac{1}{3}\% \text{ of 300.}$$

Now you're ready to try a few on your own. See if you can use both methods in the following problem set.

Problem Set:

1) What number is 62% of 900?

2) 180 is 30% of what number?

3) What percent of 480 is 192?

4) What number is $\frac{1}{2}$% of 500?

5) What percent of 500 is 2.5?

1) Translation Method

$x = 0.62(900) = 558$

Proportion Method

$$\frac{62}{100} = \frac{x}{900}$$

$62(900) = 100x$

$55,800 = 100x$

$558 = x$

2) Translation method

$180 = 0.3x$

$x = 600$

Proportion Method

$$\frac{180}{x} = \frac{3}{10}$$

$180(10) = 3x$

$$\frac{1800}{3} = \frac{3x}{3}$$

$600 = x$

3) Translation Method

$x(480) = 192$

$x = 0.4 = 0.40 = 40\%$

Proportion Method

$$\frac{x}{100} = \frac{192}{480}$$

$$\frac{x}{100} = \frac{2}{5}$$

$5x = 200$

$x = 40$

4) Translation Method $\quad \frac{1}{2}\% = 0.5\% = 0.005 \quad$ or $\quad \dfrac{\frac{1}{2}}{100} = \dfrac{\frac{1}{2}}{\frac{100}{1}} = \frac{1}{2}\left(\frac{1}{100}\right) = \frac{1}{200}$

$x = 0.005(500)$

$x = 2.5$

Proportion Method

$$\frac{\frac{1}{2}}{100} = \frac{x}{500}$$

$$\frac{1}{2}(500) = 100x$$

$$250 = 100x$$

$$\frac{250}{100} = x$$

$$2.5 = x$$

5) Translation Method

$$x(500) = 2.5$$

$$x = \frac{2.5}{500}$$

$$x = 0.005 = 0.5\% \text{ or } \frac{1}{2}\%$$

Proportion Method

$$\frac{x}{100} = \frac{2.5}{500}$$

$$500x = 2.5(100)$$

$$500x = 250$$

$$x = \frac{250}{500} = \frac{1}{2}\%$$

1) What number is 6% of 400?

2) What number is 7% of 300?

3) What percent of 400 is 24?

4) What percent of 300 is 21?

5) 24 is 6% of what number?

6) 24 is 7% of what number?

7) What percent of 400 is 240?

8) What percent of 300 is 210?

9) What percent of 400 is 2,400?

10) What percent of 300 is 2,100?

11) What number is 0.6% of 400?

12) What number is 0.7% of 300?

13) 324 is 36% of what number?

14) 1,200 is 15% of what number?

15) What percent of 1250 is 187.5?   16) What percent of 350 is 21.7?

17) What number is $\frac{1}{4}$% of 25?   18) What number is $\frac{2}{5}$% of 400?

19) 0.0625 is what percent of 25?   20) 1.6 is what percent of 400?

21) What number is $\frac{1}{3}$% of 900?   22) What number is $\frac{1}{6}$% of 1200?

23) What percent of 450 is 50? Round your answer to the nearest tenth of a percent.

24) What percent of 30 is 28? Round your answer to the nearest tenth of a percent.

7.3   Solving Percent Word Problems

In this section we'll apply what we learned in the previous section to solve word problems involving percents. This time we won't be given a sentence to use the translation method or the proportion method to solve the problem. All the examples in the last section were already set up for us. This time we'll have to read a problem and decide what the problem asks us to find, the percent, the portion, or the total. Setting up the problem is more difficult than solving a problem that has already been set up.

Example 1:   A basketball player made 10 out of 17 shots attempted. What percent of the shots taken did she make? What percent did she miss? Round your answer to the nearest percent.

The easiest way to find a percent is to divide the portion by the total and move the decimal point two places to the right.

$$total\overline{)portion}$$

SOLUTION: A percent is a portion of a total. The portion is 10 and the total is 17.

In a fraction form it would look like this: $\frac{10}{17}$. We'll use a division

box to convert the fraction to a decimal. $17\overline{)10}$ = 0.588... The directions instructed us to round to the nearest percent, so we'll stop dividing this continuous decimal at the thousandths' place and round to 0.59. To convert a decimal to a percent, we move the decimal point two places to the right. The basketball player made 59% of her shots. She missed 41% (100% - 59%) of her shots.

Example 2:   15% of the 200 employees of Marlene's Marvelous Makeup Mart are senior citizens. How many senior citizens work at the factory?

SOLUTION: The percent is 15, the total is 200, and we're asked to find the portion. We can use the translation method by using the following statement:

15% of the employees are senior citizens.

0.15 (200) = x   "Of" means multiply. Use 200 to represent
30 = x   the employees.

Another way to solve this problem is to use the proportion method.

$\frac{15}{100} = \frac{x}{200}$ Reduce the fraction.

$\frac{3}{20} = \frac{x}{200}$ Cross multiply.

3(200) = 20x   Multiply.
600 = 20x   Divide both sides by 20.
30 = x   30 senior citizens work at the factory.

We're not going to use both methods on the following examples. You decide which method you prefer to use.

Example 3: 15% of the employees of Marlene's Marvelous Makeup Mart are senior citizens. If there are 30 senior citizens, how many employees are there?

SOLUTION: 15% of the employees are senior citizens.

$0.15\, x = 30$ Divide both sides by 0.15

$$\frac{0.15x}{0.15} = \frac{30}{0.15}$$

$x = 200$ There are 200 employees.

Example 4: 30 of the 200 employees of Marlene's Marvelous Makeup Mart are senior citizens. What percent of the employees are senior citizens?

SOLUTION: To find a percent, just divide the portion by the total and move the decimal point two places to the right.

$$\frac{30}{200} = \frac{15}{100} = 0.15 = 15\%$$

Examples 2, 3, and 4 are the same problem in a different order. In example 2 we were asked to find the portion, in example 3 we were asked to find the total and in example 4 we were asked to find the percent.

Example 5: If a person whose yearly income is $67,892 negotiates a 6.5% increase in salary, how much is the person's new salary?

SOLUTION: To find the amount of the increase we multiply $67,892 by 0.065. The increase is $4,412.98. We'll calculate the new salary by adding the amount of the increase $4,412.98 to the old salary $67,892 to find the new salary of $72,304.98.

Another way to find the new salary is to multiply the old salary by 1.065.

The new salary is $67,892 (1.065) = $72,304.98. We multiplied by 1.065 because we're keeping 100% (1) of the salary and adding 6.5% (0.065) to the salary.

Problem Set:

1) On a 120-question test, a student answered 90 questions correctly. What percent of the problems did the student answer correctly?

2) Of the 4,200 students enrolled in calculus, only 25% of them are over 25 years old. How many calculus students are over 25 years old?

3) 30% of the patrons of a particular ice cream store bought vanilla ice cream. If 120 patrons bought vanilla ice cream during a particular week, what's the total number of patrons who were at the store that week?

4) A sweater that normally sells for $68 was marked up 25%. What's the new price of the sweater?

1) $\dfrac{90}{120} = \dfrac{3}{4} = 0.75 = 75\%$

2) 25% of the students are over 25 years old.

$0.25\ (4200) = x$

1,050 students are over 25 years old.

3) 30% of the patrons bought vanilla ice cream.

$0.3\,x = 120$

$\dfrac{0.3x}{0.3} = \dfrac{120}{0.3}$

$x = 400$ 400 patrons were at the store that week.

4) $68(0.25) = 17\ \ 68 + 17 = \$85$

Or we could take a short-cut.

$68\ (1.25) = \$85$ The new price of the sweater is $85.

Percent Increase / Percent Decrease

Now we're going to work on a different type of percent problem. This type will allow us to find the percent of increase or decrease. To find the percent of increase or decrease all you have to do is divide the amount of the increase or decrease by the original amount

$$\text{Percent of increase or decrease} = \frac{\textit{the amount of the increase or decrease}}{\textit{the original amount}} \times 100$$

Don't forget to move the decimal point two places to the right to convert the decimal to percent form.

Example 6:   **What's the percent of increase for a person who was making $16.00 an hour and is now making $24.00 an hour?**

SOLUTION: The amount of the increase is $8.00 an hour. The original salary is $16.00 an hour. To find the percent of increase divide 8 by 16 which is 0.5 or 50%. This person received a 50% increase in their salary.

Example 7:   **A sweater that normally sells for $85 has been reduced to $68. What's the percent of decrease?**

SOLUTION: The amount of the decrease is $17. To find the percent of the decrease, divide 17 by 85 which is 0.20 or 20%. There was a 20% decrease in the price of the sweater.

Example 8: Of the 25 students in Professor Magliano's 1991 Calculus I course, 20 passed the course. Of the 20 students in Professor Magliano's 2001 Calculus I course, 12 of them passed the course. What's the change in the percents of the number of students who passed his course?

SOLUTION: Notice this problem is a little different, it doesn't ask for the percent of increase or decrease — just the change in the percents.

First we have to find the percent who passed the courses. To find the percent divide the portion by the total. In 1991 the percent who passed the course is $\frac{20}{25} = 0.8 = 0.80 = 80\%$. In 2001 the percent who passed the course is $\frac{12}{20} = 0.6 = 0.60 = 60\%$. The change in the percents is 20%. 20% more students passed the course in 1991.

Example 9: If the price of a stock is $5.00 a share in January and rises to a price of $20.00 a share in June, what's the percent of increase in the value of the stock?

SOLUTION: The amount of the increase is $15.00. To find the percent of increase in the price of the stock, divide the amount of the increase $15 by the original price of the stock, $5. $15 \div 5 = 3 = 3.00 = 300\%$ increase in the price of the stock.

We'd like to buy some of this stock and retire.

Problem Set: Round your answers to the nearest percent.

1) The average number of books checked out of the Bayonne Public Library on a weekly basis in 1976 was 572. The average number of books checked out of the Bayonne Public Library on a weekly basis in 2001 is 355. Find the percent decrease in the number of books checked out of the library.

2) The number of students who failed the basic skills exam in 1990 at Genowa City College is 1,240. The number of students who failed the basic skills exam in 2001 is 1,456. Find the percent increase in the number of students who failed the basic skills exam.

3) On New Year's Eve, you made a resolution to lose 30 pounds by the end of March. If your weight dropped from 140 to 110, what's the percent of decrease in your weight?

4) If Jessica Storey's real estate taxes rose from $6,000 to $22,000 over a ten year period, what's the percent of increase in her real estate taxes?

SOLUTIONS TO PROBLEM SET:

1) The amount of the decrease is $572 - 355 = 217$. Divide the amount of the decrease, 217, by the original amount, 572, round off to two decimal places, and move the decimal point two places to the right.

$217 \div 572 = 0.38 = 38\%$ decrease.

2) The amount of change is $1,456 - 1,240 = 216$. Divide 216 by the original amount 1,240, round to two decimal places, and move the decimal point two places to the right.

$216 \div 1,240 = 0.17 = 17\%$ increase.

3) The amount of the decrease is 30. $30 \div 140 = 0.21 = 21\%$ decrease.

4) The amount of the increase is $22,000 - 6,000 = 16,000$.

$16,000 \div 6,000 = 2.67 = 267\%$ increase.

7.3 Workbook

Round your percent increase or decrease to the nearest percent.

1) Find the percent change when you go from 150 to

    a) 30   b) 50   c) 150   d) 200   e) 450

2) Find the percent change when you go from 60 to

    a) 20   b) 50   c) 60   d) 80   e) 120

3) A bike that normally sells for $400 is reduced to $300. What's the percent of decrease?

4) Juan received a raise for all the good work he's done. His salary went from $67,000 a year to $73,000 a year. What's the percent increase in his annual salary?

5) The annual dues for the P.T.A. increased from $4.00 to $4.50. What's the percent of increase in the dues?

6) The toll at the Washington Bridge rose from $7.00 to $9.00. What's the percent of increase in the toll?

7) A factory that produces cans of tuna fish installed all new machines. The old machines produced 23,000 cans per day; the new machines produce 34,000 cans a day. What's the percent increase in the number of cans produced a day?

8) In 1989 90 students were enrolled in kindergarten at the Horace Mann School. In 2001 there were 82 students enrolled in kindergarten at the Horace Mann School. What's the percent of decrease in the number of kindergartners enrolled at the Horace Mann School?

9) If Bobby the pool boy caught 24 blue fish on Monday and 36 blue fish on Tuesday, what's the percent increase in the number of blue fish he caught?

10) If Karen danced with 12 different men on Friday and 8 different men on Saturday, what's the percent decrease in the number of men she danced with?

11) If John won 6 pool tournaments in 2000 and 13 tournaments in 2001, what's the percent of increase in the number of pool tournaments he won?

12) If Larry the lawyer won 8 cases in January and 10 in February, what's the percent of increase in the number of cases he won?

13) Juris sold 450 pairs of shoes in March and 660 pairs in November; what's the percent increase in the number of pairs of shoes he sold?

14) If our book sold 12,235 copies in 2001 and 24,567 copies in 2002, what was the percent of increase in the number of copies sold?

15) If a class's average grade on exam 1 was 78 and the average grade on exam 2 was 68, what's the percent of decrease?

CHALLENGE PROBLEMS

16) In 1999 1,032 people graduated from Towson State University. In 2000 there were 1,079 graduates, and in 2001 there were 1,213 graduates. By what percentage did the number of graduates rise between 1999 and 2001?

17) Larry's commissions fell from $76,500 to $68,700. Samantha's commissions fell from $93,100 to $84,600. Whose commissions fell by a greater percentage?

7.4   Applications

Sales Tax

Let's put all of the skills you've developed with percentages to some practical use. We'll start by figuring out sales tax, and then move on to even more wonderful applications.

Example 1:   The sales tax rate in New Jersey is 5%. How much tax is charged on the purchase of a New Jersey Nets ticket that sells for $50? And what is the total price that you have to pay?

SOLUTION: To find the amount of the tax multiply 50 by .05, which gives us $2.50 tax. Add the $2.50 tax to the price of the ticket $50. $52.50 is the total price you have to pay.

When you go shopping, the prices you see do not include sales tax. The cashier will add that to most of your purchases. The sales tax rate varies from state to state, and even within most states. In fact, some states don't even charge sales tax. So when you're there, you won't have to work out any of these problems.

Example 2:   The sales tax is $28 on the purchase of a desk that sells for $400. What is the sales tax rate?

SOLUTION: To find the rate all you have to do is divide the amount of tax by the selling price and move the decimal point two places to the right to convert to a percent. (The selling price doesn't include the tax.)

$$Sales\,tax\,rate = \frac{amount\,of\,tax}{selling\,price} \times 100 \left(\frac{\$28}{\$400} \right) \times 100 = 7\%$$

That's it for sales tax. Please complete this problem set and then go on to sales commission.

Problem Set:

1) If the sales tax rate is 4%, how much is the tax charged on a purchase of $239.99? What is the total price?

2) If the sales tax rate is 6.5%, how much is the tax charged on a purchase of $529.50? What is the total price?

3) In New York City, the sales tax rate is 8.25%. If a New York Mets ticket sells for $28, how much tax is charged? What is the total purchase price?

4) If the sales tax is $27.00 on the purchase of an item that sells for $540, what is the sales tax rate?

5) The sales tax on a new car is $1,125. If the selling price of the car is $15,000, find the sales tax rate.

1) 239.99 (0.04) = 9.5996 which rounds to $9.60. Remember, money automatically rounds to two decimal places.

$$\begin{array}{r} \$239.99 \\ +\$\ \ \ 9.60 \\ \hline \$249.59 \end{array}$$

2) 529.50 (0.065) = $34.42.

$$\begin{array}{r} \$529.50 \\ +\$\ \ 34.42 \\ \hline \$563.92 \end{array}$$

3) 28 (0.0825) = $2.31

$$\begin{array}{r} \$\ \ 28.00 \\ +\$\ \ \ \ 2.31 \\ \hline \$\ \ 30.31 \end{array}$$

4) $Sales\ tax\ rate = \dfrac{amount\ of\ tax}{selling\ price} = \left(\dfrac{\$27}{\$540}\right) \times 100 = 5\%$

5) $Sales\ tax\ rate = \dfrac{amount\ of\ tax}{selling\ price} = \left(\dfrac{\$1,125}{\$15,000}\right) \times 100 = 7.5\%$

Commission

Have you ever worked on commission? The more you sell, the more you earn. And you know what you earn when you sell nothing.

A salesperson usually earns a certain percent of sales. The following formula will help you calculate the amount of commission a salesperson earns.

$$Amount\ of\ Commission = (Amount\ of\ sales) \times (Commission\ rate)$$

Example 3: **If you were paid a 15% commission and had sales of $10,000, what would your commission be?**

SOLUTION: Your commission would be ($10,000 × 0.15) = $1,500.

Example 4: **How much commission would you earn if you were paid 7.5% on sales of $43,150?**

SOLUTION $43,150 (0.075) = $3,236.25

Sometimes you'll be asked to find the commission rate. The following formula will help you with this type of problem.

$$Commission\ Rate = \dfrac{amount\ of\ commission}{amount\ of\ sales} \times 100$$

Example 5: A salesperson earns a commission of $2,000 selling $50,000 worth of appliances. What is her commission rate?

$$Commission\ rate = \frac{amount\ of\ commission}{amount\ of\ sales} = \times 100 = \left(\frac{\$2,000}{\$50,000}\right) \times 100 = 4\%$$

Did you notice that figuring out the commission rate is almost the same as figuring out the sales tax rate, which we did in the last section?

Problem Set

1) How much commission is earned by someone paid a rate of 25% on sales of $9,433?

2) A salesperson's commission rate is 4.5%. What is the commission from a sale of $42,150?

3) What is the commission rate of a salesperson who earns a commission of $1,500 on sales of $20,000?

4) What is the commission rate of a salesperson who earns a commission of $240 on sales of $7,200?

SOLUTIONS TO PROBLEM SET:

1) $9,433 (0.25) = $2,358.25

2) $42,150 (0.045) = $1,896.75

3) $Commission\ rate = \dfrac{amount\ of\ commission}{amount\ of\ sales} \times 100 = \dfrac{\$1,500}{\$20,000} \times 100 = 0.0750 \times 100 = 7.5\%$

4) $Commission\ rate = \dfrac{amount\ of\ commission}{amount\ of\ sales} \times 100 = \dfrac{\$240}{\$7200} \times 100 = 0.033 \times 100 = 3.3\%\ or\ 33\frac{1}{3}\%$

Sales Discounts

Everybody likes a bargain. So in this section everything is on sale. We've got a dress with a *marked price* of $80, but for today only, you may have a discount of 40%. That's right, take 40% off. So, how much is the discount and how much is the selling price of the dress?

Selling Price = Marked Price – Amount of the Discount.

SOLUTION: We'll multiply the price ($80) by the rate (0.4) to find the amount of the discount. 80 (0.4) = $32. To find the selling price, subtract the amount of the discount from the marked price. $80 – $32 is a selling price of $48.

Would you like to see a shortcut? If all you need to find is the marked down selling price, not the amount of the discount, we can save you a step. If the discount is 40% of the marked price, the selling price is 60% (100% − 40%) of the marked price. If you save 40%, you pay 60%. Instead of multiplying $80 by 0.4, multiply $80 by 0.6 to get the selling price directly.

$$\$80\ (0.6) = \$48$$

Example 6: A sofa is marked $599 and is on sale at 35% off. How much is the discount? How much is the sale price?

SOLUTION: The amount of the discount is $599 (0.35) = $209.65. The sale price is

$$\begin{array}{r} \$599.00 \\ -\$209.65 \\ \hline \$389.35 \end{array}$$

Let's find the selling price here using the shortcut method. Instead of multiplying $599 by 0.35 to find the discount, let's multiply $599 by 0.65 (100% − 35% = 65%).

$599 (0.65) = $389.35.

Problem Set:

1) A suit is marked $199 and is on sale at 40% off. How much is the discount? How much is the sale price?

2) A VCR has a marked price of $250, but there is a storewide sale with 25% off every item. How much is the discount? How much is the sale price?

3) A telephone answering machine is marked at $25.99 and is on sale at 20% off. How much is the discount? How much is the sale price?

4) A living room set is marked at $1,199 and is on sale at 25% off. How much is the sale price?

SOLUTIONS TO PROBLEM SET:

1) $199 (0.4) = $79.60

$$\begin{array}{r} \$199.00 \\ -\$\ 79.60 \\ \hline \$199.40 \end{array}$$

2) $250 (0.25) = $62.50

$$\begin{array}{r} \$250.00 \\ -\$\ 62.50 \\ \hline \$187.50 \end{array}$$

3) $25.99 (0.2) = $5.20

$$\begin{array}{r} \$25.99 \\ -\$\ 5.20 \\ \hline \$20.79 \end{array}$$

4)

$$\begin{array}{r} 1199 \\ \times\quad .75 \\ \hline 5995 \\ 8393\quad \\ \hline \$899.25 \end{array}$$

Simple Interest

When you borrow money, you usually have to pay interest. And when you put money in the bank, you are usually paid interest. The formula to calculate simple interest is:

Interest = Principle × Rate × Time (measured in years)

$I = P \times R \times T$

Example 7: If you put $100 in your savings account for one year and get paid 4% interest, how much interest would you earn after one year?

SOLUTION: The principal is $100, the rate 4% or 0.04, and the time is 1 year.

$$I = P \times R \times T$$

Interest = 100 (0.04) (1) = $4.00

Example 8: Suppose you received 5% interest on a principal of $1,000. At the end of two years, how much would you earn in interest? Assuming you didn't take out any money, how much would your principal be at the end of two years?

SOLUTION: The principal is $1,000, the interest rate is 5% or 0.05, and the time is 2 years.

$$I = P \times R \times T$$

Interest = 1,000 (0.05) (2) = $100

New Principal = $1,000 + $100 = $1,100

Example 9: If you borrow $12,000 for six months at an annual interest rate of 8%, how much money do you have to pay back at the end of six months?

SOLUTION: The principal is 12,000, the rate is 8% or 0.08, and the time is 0.5. Remember, the time is in terms of years, and six months is half a year, which is 0.5. If you mistakenly used a 6 for time, it would mean the loan was for 6 years, not 6 months.

$$I = P \times R \times T$$

Interest = 12,000 (0.08) (0.5) = $480

Money owed at the end of six months = $12,000 + $480 = $12,480.

Example 10: How much money would you owe if you borrowed $2,000 for 480 days at an annual interest rate of 8.75%?

SOLUTION: The principal is $2,000, the rate is 8.75% or 0.0875. In this example the time is labeled in terms of days. When performing calculations of this type, we assume a year is 360 days. $\frac{480}{360} \approx 1.33$ years

$$I = P \times R \times T$$

Interest = $2,000 (0.0875) (1.33) = $232.75

Amount of money owed = $2,000 + $232.75 = $2,232.75

Problem Set:

1) How much interest would you receive if you left $2,000 in the bank for one year at 3% interest?

2) How much interest would you have to pay on a five-year $5,000 loan at a 9% interest rate?

3) If you deposited $4,000 in the bank and received 4% interest, how much would your principal be in 3 months?

4) Bill Shea borrowed $7,000 from a bank for 90 days at 7% interest. How much money would he have to repay the bank at the end of the 90 days.

SOLUTIONS TO PROBLEM SET:

$I = P \times R \times T$

1) The principal is 2,000, the rate is 0.03, and the time is 1 year.

   Interest $= 2,000\,(0.03)\,(1) = \$60.00$

2) The principal is $5,000, the rate is 0.09, and the time is 5 years.

   Interest $= 5,000\,(0.09)\,(5) = \$2,250$

3) The principal is $4,000, the rate is 0.04, and the time is $\dfrac{3 \text{ months}}{12 \text{ months}} = 0.25$ years

   Interest $= 4,000\,(0.04)\,(0.25) = \$40$

   New principal $= \$4,000 + \$40 = \$4,040$

4) The principal is $7,000, the rate is 0.07, and the time is $\dfrac{90 \text{ days}}{360 \text{ days}} = 0.25$ years

   Interest $= 7,000\,(0.07)\,(0.25) = \$122.50$

   He would have to repay $\$7,000 + \$122.50 = \$7,122.50$

Sometimes it's necessary to figure out the interest rate. Don't worry: there's a nice easy formula for that.

$$Interest\ Rate = \frac{amount\ of\ interest}{amount\ borrowed \times number\ of\ years} \times 100$$

Example 11: If you lent your friend $500 and were paid back $530 (including interest) one year later, what interest rate did you charge?

SOLUTION:

$$Interest\ rate = \frac{30}{500 \times 1} \times 100 = 6\%$$

Problem Set:

1) Ms. Lang lent $8,000 to Mr. Gray. One year later he repaid the loan with $400 interest. What was the interest rate on the loan?

2) Mr. Jones borrowed $10,000 from the bank for one year. He paid the bank $10,700 (including interest). What was the interest rate on the loan?

1) $Interest\ rate = \dfrac{amount\ of\ interest}{amount\ borrowed} \times 100 = \dfrac{400}{8,000} \times 100 = 5\%$

2) $Interest\ rate = \dfrac{amount\ of\ interest}{amount\ borrowed} \times 100 = \dfrac{700}{10,000} \times 100 = 7\%$

1) If the sales tax is 5%, how much is the tax charged on a purchase of $349.95? What is the total price?

2) If the sales tax is 7.5%, how much is the tax charged on a purchase of $450? What is the total price?

3) If the sales tax in Missouri is 6%, how much tax is charged on a $45 ticket to a St. Louis Cardinal game? What is the total purchase price?

4) If the sales tax is $49.00 on the purchase of an item that sells for $980, what is the sales tax rate?

5) The sales tax on a new car is $550. If the selling price of the car is $16,500, find the sales tax rate.

6) How much commission is earned by someone paid a rate of 25% on sales of $8,417?

7) A salesperson's commission rate is 6.5%. What is the commission from a sale of $38,150?

8) What is the commission rate of a salesperson who earns a commission of $4,500 on sales of $50,000?

9) What is the commission rate of a person who earns a commission of $360 on sales of $7,200?

10) A pair of shoes marked $99 is on sale at 60% off. How much is the discount? How much is the sale price?

11) Everything at Bargain City is on sale at 45% off. If a table is marked at $50, how much is the discount? How much is the sale price?

12) How much interest would you receive if you left $3,000 in the bank for one year at 3%?

13) How much interest would you have to pay on a two-year $2,000 loan at a 9% interest rate?

14) If you deposited $5,000 in the bank and received 4% interest, how much would your principal be in 9 months?

15) Michelle Chavez borrowed $6,000 for 180 days at an 8% interest rate. How much money did she have to repay?

16) Ms. Green lent $8,000 to Mr. Blue. One year later he repaid the loan with $800 interest. What was the interest rate on the loan?

17) Charles DeGaulle borrowed $5,000 from the bank for two years. He repaid the bank $6,600 (including interest). How much was the interest rate on the loan?

CHALLENGE PROBLEMS

18) Charlie and Pam are sales reps. Charlie earned a 12.5% commission on sales of $14,300. Pam earned a 9.5% commission on sales of $18,600. Who earned the higher commission?

19) A dress that was originally $19 is marked down by 30%. How much was its sale price? When the dress was still unsold, the store offered an additional 20% off its sale price. What was its new low price?

Chapter 7 Review

Convert each of the following fractions to percents. Round to the nearest percent.

1) $\dfrac{4}{5}$

2) $\dfrac{6}{12}$

3) $\dfrac{12}{14}$

Convert each of the following percents to fractions.

4) 24%

5) 124%

6) 2.4%

7) 5.6%

8) $2\dfrac{2}{3}\%$

9) $\dfrac{5}{6}\%$

Convert each of the following decimals to percents.

10) 0.05

11) 5.5

12) 1

Convert each of the following percents to decimals.

13) 10%

14) 0.1%

15) 100%

How much is 29% of

16) 100?

17) 562?

18) 901

19) How much soda is in a 64-ounce bottle that is 40% full?

20) Mr. Gambino received a pay increase of 12%. If his original weekly salary was $625,
a) how much was his pay increase?   b) how much is his new weekly salary?

21) Of 200 students, 75 are female; what percent of the students are female?

22) 30% of a store's customers paid by cash. If 45 customers paid by cash, how many customers did the store have?

23) At the beginning of the month your checking account balance was $2,295.03. At the end of the month it was $1,840.67. What was the percent change in your balance?

24) Last year you earned $3,050 a month. This year you earn $3,475 a month. What is the percent increase in your monthly earnings?

25) If the sales tax is 5.5%, how much tax is charged on a purchase of $1,490? What is the total price?

26) If the sales tax is 7%, how much tax is charged on a $25.99 shirt? What is the total purchase price?

27) The sales tax on a car is $800. If the selling price of the car is $16,000, find the sales tax rate.

28) What is the commission rate of a salesman who earns a commission of $600 on a sale of $12,000?

29) How much interest would you receive if you left $6,000 in the bank for a year at 3% interest?

30) If you deposited $3,000 in the bank and received 4% interest, how much would your principal be in 390 days?

31) Michael Jackson borrowed $10,000 from the bank for three years. He repaid the bank $15,000 (including interest). How much was the interest rate on the loan?

Chapter 7 • Test 1

Convert each of the following fractions to percents. Round to the nearest percent.

1) $\dfrac{7}{8}$

2) $\dfrac{8}{16}$

3) $\dfrac{3}{5}$

Convert each of the following percents to fractions.

4) 74%

5) 151%

6) 4.7%

7) 15.2%

8) $10\dfrac{3}{4}\%$

9) $\dfrac{4}{5}\%$

Convert each of the following decimals to percents.

10) 0.8

11) 13.4

12) 4

Convert each of the following percents to decimals.

13) 50%

14) 0.03%

15) 250%

How much is 37% of

16) 130?   17) 396?   18) 25?

19) How much juice is in a 32-ounce container that is 60% full?

20) Ms. O'Toole received a pay increase of 6%. If her original monthly salary was $4,150, a) how much was her pay increase? b) how much is her new monthly salary?

21) Of 300 guests at the ambassador's Fourth of July party, 122 were Americans. What percent of the guests were Americans?

22) At the beginning of the month your checking account balance was $5,046.55. At the end of the month it was $6,220.13. What was the percent change in your balance?

23) 40% of a store's customers paid by credit card. If 30 customers paid by credit card, how many customers did the store have?

24) If the tax on gasoline is 12.4%, how much tax is charged on a purchase of $21.33? What is the total price?

25) If sales tax is 7.5%, how much tax is charged on a $89.95 dress? What is the total purchase price?

26) The sales tax on a pickup truck is $900. If the selling price of the truck is $18,000, find the sales tax rate.

27) What is the commission rate of a saleswoman who earns a commission of $2,500 on a sale of $75,000?

28) How much interest would you receive if you left $12,000 in the bank for a year at 5% interest?

29) If you deposited $8,000 in the bank and received 6% interest, how much would you have in the bank in 480 days?

Chapter 7 • Test 2

Convert each of the following fractions to percents. Round to the nearest percent.

1) $\dfrac{5}{6}$

2) $\dfrac{9}{18}$

3) $\dfrac{3}{4}$

Convert each of the following percents to fractions.

4) 47%

5) 2.1%

6) 111%

7) 17.5%

8) $29\dfrac{1}{2}$ %

9) $\dfrac{9}{10}$ %

Convert each of the following decimals to percents.

10) 0.7

11) 5.3

12) 6

Convert each of the following percents to decimals.

13) 35%

14) 0.08%

15) 130%

How much is 43% of

16) 180?

17) 287?

18) 45?

19)   How much water is in a 64-ounce container that is 30% full?

20) Mr. Jefferson received a pay raise of 9%. If his original monthly salary was $3,925, a) how much was his pay raise? b) how much is his new monthly salary?

21) Of 1,450 visitors to the Brooklyn Museum of Art, 550 were members. What percent of the visitors were members?

22) At the beginning of the month your checking account balance was $1,308.12. At the end of the month it was $778.29. What was the percent change in your balance?

23) 60% of a store's customers pay with cash. If 90 customers paid with cash, how many customers did the store have?

24) If the sales tax is 8.25%, how much tax is charged on a purchase of $231.90? What is the total price?

25) If the sales tax is 6.5%, how much tax is charged on a $239.99 suit? What is the total purchase price?

26) The sales tax on a SUV is $1,200. If the selling price of the SUV is $24,000, find the sales tax rate.

27) What is the commission rate of a salesperson who earns a commission of $450 on a sale of $5,000?

28) How much interest would you receive if you left $9,000 in the bank for two years at 4% interest?

29) If you deposited $6,000 in the bank and received 4% interest, how much would you have in the bank in 180 days?

Chapter 8.   Signed Numbers

Signed numbers are divided into two sets. Positive and negative numbers. 1 is a positive number and −1 is a negative number. The number 0 is non-positive and non-negative. We'll just think of 0 as the dividing line between the positive and negative numbers. Notice on the number line below that any number to the right of 0 is a positive number and any number to the left of 0 is a negative number. A negative symbol in front of a number means "the opposite of". For example −3 means the opposite of 3. 3 is three units to the right of the zero; −3 is three units to the left of zero. So −3 is on the opposite side of 3 on the number line. Keep in mind negative means opposite of. We could say − (−3), which means the opposite of −3 which is positive 3. We'll discuss this in greater detail later in the chapter. For now just keep in mind that negative means opposite of.

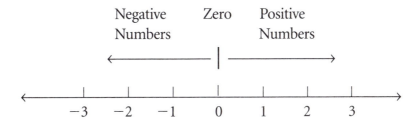

Most people don't realize we use negative numbers in our lives, but we do all the time. We use negative numbers when we bounce a check. If we have $100 in our checking account, but we write a check for $300, we would have bounced that check by $200. We would have −$200 in our account. When we read our credit card statements, we see negative and positive numbers. When we make a bad investment and lose money, we show a negative return on our investment. Sometimes on a very cold day the temperature is a negative number of degrees.

There are two sets of rules that apply to arithmetic operations for signed numbers. There's one set of rules for addition and subtraction, and a different set of rules for multiplication and division. In section 8.2 you'll learn the rules for addition and subtraction of signed numbers. We'll begin this chapter by teaching you the rules for multiplication and division. We'll state these rules in words and symbols and give examples of each rule. The + sign means positive and the − sign means negative. When we want to write a positive number such as 5, it's not necessary to write the + next to the five. If there isn't a + or − sign in front of a number, assume the number is positive.

1.1   Multiplication and Division of Signed Numbers

Multiplication Rules in:

Symbols	Words	Examples
$(+)(+) = +$	A positive times a positive is a positive.	$2(3) = 6$
$(+)(-) = -$	A positive times a negative is a negative.	$2(-3) = -6$
$(-)(+) = -$	A negative times a positive is a negative.	$-2(3) = -6$
$(-)(-) = +$	A negative times a negative is a positive.	$-2(-3) = 6$

Notice that when we have the same signs multiplied together, $(+)(+) = +$ and $(-)(-) = +$, the product is positive. When we have different signs multiplied together, $(-)(+)$, and $(+)(-)$, the product is negative. Remember, same signs positive, different signs negative.

Let's try applying these rules to the following problems.

Example 1: $4\,(-2)$

SOLUTION: $4\,(-2)$ The 4 is a positive number and the 2 is a negative number.
                        $4\,(-2) = -8$ A positive times a negative is a negative.
                                       Different signs multiplied together is negative.

Example 2: $-4\,(-2)$

SOLUTION $-4\,(-2)$ The 4 is a negative number and the 2 is a negative number.
                      $-4\,(-2) = 8$ A negative times a negative is a positive.
                                        Same signs multiplied together is positive.

Example 3: $5\,(7) = 35$

SOLUTION: $5\,(7)$ The 5 is a positive number and the 7 is a positive number.
                      $5\,(7) = 35$ A positive times a positive is a positive.
                                        Same signs multiplied together is positive.

So far we've multiplied only two numbers together. Let's try a few more challenging problems.

Example 4: $-1\,(-9)\,(-4)$

        SOLUTION: $-1(-9)(-4)$ The 1 is a negative, the 9 is a negative, and the 4 is a
                                      negative. We'll work from left to right.
$-1\,(-9)\,(-4)$
     $9\,(-4)$ $-1\,(-9) = 9$ Same signs are positive.
       -36 $9\,(-4) = -36$ Different signs are negative.

Notice we had an odd number (three) of negative signs in this multiplication problem and the product is negative.

Example 5: $2\,(-1)\,(3)\,(-5)$

SOLUTION: $2\,(-1)\,(3)\,(-5)$ We'll work from left to right.
              $-2\,(3)\,(-5)$ $2\,(-1) = -2$ Different signs are negative.
                $-6\,(-5)$ $-2\,(3) = -6$ Different signs are negative.
                   30 $-6\,(-5) = 30$ Same signs are positive.

Notice this multiplication problem has an even number (two) of negative signs and our answer is positive.

Example 6: $-1\,(-2)\,(3)\,(4)\,(-2)\,(-3)\,(5)$

    SOLUTION: If we work this one out step by step like the last two examples, it would look like this. We'll multiply left to right.

$-1\,(-2)\,(3)\,(4)\,(-2)\,(-3)\,(5)$	$(-1)(-2) = 2$
$2\,(3)\,(4)\,(-2)\,(-3)\,(5)$	$2(3) = 6$
$6\,(4)\,(-2)\,(-3)\,(5)$	$6(4) = 24$
$24\,(-2)\,(-3)\,(5)$	$24(-2) = -48$
$-48\,(-3)\,(5)$	$-48(-3) = 144$
$144\,(5)$	$144(5) = 720$
720	

Notice this example has an even number (four) of negatives and the answer is positive.

Working out this problem step-by-step was too long, so we're going to give you a shortcut to use to find the sign of the answer without working out the problem one step at a time.

The product of an even amount of negative numbers is positive.

The product of an odd amount of negative numbers is negative.

Now that you know these shortcut rules, all you have to do is multiply the numbers, then count the amount of negative numbers in the problem. If the problem has an even amount of negative numbers multiplied together, the product is positive. If the problem has an odd amount of negative numbers multiplied together, the product is negative. By the way, the faster way to multiply the numbers together is to group them together instead of multiplying them one at a time, left to right. You don't have to group them the exact same way we do. You'll get the same product no matter how you group the numbers because multiplication is commutative. In this problem we grouped the $-1(-2)(3)$ together, then we grouped $4(-2)$ together, and lastly we grouped $(-3)(5)$ together.

$$-1(-2)(3) = 6 \qquad 4(-2) = -8 \qquad (-3)(5) = -15$$

$$-1\,(-2)\,(3)\,(4)\,(-2)\,(-3)\,(5)$$
$$6\,(-8)(-15)$$
$$-48\,(-15)$$
$$720$$

This problem has an even amount (four) of negative numbers multiplied together; therefore the answer is positive.

Try applying these rules to the following problems.

Example 7: $-8\,(2)\,(-3)\,(2)\,(-1)\,(-1)\,(4)\,(-10)$

    SOLUTION: This problem has an odd amount (five) of negative numbers multiplied together; therefore the product is negative. Now all we have to do is multiply. We're not going to work this out left to right. We'll save some time by grouping numbers together. You don't have to group the numbers in the same way we did. You can group them in any order and still get the same answer.

$$-8\,(2)\,(-3)\,(2)\,(-1)\,(-1)\,(4)\,(-10) \qquad -8(2) = -16, \ (-3)(2)(-1)(-1) = -6, \ (4)(-10) = -40$$
$$-16\,(-6)(-40) \qquad -16(-6) = 96$$
$$96\,(-40) \qquad 96(-40) = -3{,}840$$
$$-3840$$

Now it's your turn to try a few problems.

Problem Set: Perform the following operations.

1) $12\,(-2)$

2) $-12\,(-2)$

3) $-3\,(-1)\,(9)$

4) $-3\,(-1)\,(-9)$

5) $6\,(-11)\,(-1)\,(9)\,(8)$

6) $-6\,(-11)\,(-1)\,(9)\,(8)$

SOLUTIONS TO PROBLEM SET:

1) -24

2) 24

3) $3(9) = 27$

4) $3(-9) = -27$

5) $-66(-72) = 4{,}752$

6) $66(-72) = -4{,}752$

In chapter 2 we studied exponents, which is the repeated multiplication of a number times itself. Let's take a look at some problems involving signed numbers and exponents.

Example 8:   Simplify the following example. How would we write this problem using exponents?

$$(-2)\,(-2)\,(-2)\,(-2) =$$

SOLUTION: There's an even number of negatives multiplied together, so the answer is positive 16. We would write this problem using exponents in the following way: $(-2)^4$ Notice we put parentheses around the 2 **and** the negative sign to show the number to the fourth power is -2, not 2. When a negative sign is in parentheses to a power, the negative also goes to the power. The base is -2; the power is 4. This problem is read negative two to the fourth power.

$$(-2)^4 = (-2)\,(-2)\,(-2)\,(-2) = 16$$

Example 9:   Simplify the following problem. How would we write this problem using exponents? Hint: Even though this problem looks similar to example 8, it's not the same.

$$-(2)(2)(2)(2)$$

SOLUTION:
$-2^4 =$

Notice positive 2, not negative 2, is written four times. The negative sign is written only once. In example 8 it was written four times.

This problem is read the opposite of two to the fourth power. 2 is the base and 4 is the exponent. Two to the fourth power is 16, so the opposite of two to the fourth power is negative 16. We would write that as -2^4. The negative doesn't belong in parentheses because it's not part of the base; it means the opposite of two to the fourth power.

$$-2^4 = -(2)\,(2)\,(2)\,(2) = -16$$

Now that we've worked out examples 8 and 9, you should have noticed that it's very important to observe whether the negative is inside parentheses or not. If the negative is inside parentheses raised to a power, the negative is part of the base and goes to the power. If the negative is not inside parentheses, the negative is not part of the base and does not go to the power. Just think of it as the opposite of the base to the power. Remember the order of operations and raise the base to the power first, then take the opposite of that answer. Let's try a few more examples and then you can try a problem set.

Example 10: Simplify the following.

    a) -5^2 b) $(-5)^2$

    SOLUTION:

        a) $-5^2 = -(5)(5) = -25$ This is read the opposite of five squared. The base is 5, the exponent is 2, The negative does not go to the power and is not part of the base because it's not in parentheses.

        b) $(-5)^2 = (-5)(-5) = 25$ This is read negative five squared. The base is -5, the exponent is 2. The negative goes to the power and is part of the base because it's in parentheses.

Example 11: Simplify the following.

    a) $(-5)^3$ b) -5^3

    SOLUTION:

    a) $(-5)^3 = (-5)(-5)(-5) = -125$ This is read -5 cubed. The base is -5, the exponent is 3. The negative goes to the power and is part of the base because it's in parentheses.

    b) $-5^3 = -(5)(5)(5) = -125$ This is read the opposite of five cubed. The base is 5, the exponent is 3. The negative does not go to the power and is not part of the base because it's not in parentheses.

Notice that when the exponent is an odd number, whether the negative is inside or outside the parentheses, it doesn't change the answer.

We're going to state two basic rules for exponents involving negative numbers.

A negative number raised to an even power is positive.

A negative number raised to an odd power is negative.

Here are some examples of these rules:

$(-1)^2 = 1$ $(-1)^3 = -1$ $(-2)^5 = -32$ $(-2)^6 = 64$ $(-1)^{200} = 1$

Problem Set: Simplify the following.

1) $(-3)^3$ 2) -3^3 3) $(-3)^4$ 4) -3^4 5) $(-1)^{31}$

1) $(-3)^3 = (-3)(-3)(-3) = -27$ A negative to an odd power is negative.

2) $-3^3 = -(3)(3)(3) = -27$ The opposite of three cubed is -27.

3) $(-3)^4 = (-3)(-3)(-3)(-3) = 81$ A negative to an even power is positive.

4) $-3^4 = -(3)(3)(3)(3) = -81$ The opposite of three to the fourth power is -81.

5) $(-1)^{31} = -1$ A negative to an odd power is negative.
                                        1 to any power is 1.

Now that you know everything there is to know about multiplication of signed numbers, let's get started on division of signed numbers.

Division Rules in:

Symbols	Words	Examples
$\dfrac{+}{+} = +$	A positive divided by a positive is a positive	$\dfrac{4}{2} = 2$
$\dfrac{-}{-} = +$	A negative divided by a negative is a positive.	$\dfrac{-4}{-2} = 2$
$\dfrac{+}{-} = -$	A positive divided by a negative is a negative.	$\dfrac{4}{-2} = -2$
$\dfrac{-}{+} = -$	A negative divided by a positive is a negative.	$\dfrac{-4}{2} = -2$

Notice that when we have the same signs divided together, $\dfrac{+}{+} = +$ and $\dfrac{-}{-} = +$ the quotient is positive.

When we have different signs divided together, $\dfrac{-}{+} = -$ and $\dfrac{+}{-} = -$, the quotient is negative. Remember: same signs positive, different signs negative: These are the same as the rules of multiplication.

Example 13: **Simplify the following:**

     a) $\dfrac{-16}{8}$ b) $\dfrac{16}{8}$ c) $\dfrac{16}{-8}$ d) $\dfrac{-16}{-8}$

     SOLUTION:

     a) $\dfrac{-16}{8} = -2$ A negative number divided by a positive number is negative.

                                      Different signs, the answer is negative.

     b) $\dfrac{16}{8} = 2$ A positive number divided by a positive number is positive.

                                      Same signs, the answer is positive.

c) $\dfrac{16}{-8} = -2$ A positive number divided by a negative number is negative.

Different signs, the answer is negative.

d) $\dfrac{-16}{-8} = 2$ A negative number divided by a negative number is positive.

Same signs, the answer is positive.

Example 14: **Simplify the following.**

a) $36 \div (-6) \div (-1)$ b) $(-36) \div [(-6) \div 6]$

SOLUTIONS:

a) We'll follow the order of operations; since we have two divisions, we'll work left to right.

$36 \div (-6) \div (-1)$ $36 \div (-6) = -6$

$(-6) \div (-1)$

6

b) We'll follow the order of operations and simplify inside the grouping first.

$(-36) \div [(-6) \div 6]$ $(-6) \div 6 = -1$

$(-36) \div [-1]$

36

Example 15: **Simplify the following**

a) $\dfrac{-2(15)}{-30}$ b) $\dfrac{-81}{(-3)^4}$

SOLUTIONS:

a) Simplify the numerator first.

$\dfrac{-2(15)}{-30}$ $-2(15) = -30$

$\dfrac{-30}{-30}$ A negative divided by a negative is positive.

1

b) Simplify the exponent first.

$\dfrac{-81}{(-3)^4}$ $(-3)^4 = (-3)(-3)(-3)(-3) = 81$

$\dfrac{-81}{81}$ A negative divided by a positive is negative.

-1

Problem Set: Perform the following operations.

1) $\dfrac{-25}{-5}$

2) $-50 \div -25$

3) $-100 \div 4$

4) $\dfrac{4}{-100}$

5) $0 \div -3$

6) $\dfrac{-3}{0}$

SOLUTIONS TO PROBLEM SET:

1) $\dfrac{-25}{-5} = 5$ A negative divided by a negative is positive.

2) $\dfrac{-50}{-25} = 2$ A negative divided by a negative is positive.

3) $-100 \div 4 = -25$ A negative divided by a positive is negative.

4) $\dfrac{4}{-100} = -\dfrac{1}{25}$ A positive divided by a negative is negative.

5) $0 \div -3 = 0$ Zero divided by a positive or a negative is zero.

6) $-3 \div 0$ is undefined   Division by zero is undefined.

Workbook

Perform the following operations.

1) $-3(-4)$

2) $-2(-5)$

3) $3(-4)$

4) $2(-5)$

5) $-1(9)$

6) $-4(6)$

7) $\dfrac{-9}{3}$

8) $\dfrac{8}{-2}$

9) $5(1)(-3)$

10) $-3(4)(-3)$

11) $\dfrac{-15}{-3}$

12) $\dfrac{(-4)(2)}{-8}$

13) $-2(-4)(3)(-2)$

14) $49 \div 7 \div (-7)$

15) $\dfrac{4(10)}{-2(-10)}$

16) $\left[49 \div \left(7 \div -7\right)\right]$

17) $-3(-3)(-3)(-3)$

18) $-5(-5)(-5)(-5)$

19) $10(-1)(5)(-4)$

20) $-7(3)(-1)(-1)(4)$

21) $-20(30)(-2)(-1)(-4)(2)$

22) $5(-4)(-1)(-1)(3)(90)$

23) $\dfrac{-7(14)(-2)}{-49(2)}$

24) $-50 \div (-5)^2$

25) $-125 \div (-5)^3$

26) $\dfrac{-12}{-2(2)}$

27) $\left(-4\right)^2$

28) $\left(-8\right)^2$

29) -4^2

30) -8^2

31) -4^3

32) -6^3

33) $\left(-4\right)^3$

34) $(-6)^3$

35) $\left(-1\right)^{100}$

36) -1^{100}

37) -1^{101}

38) $\left(-1\right)^{101}$

8.2   Addition and Subtraction of Signed Numbers

There are three basic rules for addition and subtraction of signed numbers. Before we go over them, it's important to make sure you can identify which numbers are positive and which numbers are negative. A negative number represents a reduction or a "taking away" of a quantity of some item. If we have the problem $2 + -3$ we can easily see that the 2 is positive and the 3 is negative. It means we have 2 of something and we want to take away 3 of that same item. For example, suppose we have two dollars and we would like to buy something that costs three dollars. We would be spending three of our two dollars, which we know can't be done, because we're one dollar short. Another way to state the same problem is $2 - 3$; again the 2 is positive and the 3 is negative. This shows that we want to take three away from two. We will consider any number that immediately follows a subtraction sign to be a negative number. Subtraction decreases the value of a number and is therefore considered to be a negative symbol. In the following examples we will identify which number is positive and which number is negative.

Example 1:	$-5 + 7$	The 5 is negative and the 7 is positive.
Example 2:	$11 - 25$	The 11 is positive and the 25 is negative.
Example 3:	$-1 - 4$	Both the 1 and the 4 are negative.
Example 4:	$-3 + 6 - 4$	The 6 is positive, the 3 and 4 are negative.

Now that you can identify numbers as positive or negative, you're ready to look at the first rule for addition and subtraction of signed numbers.

Rule 1:   If the numbers have the same sign, keep the sign and add the numbers.

Example 5:	$-3 + -4 = -7$	The 3 and the 4 are both negative, so keep the negative sign and add the numbers.
Example 6:	$-6 - 9 = -15$	The 6 and the 9 are both negative, so keep the negative sign and add the numbers.
Example 7:	$4 + 8 = 12$	The 4 and the 8 are both positive, add the numbers, keep the positive sign.
Example 8:	$-2 - 4 - 5 = -11$	All the numbers are negative, so keep the negative sign and add the numbers.

Try the following problem set and see how you do.

Problem Set: Perform the following operations.

1) $-1 - 2$	2) $2 + 7$	3) $3 + 6$	4) $-10 - 25$
5) $--1 + -5$	6) $-2 - 7$	7) $9 + 11 + 3$	8) $-8 + -7 - 2$

1) $-1 - 2 = -3$ 2) $2 + 7 = 9$ 3) $3 + 6 = 9$

4) $-10 - 25 = -35$ 5) $-1 + -5 = -6$ 6) $-2 - 7 = -9$

7) $9 + 11 + 3 = 23$ 8) $-8 + -7 - 2 = -17$

Now that you've mastered the first rule, here comes the second rule.

Rule 2: If the numbers have different signs, keep the sign of the larger number and subtract the numbers.

Example 9: $-2 + 3 = 1$ The 2 is negative and the 3 is positive, so the signs are different. We'll keep the sign of the larger number, positive three, and subtract the numbers to find the answer, 1.

Example 10: $4 - 7 = -3$ The 4 is positive and the 7 is negative, so the signs are different. We'll keep the sign of the larger number, negative seven, and subtract the numbers to find the answer, -3.

Example 11: $-5 + 5 = 0$ We have different signs, so we must subtract the numbers which gives us the answer, 0. Remember, 0 is not positive or negative.

Example 12: $-11 + 2 + 3$
$-11 + 5$
-6

We have three numbers, so we'll use the first rule to add the 2 and the 3 together because they're the same sign. This gives use $-11 + 5$ which requires us to use the second rule because -11 and 5 are different signs. We'll keep the sign of the larger number, -11, and subtract the numbers to get our answer of -6.

In this problem we combined the positives first, $2 + 3 = 5$. Then we added the -11 to the 5. As the problems get longer, we'll combine the numbers that have the same signs first. It makes the problem shorter.

This same problem can also be simplified by adding left to right.

$-11 + 2 + 3$
$-9 + 3$
-6

-11 and 2 are different signs, so we'll keep the sign of the larger number, -11, and subtract the numbers. -9 and 3 are different signs, so we'll keep the sign of the larger number, -9, and subtract the numbers.

Example 13: $200 - 300$
-100

The 200 is positive and the 300 is negative. Keep the sign of the larger number, -300, and subtract the numbers.

If you get confused, think of money. Suppose we have $200 in our bank account and we write a check for $300. Wouldn't we bounce our check by $100?

Example 14: $-5 + 4 - 2 + 4 - 10 - 6$

$-5 - 2 - 10 - 6 + 4 + 4$ We could work this problem out from left to right, but it's faster if we group all the negatives together $(-5 - 2 - 10 - 6 = -23)$, and all the positives together $(4 + 4 = 8)$. Keep the sign of the larger number -23 and subtract the numbers.

$-23 + 8$

-15

Now it's time for you to try the following problem set.

Problem Set: Perform the following operations.

1) $-3 + 8$ 2) $9 - 11$ 3) $-2 + 2$ 4) $-10 + 12$

5) $8 - 9$ 6) $-8 - 1 + 2$ 7) $-7 + 4 + 5$ 8) $-1 - 1 - 1 + 3$

SOLUTIONS TO PROBLEM SET:

1) 5 2) -2 3) 0 4) 2

5) -1 6) $-9 + 2 = -7$ 7) $-7 + 9 = 2$ 8) $-3 + 3 = 0$

Rule 3: Minus a negative becomes plus a positive.

If a subtraction sign is immediately followed by a negative sign with no number between them, you must use the third rule. Minus a negative means the opposite of the negative number, which would be a positive number. Minus a negative becomes plus a positive.

Example 15: $2 - (-3)$ The opposite of -3 is 3.
Minus a negative becomes plus a positive.

$2 - (-3) = 2 + 3 = 5$

Example 16: $3 - (-5)$ The opposite of -5 is 5.
Minus a negative becomes plus a positive.

$3 - (-5) = 3 + 5 = 8$

Example 17: $-3 - (-8)$ The opposite of -8 is 8.
Minus a negative becomes plus a positive.

$-3 - (-8) = -3 + 8 = 5$ We have different signs, so we kept the sign of the larger number and subtract the numbers.

Problem Set: Perform the following operations.

1) $-6 - (-7)$ 2) $6 - (-7)$ 3) $14 - (-9)$ 4) $-14 - (-9)$

5) $2 - (-5)$ 6) $3 - (-2) - 4$ 7) $7 - (-7)$ 8) $-7 - (-7)$

SOLUTIONS TO PROBLEM SET:

1) $-6 + 7 = 1$ 2) $6 - (-7) = 6 + 7 = 13$

3) $14 - (-9) = 14 + 9 = 23$ 4) $-14 - (-9) = -14 + 9 = -5$

5) $2 - (-5) = 2 + 5 = 7$

6) $3 - (-2) - 4 = 3 + 2 - 4 = 5 - 4 = 1$

7) $7 - (-7) = 7 + 7 = 14$

8) $-7 - (-7) = -7 + 7 = 0$

Let's look at a few examples that combine everything we've learned so far in this chapter.

Example 18: Simplify $-4- 5\left(-70 \div -10\right)$

SOLUTION: Before we begin this problem, let's identify which numbers are positive and which are negative. The 4, 5, 70, and 10 are negative. Remember, the 5 directly follows a subtraction sign, so we'll consider it a negative.

$-4 - 5\left(-70 \div -10\right)$ Begin inside the parentheses first. $-70 \div -10 = 7$.

$-4 - 5\left(7\right)$ Multiply next. The five is a negative. $-5(7) = -35$.

$-4 - 35$ Use the rules for addition and subtraction. The signs

-39 are the same; add the numbers and keep the negative sign.

Example 19: Simplify $\dfrac{(1-3)(-2-2)}{-1-1}$

SOLUTION: Use the rules for addition and subtraction to simplify inside the parentheses first.

$\dfrac{(1-3)(-2-2)}{-1-1}$ $1 - 3 = -2, \;\; -2 - 2 = -4, \;\; -1 - 1 = -2$

$\dfrac{(-2)(-4)}{-2}$ Multiply $-2(-4) = 8$

$\dfrac{8}{-2}$ A positive divided by a negative is negative.

-4

Example 20: Simplify $\left[-5\left(-4 \div (-2)^2\right)+-1\right]$

SOLUTION: $\left[-5\left(-4 \div (-2)^2\right)+-1\right]$ Inside groupings first. Start with $\left(-2\right)^2 = 4$.

$\left[-5\left(-4 \div 4\right)+-1\right]$ Now divide $-4 \div 4 = -1$.

$\left[-5\left(-1\right)+-1\right]$ Multiply $-5(-1) = 5$.

$\left[\, 5 + -1 \,\right]$ Add $5 + -1 = 4$

4

The following problem set combines addition, subtraction, multiplication, and division of signed numbers. Remember to follow the order of operations and the appropriate set of rules for addition/subtraction or multiplication/division.

Problem Set: Simplify the following.

1) $12 + \dfrac{-2-2-2-2}{(-2)^3}$

2) $\dfrac{-64 \div -8}{-4+6}$

3) $(-50 \div 5) \div (-1-1) \div (-4-1)$

4) $\{-2(4-5) \div [-8-(-8)]\}$

SOLUTIONS TO PROBLEM SET:

1) $12 + \dfrac{-2-2-2-2}{(-2)^3}$ Combining the $-2-2-2-2$ gives us -8.

$12 + \dfrac{-8}{-8}$ A negative to an odd power is negative. $(-2)^3 = -8$.

$12 + 1$ A negative divided by a negative is positive. $\dfrac{-8}{-8} = 1$.

13

2) $\dfrac{-64 \div -8}{-4+6}$ A negative divided by a negative is positive. $-64 \div -8 = 8$.

Adding the -4 and 6 gives us 2.

$\dfrac{8}{2}$ Divide.

4

3)   Simplify inside parentheses first.

$(-50 \div 5) \div (-1-1) \div (-4-1)$ $-50 \div 5 = -10$, $-1-1 = -2$, $-4-1 = -5$

$(-10) \div (-2) \div (-5)$ Divide left to right. $-10 \div (-2) = 5$,

$5 \div -5$ $5 \div -5 = -1$

-1 A positive divided by a negative is negative.

4) $\{-2(4-5) \div [-8-(-8)]\}$ Begin inside the groupings $4-5 = -1$.

$\{-2(-1) \div [-8+8]\}$ $-8-(-8) = -8+8. -8+8 = 0. -2(-1) = 2.$

$2 \div 0$ Division by zero is undefined.

undefined

Simplify the following.

1) $-11 - 5$

2) $11 - 5$

3) $-14 + 15$

4) $14 + -21$

5) $20 - 39$

6) $34 - 99$

7) $-5 + 5$

8) $-2 + 2$

9) $0 - 10$

10) $0 - 4$

11) $-8 + 4 - 1$

12) $-4 - 7 - 19$

13) $-5 - (-15)$

14) $-23 - (-34)$

15) $17 - (-17)$

16) $23 - (-23)$

17) $-8 - (-5) - 4$

18) $-1 - 5 + 4$

19) $-5 + 2 - 3 - 4 + 5$

20) $6 - 7 - 8 + 12 - 6$

21) $-9(2 - 3)$

22) $-7(11 - 5 - 7)$

23) $-2(12-11-4)$

24) $6-(5-8-2)$

25) $-2+32 \div -16$

26) $-10-2 \div (-2)$

27) $\dfrac{-6(-1-1)}{3(-4)}$

28) $\dfrac{12-4(-3)}{-2}$

29) $\dfrac{-11-200}{-3-(-2-1)}$

30) $\dfrac{-2(4-8)}{(-2)^2}$

CHALLENGE PROBLEMS:

31) $[-9-(5-7)(-2-2-3)]$

32) $-8-[4-5(-3)-5]$

33) $\left[-33-\left(-64 \div (-4)^3\right)\right]$

34) $[-20-(-500) \div (-10)^2]$

8.3 Applications

Business owners usually make profits, but sometimes they take losses. Profits are stated as positive numbers, while losses are stated as negative numbers. So a business firm might make a profit of $60,000 or $250,000 or perhaps $10,000,000. Suppose a business lost $75,000. It would be a profit of −$75,000 or a loss of $75,000.

Example 1: You begin the day with $120 and spend $185 (luckily your credit is good). How much money do you really have now?

SOLUTION: The answer is −$65. In other words, you started out with $120, but after spending $185, you now owe $65. $120 − $185 = −$65

Example 2: If you owe $157 and pay back $99, how much do you still owe?

SOLUTION: −$157 + $99 = −$58. You still owe $58.

Example 3: The temperature in Fairbanks, Alaska, dropped from 14 degrees at noon to −19 degrees at midnight. By how many degrees did the temperature drop?

SOLUTION: 14 degrees − (−19 degrees) = 14 degrees + 19 degrees = 33 degrees.

Example 4: Everything you have in the world is in your back pocket. It's four IOUs for $12 each. How much money are you worth?

SOLUTION: −$12 (4) = −$48

Example 5: A father leaves a debt of $850 to five children, which they share equally. How much money does each child inherit?

SOLUTION: $\dfrac{-\$850}{5} = -\$170$

Problem Set:

1) The temperature is −4°F. If it falls another 8°, how much is the temperature?

2) A business lost $100,000 last year (profit = −$100,000). This year they lost $15,000 more than last year. State their profit this year.

3) Jones owes $113. He pays back $64. How much does he still owe?

4) An estate with a net worth of −$4,200 is divided equally among four heirs. How much does each heir inherit?

5) A bus with 12 people stopped at Avenue A. Eight people got off and three people got on. At Avenue B six people got off and nine people got on. At Avenue C five people got off the bus and seven people got on. How many people were now on the bus?

SOLUTIONS TO PROBLEM SET:

1) −4 − 8 = −12°

2) −$100,000 + −$15,000 = −$115,000

3) −$113 + $64 = −$49

4) $\dfrac{-\$4,200}{4} = -\$1,050$

5) 12 − 8 + 3 − 6 + 9 − 5 + 7 = 12

1) Jessica had $2,011 in her bank account on September 1. She deposited $396 and $1,135 during the month, and wrote checks for $214, $481, and $459 during September. What was her balance at the end of the month?

2) A bus with twenty-two people stopped at Ave. A. Six people got on and ten people got off. At Avenue B eleven people got off and five people got on. At Avenue C five people got on and two people got off. How many people were now on the bus?

3) Mr. Reilly owes $49. He pays back $17. How much does he still owe?

4) The temperature on Mount Everest rose from -15^0 F to 8^0 F. How much was the rise in temperature?

5) If you had no money in your checking account and wrote five checks for $100 each, what would your new checking account balance be?

6) A business with a net worth of $-$300$ is divided equally among four partners. What is the value of each partner's share?

7) You owe your friend $34 and borrow another $13. How much do you now owe?

8) Frank owned four stocks. This year his holdings in SBC stock went up $2,350, in AT&T it went up $3,900, in Texaco it went down $4,125, and in IBM it went down $3,875. How much less are his holdings worth at the end of the year compared to the beginning of the year?

9) On Saturday the temperature reached a high of 14^0 above 0 and a low of 17^0 below 0. What is the difference between the high and low temperatures for Saturday?

10) A company made a profit of $18,700 in year 1, a loss of $12,950 in year 2, and a loss of $9,425 in year 3. How much was the company's total profit over this three-year period?

11) A submarine dives to a depth of 450 feet below the ocean's surface. It then dives another 173 feet before climbing 120 feet. What is the depth of the submarine?

12) Mrs. Smith has a bank balance of $5,615 at the beginning of March. She deposits $1,903 and $2,490 during the month, and writes checks for $750 and $2,755. How much money is in her account at the end of the month?

13) A helicopter is 485 feet above sea level and a submarine directly below it is 540 feet below sea level. How far apart are they?

CHALLENGE PROBLEMS

14) A sailboat's mast was 60 feet above sea level. If the boat was sinking at a rate of 18 feet per minute, where would the mast be in 6 minutes?

15) A company has a net worth of $100,000. If it loses $40,000 a quarter for six consecutive quarters, find the net worth of the company.

16) You own a house worth $350,000 with a mortgage of $400,000. You have $10,000 in the bank and owe $35,000 on your credit cards. What is your net worth?

17) Your bank allows you to write overdraft checks (that means, you may write checks even if you don't have enough money in your account to cover these checks). You begin the day with $250 in your account. You write checks for $85, $270, and $900. You also deposit $335 in your account. How much is your account balance?

Chapter 8 Review:

Perform the following operations.

1) $12(-5)$

2) $12 - 5$

3) $5 - 12$

4) $-12(-5)$

5) $\dfrac{-24}{-4}$

6) $\dfrac{24}{-4}$

7) $\dfrac{-2(3-4)}{-1(-1)}$

8) $(-6)^2$

9) -6^2

10) Find the sum of -10 and -11.

11) $-3 + 4 + 60 - 72$

12) Find the difference of -4 and -7.

13) $(-8)(2)(-1)(-3)$

14) $2 - 6(-4)^3$

15) You have a net worth of $200. You borrow $500 from your friend and receive a tax refund of $150. How much is your net worth now?

16) Carlos spends $470 on lottery tickets and wins prizes of $50 and $175. How far ahead or behind is he now?

17) $6 - (-3)$

18) $-6 - (-3)$

19) If you have $2,000 in the bank and write checks for $450, $1,234, and $1,345, what's your new balance?

20) $-3 - (4-7)$

21) $\dfrac{-9-(6-9)\div-1}{-2^2}$

22) $11 + \left[40 \div \left(-3-5\right) \div (2-1)\right]$

23) $\left(-4\right)^2 \div \left(-2\right)^0$

24) $\left(-10\right)^0$

25) $\left(5-7\right)^2 \div \left(-3+3\right)$

26) The temperature in Livingston, Montana rose from $-22^0\,$F to $13^0\,$F. How much was the rise in temperature?

27) Your bank allows overdrafts (that means, you may write checks even if you don't have enough money in your account to cover these checks). You begin the day with $129 in your account. You write checks for $150 and $215. How much is your account balance?

Chapter 8 • Test 1

Simplify the following problems.

1) $15(-6)$

2) $-14 - 8$

3) $8 - 14$

4) $-18(-4)$

5) $\dfrac{36}{-9}$

6) $\dfrac{-42}{-7}$

7) $\dfrac{-82}{12}$

8) $\dfrac{-3(5-8)}{(3)(-3)}$

9) -5^3

10) $(-2)^5$

11) Find the sum of -12 and -4.

12) $-8 + 5 - 30 + 6$

13) Find the difference of -16 and -4.

14) $90 - 5(-5)^2$

15) $(-6)(4)(-1)(-2)$

16) Joseph has a net worth of $315. He borrows $600 from a friend and receives a gift of $250 from another friend. How much is his net worth now?

17) Jessica spends $10 a week on lottery tickets for one year. During that time she wins prizes of $100 and $300. How much is she ahead or behind for the year?

18) $24 - (-5)$

19) $-18 + (-6)$

20) If you have $1,000 in the bank and write checks for $175, $408, and $1,250, what's your new balance?

21) $-5 - (9 - 13)$

22) $\dfrac{-7 - (4 - 10) \div 2}{(-1)^3}$

23) $15 + \{64 \div (4 - 20) \div (-2 - 2)\}$

24) $(-5)^2 - (-17)^0$

25) $(-12)^0$

26) $(10 - 15)^2 \div (-6 + 6)$

28) The temperature at the North Pole drops from $4^0\,$F to $-31^0\,$F. By how many degrees did the temperature fall?

Chapter 8 • Test 2

Perform the following operations.

1) $8 + 9 - 15$

2) $-30 - 10$

3) $9(-6)$

4) $-6(-7)$

5) $10 + (-2)(4)$

6) $-7 - (9)(-2)$

7) $-9 + 5 - 10 - 8$

8) $-4(-3)(-1)$

9) A company made a profit of \$6,200 in 2005. It lost \$7,500 in 2006. In 2007 it made a profit of \$900. How much was the company's total profit or total lost over this three year period?

10) A helicopter is 730 feet above sea level and a submarine directly below it is 380 feet below sea level. How far apart are they?

11) $\dfrac{-11 - 200}{-3 - (-2 - 1)}$

12) $(-9)^2$

13) -7^2

14) Find the sum of -12 and -7.

15) Find the difference between -3 and -9.

16) $(-6)(-2)(5)(-3)$

17) $-13 - (-8)$

18) $14 - (-10)$

19) $\dfrac{-8 - (-42 - 6) - (-2)^2}{9 \div (-1)}$

20) $40 - [64 \div (-7 - 9) \div (-2 + 6)]$

21) You have a net worth of $300. You borrow $750 from your friend and receive a tax refund of $275. How much is your net worth now?

22) A business firm has $1,000 in cash and owes $3,000. If the firm has four partners, how much is each partner's share of the firm worth?

23) $(-16)^0$

24) $-5 - (7-15) \div \{(-3)^2 - (14-5)\}$

25) $-8 - 8\,(-2)^3$

26) $-5^2 - (-12)^0$

27) You began the month with a checking account balance of $455. During the month you deposit $500 and write checks for $225, $820, and $650. How much is your balance at the end of the month?

28) $-4(-2) + 7(-6) \div 3 - 5^2$

Chapter 9. Introduction to Geometry

Suppose you wanted to order enough fencing to enclose a pool. You would have to measure the distance around the pool. The distance around an enclosed area is called its perimeter. Suppose you wanted to calculate the amount of ground the pool covers. The amount of ground the pool covers is called its area. And what if you wanted to calculate how much water is needed to fill the pool? The part of the pool that is filled with water is called the volume. In this chapter you'll learn how to find the perimeter, the area, and the volume of a geometric figure.

9.1 Perimeter and Area

The boundary of a figure is its *perimeter*. To find the perimeter of a figure, just find the sum of its sides. We'll look at perimeters of squares, rectangles, triangles, and other geometric figures. We'll use P to represent perimeter.

Squares

A square contains four right angles and has four sides of equal length. **The formula for the perimeter of a square is four times the length of one of its sides.**

$$P = 4s$$

Example 1: Find the perimeter of the following square.

$s = 2$ feet

SOLUTION: To find the perimeter, just add up the sides. $2 + 2 + 2 + 2 = 8$ feet. Since a square consists of four sides of equal length, a faster way to find the perimeter of a square is to multiply the length of one of its sides by 4.

$P = 4s = 4(2) = 8$ feet

Example 2: Find the perimeter of a square with sides $4\frac{1}{2}$ inches long.

SOLUTION: $P = 4s = 4(4.5) = 18$ inches

Example 3: Find the perimeter of a square with sides $2\frac{1}{3}$ yards long.

SOLUTION: $P = 4\left(2\frac{1}{3}\right) = \frac{4}{1}\left(\frac{7}{3}\right) = \frac{28}{3} = 9\frac{1}{3}$ yards

Example 4: Find the perimeter of a square with sides 3.72 feet long.

  SOLUTION: $P = 4s = 4(3.72) = 14.88$ feet

The perimeter measures the distance around the outside of the square. The area measures the amount of surface the square covers. Now let's find the area of these same squares. We'll use A to represent area.

The formula for the area of a square is $A = s^2$. To find the area of a square, just multiply the length of one side times itself.

$$A = s^2 \qquad s \;\boxed{}\; s$$

Now let's go back to examples 1 through 4 and find the area of those squares.

Example 1: Find the area of the following square.

$s = 2$ feet

  SOLUTION: $A = s^2 = 2^2 = 4$ square feet. Notice the label—feet times feet is square feet. When stating area we'll say square feet. When stating perimeter we'll say feet.

Example 2: Find the area of a square with sides $4\frac{1}{2}$ inches long.

  SOLUTION: $A = s^2 = \left(4\frac{1}{2}\right)^2 = \left(\frac{9}{2}\right)^2 = \frac{81}{4} = 20\frac{1}{4}$ square inches

Example 3: Find the area of a square with sides $2\frac{1}{3}$ yards long.

  SOLUTION: $A = s^2 = \left(2\frac{1}{3}\right)^2 = \left(\frac{7}{3}\right)^2 = \frac{49}{9} = 5\frac{4}{9}$ square yards

Example 4: Find the area of a square with sides 3.72 feet long.

  SOLUTION: $A = s^2 = (3.72)^2 = 13.8384$ square feet

Rectangles

A *rectangle* contains four right angles, a pair of equal lengths, and a pair of equal widths. The formula for the perimeter of a rectangle is, $P = 2l + 2w$.

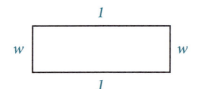

Example 5:   Find the perimeter of the following rectangle.

Width = 4

Length = 6

SOLUTION: $P = 2l + 2w = 2(6) + 2(4) = 12 + 8 = 20$

Example 6:   How much is the perimeter of a rectangle that has a length of 4 feet and a width of 2.5 feet.

SOLUTION: $P = 2l + 2w = 2(4) + 2(2.5) = 8 + 5 = 13$ feet

Now let's find the area of the same rectangles. **The area of a rectangle is the length times the width.**

$$A = lw$$

Example 7:   Find the area of the following rectangle.

Width = 4

Length = 6

SOLUTION: $l = 6, w = 4$ $A = lw = 6(4) = 24$

Example 8:   Find the area of the rectangle of length 4 feet and width 2.5 feet.

SOLUTION: $A = lw = (4)(2.5) = 10$ square feet

Problem Set:

Find the perimeter and area of a square with sides of length

1) 2 feet   2) $4\frac{3}{4}$ inches   3) 7.2 yards

Find the perimeter and area a rectangle with the following dimensions.

4) length of 9 cm. and width of 4 cm.

5) length of 12 feet and width of $4\frac{3}{5}$ feet.

6) length of 8.3 yards and width of 4.9 yards.

SOLUTIONS TO PROBLEM SET:

1) $P = 4s = 4(2) = 8$ feet

   $A = s^2 = 2^2 = 4$ square feet

2) $P = 4s = 4\left(4\frac{3}{4}\right) = 4\left(\frac{19}{4}\right) = 19$ inches

   $A = s^2 = \left(4\frac{3}{4}\right)^2 = \left(\frac{19}{4}\right)^2 = \frac{361}{16} = 22\frac{9}{16}$ square inches

3) $P = 4s = 4(7.2) = 28.8$ yards

   $A = s^2 = (7.2)^2 = 51.84$ square yards

4) $P = 2l + 2w = 2(9) + 2(4) = 18 + 8 = 26$ cm.

   $A = lw = 9(4) = 36$ square cm.

5) $P = 2l + 2w = 2(12) + 2\left(4\frac{3}{5}\right) = 24 + 2\left(\frac{23}{5}\right) = 24 + \frac{46}{5} = 24 + 9\frac{1}{5} = 33\frac{1}{5}$ feet

   $A = lw = 12\left(4\frac{3}{5}\right) = 12\left(\frac{23}{5}\right) = \frac{276}{5} = 55\frac{1}{5}$ square feet

6) $P = 2l + 2w = 2(8.3) + 2(4.9) = 16.6 + 9.8 = 26.4$ yards

   $A = lw = (8.3)(4.9) = 40.67$ square yards

Triangles

A triangle is a three-sided polygon, where all three sides are straight lines. **The formula for the perimeter of a triangle is** $P = $ side a + side b + side c. In other words, to find the perimeter of a triangle, just add up the lengths of its sides.

Example 9:   **Find the perimeter of the following triangle.**

SOLUTION: $P = a + b + c$ $P = 2 + 4 + 6 = 12$

Before we continue the study of how to find the perimeter and area of a triangle, we have to ask you one question. Would you know how to find the measure of the third side of a right triangle if the measures of only two of its sides are given? If your answer is no, don't worry; we'll show you how to use the Pythagorean Theorem to find the measure of the third side. When two of the sides of a triangle intersect and form a perfect L, a small box is written inside the triangle to show that the angle is a right angle, and the triangle is called a right triangle. *A right triangle is a triangle that has a right angle.* For those of you who already know about angles, a ninety-degree angle is called a right angle. The following triangle is a right triangle. Capital letters are used to represent angles and small letters are used to represent sides. Side a is opposite angle A. Side b is opposite angle B. And side c is opposite angle C. The hypotenuse is always the side opposite the right angle.

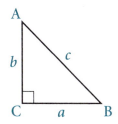

C is the right angle.

The side opposite the *right angle, "c,"* is called the *hypotenuse* of the triangle.

The *Pythagorean Theorem* states that the square of the hypotenuse is equal to the sum of the squares of the sides of a right triangle.

$$(\textbf{hypotenuse})^2 = (\textbf{side})^2 + (\textbf{side})^2$$
$$\textbf{c}^2 = \textbf{a}^2 + \textbf{b}^2$$

Example 10: **Find the measure of the hypotenuse of the following triangle.**

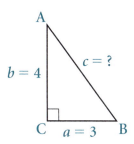

SOLUTION: By the square marked inside the triangle we can see this is a right triangle, so we can use the Pythagorean Theorem to find the length of c.

$c^2 = a^2 + b^2$ Substitute 3 for a and 4 for b in the formula.

$c^2 = 3^2 + 4^2$ Replace 3^2 by 9 and 4^2 by 16.

$c^2 = 9 + 16$

$c^2 = 25$ What number times itself is 25? $5^2 = 25$

$c = \sqrt{25}$

$c = 5$ The length of the hypotenuse is 5.

The perimeter of this triangle is $P = 3 + 4 + 5 = 12$.

Let's talk about the area of a triangle. The area of a triangle is the amount of surface the triangle covers. The area of a triangle is one half the base times the height. The formula for the area of a triangle is:

$A = \frac{1}{2}(base)(height)$ This formula applies to all triangles, not just right triangles.

height   or   height
base   base

Example 11: Find the perimeter and area of the following triangle.

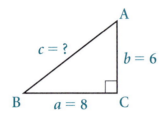

$c = ?$ A
$b = 6$
B $a = 8$ C

SOLUTION: Before we can find the perimeter or area, we have to find the length of side c. We'll begin by using the Pythagorean Theorem. The side opposite the right angle C is the hypotenuse c.

$c^2 = a^2 + b^2$ Substitute 8 for a and 6 for b.

$c^2 = 8^2 + 6^2$ $8^2 = 64, 6^2 = 36$

$c^2 = 64 + 36$

$c^2 = 100$ What number times itself is 100? $10^2 = 100$.

$c = 10$ The length of the hypotenuse is 10.

Now that we know the length of side c, it's easy to find the perimeter.

$P = a + b + c = 8 + 6 + 10 = 24$

$A = \frac{1}{2}(base)(height) = \frac{1}{2}(8)(6) = 24$

Example 12: Find the perimeter and area of the following triangle.

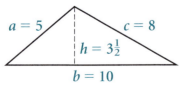

$a = 5$ $c = 8$
$h = 3\frac{1}{2}$
$b = 10$

SOLUTION: Perimeter $= 5 + 10 + 8 = 23$

$A = \frac{1}{2}(base)(height) = \frac{1}{2}(10)\left(3\frac{1}{2}\right) = \frac{1}{2}\left(\frac{\overset{5}{\cancel{10}}}{\cancel{1}}\right)\left(\frac{7}{2}\right) = \frac{35}{2} = 17\frac{1}{2}$

Example 13: **Find the perimeter and area of the following triangle.**

$a = 11$ $c = 10$ $h = 9$ $b = 8$

SOLUTION: $P = a + b + c = 11 + 8 + 10 = 29$

$$A = \tfrac{1}{2}(base)(height) = \tfrac{1}{2}\overset{4}{\cancel{(8)}}(9) = 36$$

Problem Set:

Find the perimeter and area of each of the following triangles.

1)

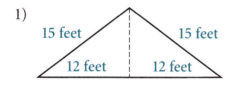

15 feet   15 feet
12 feet   12 feet

2)

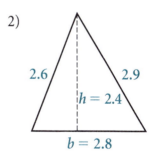

2.6   2.9 $h = 2.4$ $b = 2.8$

3)

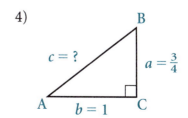

3 yards $2\tfrac{1}{2}$ yards $h = 2$ yards $b = 3\tfrac{1}{2}$ yards

4)

B $c = ?$ $a = \tfrac{3}{4}$ A $b = 1$ C

For each problem we'll use $P = a + b + c$ and $A = \frac{1}{2}(base)(height)$.

1)   This is a right triangle so we'll use the Pythagorean Theorem to find the measure of side c.

    $P = 24 + 15 + 15 = 54$ feet

    To find the measure of the height use the Pythagorean Theorem.

    $15^2 = 12^2 + h^2$

    $225 = 144 + h^2$ What number plus 144 equals 225? It's 81.

    $h^2 = 81$ What number squared is 81? It's 9.

    $h = 9$ $A = \frac{1}{2}(base)(height) = \frac{1}{\cancel{2}}(\overset{12}{\cancel{24}})(9) = 108$ square feet.

2) $P = 2.6 + 2.9 + 2.8 = 8.3$ feet.

    $A = .5(base)(height) = .5(2.8)(2.4) = 3.36$ square feet.

3) $P = 3 + 3\frac{1}{2} + 2\frac{1}{2} = 9$ yards

    $A = \frac{1}{2}(base)(height) = \frac{1}{2}\left(3\frac{1}{2}\right)\left(\frac{2}{1}\right) = 3\frac{1}{2}$ square yards

4)   This problem is a right triangle so we'll use the Pythagorean Theorem to find the measure of side c.

    $c^2 = a^2 + b^2$

    $c^2 = \left(\frac{3}{4}\right)^2 + 1^2$ $P = \frac{3}{4} + \frac{4}{4} + \frac{5}{4} = 3$

    $c^2 = \frac{9}{16} + \frac{16}{16} = \frac{25}{16}, c = \frac{5}{4}$ $A = \frac{1}{2}(base)(height) = \frac{1}{2}(1)\left(\frac{3}{4}\right) = \frac{3}{8}$

Other Geometric Shapes

We can combine squares, rectangles, and triangles into all kinds of interesting geometric shapes, and then try to find their perimeters and areas. See if you can find the perimeter and area of the geometric shape in the following example.

Example 14:

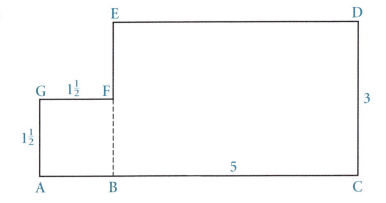

SOLUTION: BEDC is a rectangle and ABFG is a square. The perimeter of the entire shape includes only the measures of the outer border of the shape. The length of FB is not part of the perimeter. We'll begin by finding the measures of the sides that aren't labeled.

Side AB is $1\frac{1}{2}$ because FG is $1\frac{1}{2}$. Side ED is 5 because BC is 5.

Side EF is $1\frac{1}{2}$ because DC − GA = $3 - 1\frac{1}{2}$ = EF $1\frac{1}{2}$.

$$P = 1.5(AB) + 5(BC) + 3(CD) + 5(DE) + 1.5(EF) + 1.5(FG) + 1.5(GA) = 19$$

To find the area of ABCDEFG, we have to find the area of the rectangle and combine it with the area of the square.

Area of rectangle (BCDE) = *(length)(width)* = lw = 5(3) = 15.

Area of square (ABFG) = s^2 = $(1.5)^2$ = 2.25.

Area of ABCDEFG = 15 + 2.25 = 17.25.

Example 15:

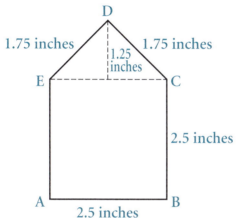

Find the perimeter and area of the geometric shape ABCDE.

SOLUTION: Perimeter = 2.5(AB) + 2.5(BC) + 1.75(DC) + 1.75(DE) + 2.5(AE) = 11 inches.

Area of the figure = area of square ABCE + area of triangle CDE.

Area of square ABCE = s^2 = $(2.5)^2$ = 6.25 square inches.

Area of triangle CDE = 0.5(base) (height) = 0.5(2.5)(1.25) = 1.5625 square inches.

Area of square ABCE + area of triangle CDE = 6.25 + 1.5625 = 7.8125 square inches.

Problem Set:

Find the perimeter and area of each of these geometric figures.

1)

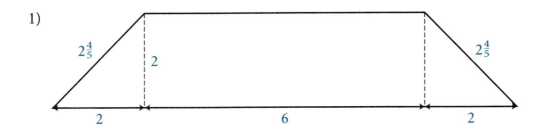

2)

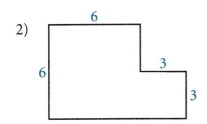

3)

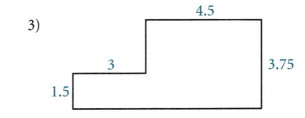

SOLUTIONS TO PROBLEM SET:

1) $P = 2 + 6 + 2 + 2\frac{4}{5} + 6 + 2\frac{4}{5} = 21\frac{3}{5}$

Area of the figure = area of rectangle + areas of the triangles

Area of rectangle = (length)(width) = 6(2) = 12

Area of one triangle = $\frac{1}{2}$(*base*)(*height*) = $\frac{1}{2}$(2)(2) = 2

Area of entire shape = area of rectangle + area of triangle + area of triangle =

12 + 2 + 2 = 16

2) Perimeter = 6 + 6 + 3 + 3 + 3 + 9 = 30

Area = area of the larger square + area of smaller square

Area of larger square = $s^2 = (6)^2 = 36$

Area of smaller square = $s^2 = (3)^2 = 9$

Area of shape = area of the larger square + area of smaller square = 36 + 9 = 45

3) P = 3 + 2.25(3.75 − 1.5) + 4.5 + 3.75 + 4.5 + 3 + 1.5 = 22.5

Area of shape = area of larger rectangle + area of smaller rectangle

Area of larger rectangle = (length)(width) = (4.5)(3.75) = 16.875

Area of smaller rectangle = (length)(width) = 3(1.5) = 4.5

Area of shape = area of larger rectangle + area of smaller rectangle

= 16.875 + 4.5 = 21.375

1) Find the perimeter and area of a square with sides:

   a) 6 feet   b) 3 inches   c) 2.8 cm.   d) $3\frac{2}{3}$ yards

2) Find the perimeter and area of a rectangle with the given measurements:

   a) length 3 inches, width 6 inches

   b) length 4.6 feet, width 7.5 feet

   c) length 3 feet, width 2 yards

   d) length $\frac{4}{5}$ meters, width $5\frac{1}{4}$ meters

3) Find the area of a triangle with the given measurements:

   a) base = 3 inches, height = 8 inches

   b) base = 10.2 inches, height = 8.5 inches

   c) base = $2\frac{2}{5}$ miles, height = $2\frac{1}{2}$ miles

4) Find the measure of the perimeter and the area of the following triangles.

a)

b)

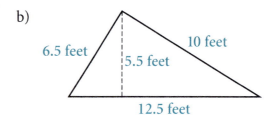

5) Find the perimeter and the area of the following figures.

a)

b)

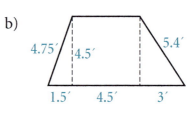

9.2 Circles

Radius, Diameter, Circumference and Area

The circle is a very familiar geometric shape. Indeed, we use the word circle in our everyday lives—e.g., traffic circle, going around in circles, let's circle around. More formally, *a circle is a planar shape consisting of a closed curve in which each point on the curve is the same distance from a fixed point called the center of the circle.*

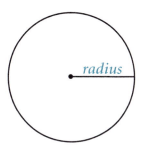

You'll notice in the circle shown above, we have a straight line called a radius. *The radius of the circle is the distance between the center and any point on the circle.* And as you can see, all radii (plural for radius) drawn for a given circle have the same length.

In the following circle, we've added a diameter. A line segment drawn through the center point with its end points on the circle is called the diameter of the circle. As you can see, the diameter is twice the radius: $d = 2r$. The radius is half the diameter: $r = \frac{1}{2}d$.

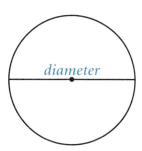

When we look at a circle, what we see is really its circumference. The perimeter of a circle is called its circumference. *The circumference of a circle is the distance around the circle.* The Greek symbol π (pronounced pie), defines the ratio between the circumference and the diameter of a circle. *π is equivalent to the circumference divided by the diameter of the circle.* The value of π is $\frac{22}{7}$, or in decimal form, approximately 3.14. The formula used to find the circumference of a circle is $C = 2\pi r$.

Example 1: **If the diameter of a circle is 18 inches, how much is its radius?**

SOLUTION: The radius is half the diameter: $r = \frac{1}{2}d = \frac{1}{2}\left(\frac{18}{1}\right) = 9$ inches.

Example 2: **If the radius of a circle is 18 inches, how much is $\frac{1}{2}$ its diameter?**

SOLUTION: The diameter is twice the radius: $d = 2r = 2(18) = 36$ inches.

Example 3: If the radius of a circle is 3 inches, how much is its circumference?

SOLUTION: $r = 3$ $C = 2\pi r = 2(3.14)(3) = 18.84$ inches

Example 4: If the diameter of a circle is 10 feet, how much is its circumference?

SOLUTION: The formula we've been using to find the circumference of a circle uses the radius, not the diameter, so we'll divide the diameter by two to find the radius.

$$d = 10 \qquad r = \tfrac{1}{2}d \qquad r = \tfrac{1}{2}\left(\tfrac{10}{1}\right) = 5$$

$$C = 2\pi r = 2(3.14)(5) = 31.4 \text{ feet}$$

Example 5: If the circumference of a circle is 314 inches, how much is its diameter?

SOLUTION: We know $C = 2\pi r$. We also know that $2r = d$. Therefore we can say $C = d\pi$. To solve for d and establish a formula for the diameter of a circle, all we have to do is divide both sides of the equation by π.

$$C = d\pi$$

$$\frac{C}{\pi} = \frac{d\pi}{\pi} \qquad \text{Reduce the common factor of } \pi.$$

$$d = \frac{C}{\pi} \qquad \begin{array}{l}\text{The diameter of a circle is equal to the}\\ \text{circumference divided by } \pi\end{array}$$

Now that we have a formula to find the diameter of a circle, we'll substitute the given value for C into the formula and find the measure of the diameter of the circle.

$$d = \frac{C}{\pi} = \frac{314}{3.14} = 100 \qquad \text{The diameter is 100 inches.}$$

Example 6: If the circumference of a circle is $1\tfrac{4}{7}$ feet, how much is its radius?

SOLUTION: We know $C = 2\pi r$. To establish a formula for the radius of a circle we divide both sides of the equation by 2π.

$$\frac{C}{2\pi} = \frac{2\pi r}{2\pi} \qquad \text{Reduce the common factor } 2\pi.$$

$$r = \frac{C}{2\pi} \qquad \text{This is the formula we'll use to find r.}$$

Since the circumference is given in fraction form ($1\tfrac{4}{7}$), we'll use the fraction form of π ($3\tfrac{1}{7}$) to find the value of r.

$$r = \frac{1\tfrac{4}{7}}{\tfrac{2}{1}\left(3\tfrac{1}{7}\right)} = \frac{\tfrac{11}{7}}{\tfrac{2}{1}\left(\tfrac{22}{7}\right)} = \frac{\cancel{11}^{\,1}\cancel{7}^{\,1}}{\cancel{7}_{1}\,\cancel{44}_{4}} = \tfrac{1}{4} \text{ foot.}$$

page 454 • Basic Mathematics

Area

The formula for the *area of a circle is $A = \pi r^2$*.

Example 7: **What is the area of the following circle?**

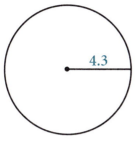

4.3

SOLUTION: $A = \pi r^2 = 3.14(4.3)^2 = 3.14(18.49) = 58.0586$

Example 8: **Find the area of a circle with a radius of $1\frac{10}{11}$ feet.**

SOLUTION: $A = \pi r^2 = \frac{22}{7}\left(\frac{21}{11}\right)^2 = \overset{2}{\cancel{\frac{22}{7}}}\,\overset{63}{\left(\frac{\cancel{441}}{\cancel{121}}\right)} = \frac{126}{11}$ square ft.

Example 9: **Find the radius of a circle with an area of 78.5 square cm.**

SOLUTION: We'll use the formula to find the area of a circle to derive a formula to find the radius of a circle.

$$A = \pi r^2$$

$$\frac{A}{\pi} = \frac{\pi r^2}{\pi}$$

$$\frac{A}{\pi} = r^2$$

$$\sqrt{\frac{A}{\pi}} = r$$

We want to isolate the r, so we'll have to divide both sides of the equation by π. Reduce the common factor π from both sides of the equation. The r is squared. The inverse of squaring a value is finding its square root, so we'll take the square root of both sides of the equation.

This is the formula we'll use to find the radius of a circle.

$$r = \sqrt{\frac{A}{\pi}}$$

Now that we have the formula, let's substitute the value for A into the equation and find the radius of the circle.

$$r = \sqrt{\frac{78.5}{3.14}} = \sqrt{25} = 5 \text{ cm}.$$

Problem Set:

1) Find the diameter of a circle with radius 5 inches.

2) Find the radius of a circle with a diameter of 20 feet.

3) Find the circumference and area of a circle with a radius of 6 inches.

4) Find the circumference and area of a circle with a radius of $2\frac{5}{22}$ meters.

5) Find the radius of a circle with a circumference of 11 yards.

1) $d = 2r$ $d = 2(5) = 10$ inches.

2) $r = \frac{d}{2}$ $r = \frac{20}{2} = 10$ feet

3) $C = 2\pi r = 2(3.14)(6) = 37.68$ inches
   $A = \pi r^2 = (3.14)(6)^2 = 113.04$ square inches

4) $C = 2\pi r = \frac{2}{1}\left(\frac{22}{7}\right)\left(2\frac{5}{22}\right) = \frac{2}{1}\left(\frac{22}{7}\right)\left(\frac{49}{22}\right) = 14$ meters

   $A = \pi r^2 = \left(\frac{22}{7}\right)\left(\frac{49}{22}\right)^2 = \left(\frac{22}{7}\right)\left(\frac{2401}{484}\right) = 15\frac{13}{22}$ meters squared

5) $r = \dfrac{C}{2\pi} = \dfrac{11}{\frac{2}{1}\left(\frac{22}{7}\right)} = \dfrac{11}{\frac{44}{7}} = \dfrac{11}{1}\left(\dfrac{7}{44}\right) = \dfrac{7}{4} = 1\frac{3}{4}$ yards

1) If the diameter of a circle is 4 yards, what's its radius?

2) If the diameter of a circle is 9 inches, what's its radius?

3) If the radius of a circle is 10 km., what's its diameter?

4) If the radius of a circle is 8 km., what's its diameter?

5) If the diameter of a circle is 3.5 yards, how much is its circumference?

6) If the diameter of a circle is $2\frac{1}{2}$ inches, how much is its circumference?

7) If the radius of a circle is 3 inches, how much is its circumference?

8) If the radius of a circle is 4.8 inches, how much is its circumference?

9) Find the area of the following figure.

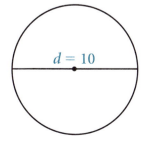

10) Find the area of the following figure.

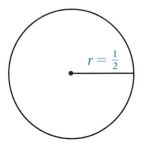

$r = \frac{1}{2}$

11) Find the radius of the circle with a circumference of 508.68 miles.

12) Find the diameter of the circle with an area of 22 square inches.

CHALLENGE PROBLEMS

13) Find the area of the following figure.

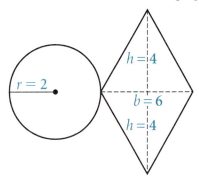

$r = 2$ $h = 4$ $b = 6$ $h = 4$

14) Find the area of the following figure.

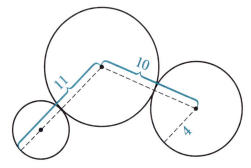

10 11 4

9.3 Volume

We've just crossed over from plane geometry to solid geometry. Now we'll be finding the volume of cubes and rectangular solids. The volume of a three dimensional figure is the amount of area inside the figure.

Cubes

A cube is a rectangular solid with its length, width, and height all equal (see the following figure). Can you think of any examples of cubes? There are sugar cubes, dice, office cubes, and even ice cubes (although most ice cubes are not really of equal length, width, and height). Back in the 1930s cubism was a very popular form of abstract art.

You'll remember that the area of a rectangle is lw. And that the area of a square is s^2, since a square's length and width are always equal. Can you guess the formula for the volume of a cube?

You could say $V = s^3$, or you could say $V = lwh$ (i.e., length times width times height). Both of these formulas would be correct. But in mathematics, we like to keep introducing new terms to keep you on your toes. One new term we're going to introduce now is *edge. An edge of a shape is the line, or one of the lines, defining the outline of that shape.* In a cube, the edges are formed where the defining surfaces meet. The formula for the volume of a cube is: *Volume = e^3*.

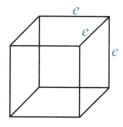

Example 1: What's the volume of a cube with an edge of 9 inches? (Be sure to state your answer in cubic inches.)

SOLUTION: Volume = e^3 = 729 cubic inches.

Example 2: Find the volume of a cube with an edge of $4\frac{1}{2}$ feet.

SOLUTION: Volume = e^3 = $(4.5)^3$ = 91.125 cubic feet.

Rectangular Solids

There are two types of rectangular solids—cubic and noncubic. Since we've just done cubes, let's do noncubic rectangular solids. The following figure is a rectangular solid.

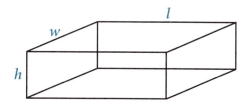

The formula for the volume of a rectangular solid is $V = lwh$ (length times width times height).

Example 3: Find the volume of a rectangular solid with a length of 4 feet, width of 6 feet, and a height of 8 feet.

SOLUTION: $V = lwh = 4(6)(8) = 192$ cubic feet.

Example 4: Find the volume of a rectangular solid with a length of $1\frac{1}{5}$ inches, a width of $4\frac{1}{6}$ inches, and a height of $7\frac{1}{5}$ inches.

SOLUTION: $V = lwh = \left(1\frac{1}{5}\right)\left(4\frac{1}{6}\right)\left(7\frac{1}{5}\right) = \left(\frac{6}{5}\right)\left(\frac{25}{6}\right)\left(\frac{36}{5}\right) = 36$ cubic inches

Problem Set:

Find the volume of the rectangular solids with these dimensions:

1) Length of 4 inches, width of 5 inches, and height of 10 inches.

2) Length of $3\frac{3}{4}$, width of $3\frac{1}{5}$, and height of $3\frac{1}{3}$.

3) A cube with edge 2.5 yards.

SOLUTIONS TO PROBLEM SET:

1) $V = lwh = 4(5)(10) = 200$ cubic inches

2) $V = lwh = \left(3\frac{3}{4}\right)\left(3\frac{1}{5}\right)\left(3\frac{1}{3}\right) = \left(\frac{15}{4}\right)\left(\frac{16}{5}\right)\left(\frac{10}{3}\right) = 40$

3) $V = e^3 = (2.5)^3 = 15.625$ cubic yards

Find the volume of the cubes with edges of:

1) $1\frac{1}{4}$ yards

2) 9.5 feet

3) 3 inches

4) $2\frac{1}{3}$ cm

Find the volume of the rectangular solids with these dimensions:

5) Length of 5 inches, width of 4 inches, and height of 5 inches.

6) Length of $5\frac{1}{2}$ feet, width of $4\frac{1}{2}$ feet, and height of 3 feet.

7) Length of 7.9 yards, width of 4.6 yards, and height of 6.2 yards.

8) Length of $11\frac{1}{3}$ inches, width of $13\frac{1}{2}$ feet, and height of $\frac{5}{8}$ inches.

9.4 Applications

Much of what you've learned in this chapter can be applied to our everyday lives. And, if nothing else, the rest can be used to solve some very interesting problems.

Perimeter and Area Applications

Example 1: A farmer gives each of his three children an equal share of a field. If the field is 150 yards long and 120 yards wide, how much land does each child get?

SOLUTION: First we have to find the total area and then divide it by three.

Total Area = $lw = (150)(120) = 18,000$ square yards.

The share of one child = $18,000 \div 3 = 6,000$ square yards.

Example 2: A town is going to put a fence around a square field that has a side of $186\frac{1}{2}$ feet. If the fencing costs $12 a foot, how much will the fence cost?

SOLUTION: First we have to find the amount of fencing needed, which is the perimeter. Then we have to multiply the perimeter by the cost of the fencing, $12 per foot.

Perimeter = $4s = 4(186.5) = 746$ feet.

Cost = *(the amount of fencing)(cost per foot)* = $746(\$12) = \$8,952$.

Example 3: A triangular piece of land with a base of 23.4 feet and a height of 17.8 feet is going to be paved over. If it costs $9.75 per square foot for paving, how much will this job cost?

SOLUTION: The paving covers the inside of the triangle; therefore we have to use the formula for the area of a triangle.

$A = .5(base)(height) = .5(23.4)(17.8) = 208.26$ square feet.

Cost = *(price per square foot)(the number of square feet)* = $(9.75)(208.26) = \$2,030.54$

Example 4: A circular stage with a radius of 30 feet is getting a new wooden floor. If it costs $24.35 per square foot to build the new floor, how much will the whole job cost?

SOLUTION: First we have to find the area, then multiply it by the cost per square foot.

Area = $\pi r^2 = (3.14)(30^2) = (3.14)(900) = 2,826$ square feet.

Cost = *(number of square feet)(cost per square foot)* = $2,826(24.35) = \$68,813.10$

Example 5: A rectangular swimming pool with a length of 40 feet and a width of 25 feet is surrounded by a walkway 3 feet wide. a) What is the perimeter of the walkway? b) How many square feet of area does the walkway cover?

SOLUTION: First, it would be very helpful to draw a picture of the pool and walkway. The length of the walkway adds 3 feet to each end of the pool.

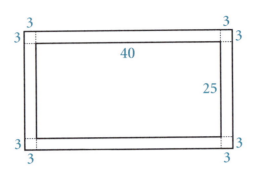

Therefore the length of the walkway is $40 + 3 + 3 = 46$ feet. The width of the walkway adds 3 feet to each side of the pool. Therefore the width of the walkway is $25 + 3 + 3 = 31$ feet.

a) Perimeter of walkway $= 2l + 2w = 2(46) + 2(31) =$
$92 + 62 = 154$ feet

b) To find the number of square feet the walkway covers, be careful to notice the length 46 includes the pool, not just the walkway. The width 31 includes the pool, not just the walkway. The product of 46 and 31, is the area 1,426 square feet, of the pool and the walkway. To find the area of just the walkway we have to subtract the area of the pool, from the area of the pool and the walkway. We already found the area of the pool and walkway is 1,426 square feet. Now let's find the area of the pool only. The length of the pool was given to be 40 feet while the width was given to be 25 feet. The area of the pool only is $A = 40(25) = 1,000$ square feet.

Area of walkway = Area of walkway and pool − Area of pool.

Area of walkway = $1.426 − 1,000 = 426$ square feet.

Problem Set:

1) A store designs a triangular sign, 22 feet across its base and 16 feet high. If the material for the sign costs $7.25 per square foot, what will the material cost?

2) How many square tiles, each 6 inches on a side, are needed to cover a floor that is 13 feet wide and 24 feet long?

Use the following figure to answers questions 3 and 4.

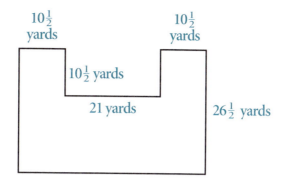

3) If fencing material costs $12 per yard, how much would it cost to fence in this area?

4) If it costs $6 per square yard to place sod on this field, how much would it cost to resod the entire field?

5) The length of a side of a square is 35 inches. If all four sides are decreased by 10 inches, how much less is the area of the new square than the area of the original square?

SOLUTIONS TO PROBLEM SET:

1) Area $= .5(base)(height) = .5(22)(16) = 176$ square feet

Cost $= (cost\ per\ square\ foot)(number\ of\ square\ feet) = (7.25)(176) = \$1,276$

2) The tiles are measured in inches and the dimensions of the floor are measured in feet. The first thing we have to do is convert the feet to inches or the inches to feet. That way all measurements are the same type. It doesn't matter which choice you make; the number of tiles needed will be the same. We'll use all feet. 6 inches is a half a foot, so we'll say the length of the tiles is 0.5 feet. The area covered by the floor is $(13)(24) = 312$ square feet. One tile covers an area of $0.5(0.5) = 0.25$ square feet. To find how many 0.25 square feet tiles it takes to cover a section of floor that is 176 square feet, divide: $312 \div 0.25 = 1,248$ tiles need to cover the floor.

3) First we have to fill in the sides of the figure that are not labeled, then add to find the perimeter of the figure. Once we know the perimeter, multiply it by $12 to find the cost of the fence.

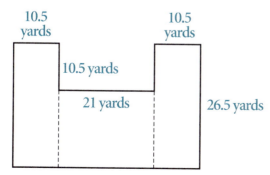

$P = 10.5 + 10.5 + 21 + 10.5 + 10.5 + 26.5 + 42 + 26.5 = 158$ yards

$C = (158)(12) = \$1,896$

4) We have to divide this figure into three rectangles, and find the area of each rectangle. Then add the areas to find the area of the figure. Lastly we have to multiply the total area by the price per square yard.

Total area = $10.5(26.5) + 21(16) + 10.5(26.5) = 278.25 + 336 + 278.25 = 892.5$ sq yds.

Cost = $6(892.5) = \$5,355$

5) The area of the original square is $35^2 = 1,225$ square inches. The new square has sides 25 inches. The area of the new square is $25^2 = 625$ square inches. The area of the new square is $(1,225 - 625)$, 600 square inches less than the original square.

1) A builder wants to construct four houses on a field with a length of 650 yards and a width of 230 yards. If each house has the same size lot, what is the area of each lot?

2) A town is going to put a fence around a square playground that has sides 563.5 feet long. If the fencing costs $8.50 per foot, how much will it cost the town to have the park fenced in?

3) The radius of a circular garden is increased from 20 yards to 30 yards. By how much is the area of the garden increased?

4) How many square tiles, each 9 inches on a side, are needed to cover a floor that is 135 feet wide and 400 feet long?

5) A triangular floor, 29 feet across the base and 17 feet high, is to be covered with linoleum. If it costs $3.75 per square foot, how much will it cost to cover the floor?

Use the following geometric figure to answer questions 6 and 7.

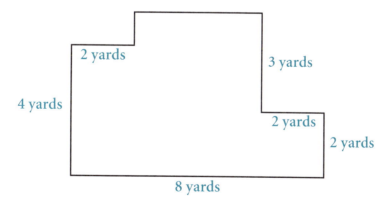

6) If fencing material costs $14.50 per yard, how much would it cost to fence in this area?

7) If it costs $7.50 per square yard to place artificial turf on this field, how much would it cost to cover the entire field?

8) A field with a length of $83\frac{1}{4}$ yards and a width of $74\frac{1}{2}$ yards is divided into three equal sections. What is the area of one section?

9) A rectangle is 38 inches long and 12 inches wide. If the length is increased by 7 inches, by how much is the area increased?

10) The radius of a circle is decreased from 4 feet to $3\frac{1}{2}$ feet. By how much is the area of the circle decreased?

11) The perimeter of a rectangle is 22. The lengths of the rectangle are 5. Find the measure of the widths.

12) The base of a triangle is 4, it's area is 6. Find its height.

13) The perimeter of a square is 32 inches. Find the length of its sides.

14) The area of a square if 25 square yards. Find the length of its sides.

15) How many cubic feet of water would it take to fill a pool 150 feet in length, 50 feet in width and 6 feet deep?

CHALLENGE PROBLEMS

16) A rectangular solid is 30 yards long, 40 feet wide, and 48 inches high. A second rectangular solid is 24 yards long, 45 feet wide, and 72 inches high.
    (a) Find the volumes of both rectangular solids.
    (b) Which one has the higher volume?
    (c) What is the difference in their volumes?

17) Helium cost $5.75 per cubic foot. How much more would it cost to fill a cube with an edge of 4 yards than a cube with an edge of 3 yards?

Chapter 9 • Review

Find the perimeter and area of a square with sides of:

1) 12

2) 5 feet

3) $6\frac{3}{4}$ inches

4) 15.3 yards

Find the perimeter and area of each of these rectangles.

5) Length of 12 and width of 9

6) Length of 14 inches and width of $8\frac{5}{6}$ inches.

7) Length of 13.4 yards and width of 10.7 yards.

8) Length of $8\frac{1}{2}$ feet, width of $4\frac{2}{3}$.

9) A triangle has sides of 9, 14 and 16. If the base is 9 and the height is 15, find the perimeter and the area.

10) Find the perimeter and the area of the following triangle.

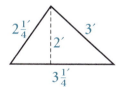

11) Find the perimeter and area of the following geometric figure. Round your answer to the nearest inch and square inch.

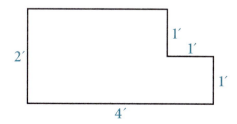

Chapter 9 • Review

Find the perimeter and the area of each of the following geometric figures.

12)

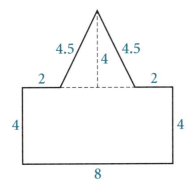

13)

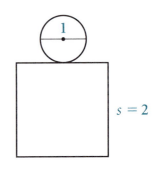

14) If the diameter of a circle is 6.4 feet, how much is its circumference?

15) If the diameter of a circle is 3.5 miles, how much is its circumference?

16) If the radius of a circle is 8, how much is its circumference?

17) If the radius of a circle is 8.3 yards, how much is its circumference?

18) If the circumference of a circle is 3.14 inches, how much is its radius?

19) If the circumference of a circle is 20.4 feet, how much is its diameter?

20) Find the area of a circle with the radius of 15.2 yards.

21) Find the area of a circle with a diameter of $12\frac{4}{7}$ feet.

22) Find the area of a circle with a diameter of 15.4 inches.

23) Find the radius of a circle with an area of 314 square yards.

Chapter 9 • Review

24) Find the diameter of a circle with a circumference of 15.7 square inches.

25) A circular path, 4 feet wide, is built around a circular garden with a radius of 15 feet. Find the area of the circular path.

26) How many square tiles, each 8 inches on a side, are needed to cover a floor that is 18 feet long and 15 feet wide.

27) A triangular floor, 39.2 feet across the base and 19.7 feet high, is to be covered with wood. If the wood costs $4.35 per square foot, how much will it cost to cover the floor?

Use the following geometric figure to answer questions 28 and 29.

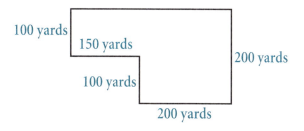

28) If fencing material costs $15.85 per yard, how much would it cost to fence in this area?

29) If it costs $8.50 per square yard to place artificial turf on this field, how much would it cost to cover the entire field?

30) A field with a length of $178\frac{2}{3}$ yards and a width of $114\frac{5}{8}$ yards is divided into three equal sections. What is the area of one section?

31) A rectangle is 43 inches long and 12 inches wide. If the length is increased by 12 inches, by how much is the area increased?

32) The radius of a circle decreased from $15\frac{3}{8}$ feet to $12\frac{1}{4}$ feet. By how much is the area of the circle decreased?

33)   If plaster costs $2 per cubic foot, how much would the plaster cost to make a cube with edges 8 feet?

34)   If it costs $2.75 per cubic inch of copper, how much would it cost to build a rectangular solid made of copper with a length of 3 inches, a width of 4 inches, and a height of 2 inches?

35)   A rectangular solid has length of 25 yards, a width of 20 yards and a height of 15 yards. A cubic solid has an edge of 20 yards. Which has greater volume? How much greater?

Chapter 9 • Test 1

1) The radius of a circle is 8 inches. What's its diameter?

2) The side of a square is 10 cm. Find its perimeter and area.

3) Find the circumference and area of a circle with a diameter of 12 feet.

Find the perimeter and area of the following figures.

4)

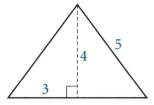

5)

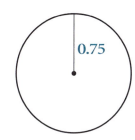

6)

$$l = 10$$

$$w = 3$$

7)

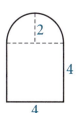

Find the volume of the following figures.

8)

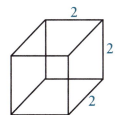

9)

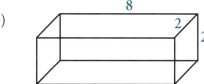

10) Find the measure of the edge of a cube with volume 27 cubic inches.

11) Find the area and the perimeter of the following triangle.

12) The length of a rectangle is 6 feet. The perimeter is 20 feet. Find the width of the rectangle.

13) Find the cost to fence in a garden which has dimensions of 20 feet by 15 feet, if the fence costs $8.50 per foot.

14) Find the cost to fill in a rectangular garden with top soil that costs $2.85 per cubic yard. The garden is 6 yards wide, 8 yards long, and 1.5 feet deep.

15) How many square tiles, each 6 inches on a side, are needed to cover a floor that is 18 feet by 10 feet?

16) The radius of a circle is 4 inches; find the area.

17) The diameter of a circle is 9.4 feet; find the area.

18) The area of a circle is 314 square yards; find the radius.

19) The circumference of a circle is 22 cm.; find the diameter.

20) Find the perimeter and area of the following geometric shape.

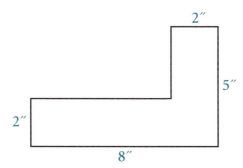

Chapter 9 • Test 2

1) The diameter of a circle is 34.8 cm; what's its radius?

2) The edge of a cube is 3 feet. Find its volume.

3) Find the circumference and area of a circle with diameter 22 feet.

Find the perimeter and area of the following figures.

4)
$s = 10$

$s = 10$

5)
$l = 4$

$w = 2$

6)
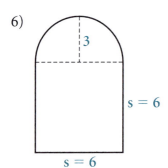

3

$s = 6$

$s = 6$

7)
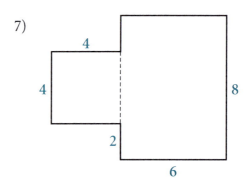

4

4

8

2

6

Chapter 9 • Test 2

Find the volume of the following figures.

8)

$l = 3$

$w = 3$

$h = 3$

9)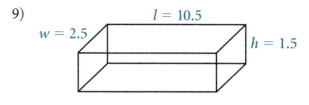

$w = 2.5$

$l = 10.5$

$h = 1.5$

10)   Find the measure of the edge of a cube with volume of 64 cubic inches.

11)   Find the area and primeter of the following triangle.

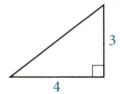

3

4

12)   The length of a rectangle is 10 inches; its area is 220. Find its width.

13)   Find the cost to fence in a circular garden with a diameter of 10 feet if the fencing costs $4.50 per foot.

Chapter 9 • Test 2

14) Find the cost to fill in a circular path 4 feet wide and 6 inches deep, with concrete. The concrete costs $5.00 per cubic foot. The path encloses a garden with a radius of 6 feet.

15) How many square tiles, each 9 inches on a side, are needed to cover a patio that is 125 feet by 45 feet?

16) Find the perimeter and area of this geometric shape.

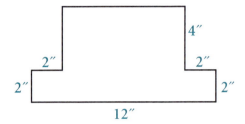

17) How much greater is the volume of a cube with an edge of 10 inches than a cube with an edge of 5 inches?

18) The circumference of a circle is 381 square meters; find the radius.

19) The area of a circle is 1,256 square feet; find the diameter.

20) The diameter of a circle is $7\frac{1}{2}$ yards; find the area.

INDEX

SOLUTIONS

Chapter 1

Solutions to Workbook 1.1

1) tens' place, 63 = 60 + 3

3) hundreds' place, 1,689 = 1,000 + 600 + 80 + 9

5) hundred thousands' place, 600,980 = 600,000 + 900 + 80

7) thousands' place, 26,897 = 20,000 + 6,000 + 800 + 90 + 7

9) ones' place, 506 = 500 + 6

11) hundreds' place, 300 = 300

13) thousands' place, 3,721 = 3,000 + 700 + 20 + 1

15) ten thousands'place, 30,721 = 30,000 + 700 + 20 + 1

17) millions'place, 3,946,125 = 3,000,000 + 900,000 + 40,000 + 6,000 + 100 + 20 + 5

19) tens'place, 930 = 900 + 30

21) sixty-three

23) six hundred three

25) six thousand, three

27) six million, six hundred sixty-three thousand, three hundred thirty

29) five hundred six

31) 823

33) 71

35) 10,300,400

Solutions to Workbook 1.2

1)	2,660	2,700	3,000
3)	350,000	350,000	350,000
5)	38,520	38,500	39,000
7)	124,800	124,800	125,000
9)	2,460,000	2,500,000	2,000,000
11)	3,900,000	3,900,000	4,000,000
13)	11,460,000	11,500,000	11,000,000
15)	9,750,000	9,800,000	10,000,000

Solutions to Workbook 1.3

1) 98
3) 78
5) 108
7) 4,672
9) 50,358
11) 81,536
13) 138,354
15) 1,809
17) 628,861

Solutions to Workbook 1.4

1) 6
3) 8
5) 13
7) 108
9) 99
11) 225
13) 94
15) 330,002
17) 808,172
19) 79,250

Solutions to Workbook 1.5

1) 644 miles
3) 88 points
5) 40,000 acres
7) 77,716 acres
9) $3,453
11) 1,165 miles
13) 107,350,000
15) 83 points
17) $87,000
19) 554

Solutions to Chapter 1 Review

1) a) ones' place, 163 = 100 + 60 + 3

 b) hundred thousands' place, 300,175 = 300,000 + 100 + 70 + 5

 c) hundreds' place, 1,394 = 1,000 + 300 + 90 + 4

 d) millions' place, 3,190,040 = 3,000,000 + 100,000 + 90,000 + 40

2) a) four hundred sixty-nine

 b) one hundred thirty-two thousand, ninety-eight

 c) forty thousand, nine hundred twenty-two

 d) one thousand, six hundred four

3) a) 4,160   4,200   4,000

 b) 82,500   82,500   83,000

 c) 1,619,240   1,619,200   1,619,000

4) a) 14,380,000   14,400,000   14,000,000

 b) 6,500,000   6,500,000   7,000,000

 c) 9,500,000   9,500,000   9,000,000

5) 79   6) 579   7) 77   8) 75   9) 1,048   10) 31,944

11) 54   12) 253   13) 3,132   14) 181   15) 1,777   16) 1,407

17) a) tens' place, 272 = 200 + 70 + 2

 b) hundreds' place, 4,719 = 4,000 + 700 + 10 + 9

 c) thousands' place, 97,460 = 90,000 + 7,000 + 400 + 60

 d) millions' place, 7,285,116 = 7,000,000 + 200,000 + 80,000 + 5,000 + 100 + 10 + 6

18) a) seven thousand, eighty-three

    b) eight hundred ninety-three

    c) four million, five hundred ten thousand, three hundred eighty-four

    d) one hundred twenty-three thousand, eighty-six

19) a) 4,380   4,400   4,000

    b) 55,500   55,500   55,000

    c) 5,700   5,700   6,000

20) a) 24,500,000   24,500,000   25,000,000

    b) 15,680,000   15,700,000   16,000,000

    c) 1,600,000   1,600,000   2,000,000

21) a) $\begin{array}{r} 50 + 6 \\ -20 - 3 \\ \hline 30 + 3 = 33 \end{array}$

    b) $\begin{array}{r} 400 + 30 + 7 \\ -200 - 10 - 4 \\ \hline 200 + 20 + 3 = 223 \end{array}$

    c) $\begin{array}{r} 5000 + 600 + 80 + 9 \\ -3000 - 200 - 50 - 3 \\ \hline 2000 + 400 + 30 + 6 = 2,436 \end{array}$

22) 81   23) 756   24) 26,646   25) 441   26) 241   27) 11,231

28) 99   29) 1,948   30) 13,878   31) $533   32) 150,000 acres

33) $35,871   34) 731 miles

Chapter 2

Solutions to Workbook 2.1

1) 100	3) 152	5) 169	**7)** 288,435
9) 563,288,999	11) 40	13) 700	15) 1,450
17) 70	19) 130	21) 800	23) 7,200
25) 36,700	27) 10,900	29) 1,299,900	31) 10,000
33) 1,000,000	35) 390	37) 3,490,000	39) 1,800
41) 27,999,600	43) 13,800	45) 200,000	

Solutions to Workbook 2.2

1) 25	3) 32	5) 8	7) 625	9) 1
11) 1	13) 36	15) 1,331	17) 100	19) 10,000

Solutions to Workbook 2.3

1) 36	3) 400	5) 900	7) 729	9) 5	11) 10
13) 9	15) 18	17) 17	19) 3	21) 5	23) 2
25) 5	27) 2	29) 1	31) 1	33) 1	

Solutions to Workbook 2.4

1) 132 $132 \times 3 = 396$ 3) 23 $23 \times 16 = 368$

5) 9 R18 $9 \times 24 = 216 + 18 = 234$ 7) 0

9) 42   42 divided by 7 = 6   11) 805 $805 \times 46 = 37,030$

13) 2004 $2,004 \times 55 = 10,220$ 15) 606 R3 $606 \times 18 + 3 = 10,908 + 3 = 10,911$

17) 51 R201 $51 \times 456 + 201 = 23,256 + 201 = 23,457$

19) 0   21) undefined

23) 0	25) 13	27) 101	29) 8	31) 1,250
33) 0	35) 1	37) 0	39) 85	41) 9
43) $67	45) 219 hours	47) $40 \div 20 = 2$	49) $80(10) = 800$	

Solutions to Workbook 2.5

1) Commutative Property of Addition

3) Commutative Property of Multiplication

5) Associative Property of Addition

7) Associative and Commutative Properties of Addition

9) Addition Property of Zero

11) Associative Property of Multiplication

13) Commutative Property of Addition

15) Commutative Property of Addition

17) Commutative and Associative Properties of Multiplication

19) Commutative Property of Addition

21) Multiplication Property of Zero

23) Multiplication Property of One

Solutions to Workbook 2.6

1) $24 + 1 = 25$

3) $4 \div 4 = 1$

5) $16 \div 1 = 16$

7) $16 \div 16 = 1$

9) $5(16) = 80$

11) $2 + 4(5 + 1) = 2 + 4(6) = 2 + 24 = 26$

13) $1 \div (4 - 4) = 1 \div 0$ is undefined.

15) $(4 - 4) \div (1) = 0 \div 1 = 0$

17) $8 \div 2 - 4 = 4 - 4 = 0$

19) $125 - 25 \div 5 = 125 - 5 = 120$

21) $12 \div 2 = 6$ 23) $7 + 1 - 3 = 5$

25) $5(81) + 16 \div 4 - 4 = 405 + 4 - 4 = 405$

27) $8^2 \div (9 - 1)^2 = 8^2 \div (8)^2 = 64 \div 64 = 1$

29) $81 + 16 \div 8 = 81 + 2 = 83$

Solutions to Workbook 2.7

1) $7 + 10 + 3 + 14 = 34$ points

3) $492(673) = \$331,116$

5) $808 + 2,166 = 2,974$ miles, $2,166 - 808 = 1,358$ miles.

7) $2,137 - 275 - 458 = \$1,404$

9) $83 - 19 = 64$

11) $15,246 \div 7 = 2,178$ VCRs

13) $601 - 382 = 219$

15) $1816 + 2003 + 1919 + 1793 = 7,531$ students

17) $13 + 19 + 12 + 8 = 52$ hours

19) $283 - 9 - 54 - 8 = \$212$

21) $19(275) = 5,225$ votes

23) $2,017 + 1,932 + 1,819 + 1,687 = 7,455$ students

25) $214 - 98 + 77 = 193$ people

27) $5,658 \div 123 = \$46$

29) $5,071 + 1,250 - 425 = \$5,896$

31) $254 \div 4 = 64$ cups

33) $16,394 - 4,199 = 12,195$

35) $36 - 14 + 19 = 41$ marbles

Solutions to Chapter 2 Review

1) 510   2) 1,296   3) 4,710   4) 15,040   5) 355,725   6) 4,220,248

7) a) 20   200   2,000

   b) 100   1,000   10,000

   c) 3,600   36,000   360,000

   d) 31,000   310,000   3,100,000

8) 63,400

9) 3,712,000

10) a) $(6)(6) = 36$

b) $(2)(2)(2)(2)(2)(2) = 64$

c) 1

d) $(3)(3)(3) = 27$

11) 128

12) 219

13) 75

14) 196 R37

15) 182 R26

16) 204

17) a) 41,600   4,160   416

b) 100,000   10,000   1,000

c) 8,200   820   82

d) 900   90   9

18) 62

19) a) Commutative Property of Multiplication

b) Associative Property of Addition

c) Commutative and Associative Properties of Multiplication

d) Multiplication Property of Zero

20) $8 + 3(5) = 8 + 15 = 23$

21) $2 + 27 = 29$

22) $20(4) = 80$

23) $(16 \div 4) = 4 - 4 = 0$

24) 34

25) 36

26) $3,974

27) $322,818

28) 241

29) 456

30) 45,570

31) 1,604,451

32) a) 640   6,400   64,000

b) 40,000   400,000   4,000,000

c) 60   600   6,000

d) 700   7,000   70,000

33) 20,640,000

34) a) $(4)(4)(4) = 64$

b) $(3)(3)(3)(3) = 81$

c) 1

d) $(10)(10)(10) = 1,000$

35) 508

36) 235 R52

37) 1,202

38) a) 3,500   350   35

b) 500   50   5

c) 434,000   43,400   4,340

39) $145,000

40) 336,400 cans

41) a) Commutative Property of Addition

b) The Addition Property of Zero

c) Associative Property of Multiplication

d) Commutative and Associative Properties of Addition

42) a) $4 + 2(4) = 4 + 8 = 12$

b) $(3)(16) - 25 = 48 - 25 = 23$

c) $(4)(4) - (3)(3) = 16 - 9 = 7$

d) $(81 - 81) \div (60 - 27) = 0 \div 33 = 0$

43) $5,122 - 1,847 = 3,275$ miles

44) $196(4,000) = 784,000$

Chapter 3

Solutions to Workbook 3.2

1) a) $\dfrac{3}{5}$ b) 3   c) 5

3) a) $\dfrac{5}{6}$ b) 5   c) 6

5) a) $\dfrac{3}{4}$ b) 3   c) 4

7) a) $\dfrac{5}{9}$ b) 5   c) 9

9) a) $\dfrac{4}{7}$ b) 4   c) 7

11) $\dfrac{4}{13}$

13) a) $\dfrac{8}{17}$ b) $\dfrac{9}{17}$

15) a) $\dfrac{12}{23}$ b) $\dfrac{9}{23}$ c) $\dfrac{2}{23}$

17) $\dfrac{5}{12}$

19) a) $\dfrac{1}{6}$ b) 1   c) 6

21) a) $\dfrac{1}{4}$ b) $\dfrac{1}{4}$ c) $\dfrac{1}{4}$ d) $\dfrac{1}{4}$

23) $\dfrac{1}{6}$

25) a) $\dfrac{9}{25}$ b) $\dfrac{2}{25}$ c) $\dfrac{6}{25}$ d) $\dfrac{8}{25}$

27) a) $\dfrac{4}{21}$ b) $\dfrac{1}{21}$

29) $\dfrac{3}{7}$

31) $\dfrac{1}{6,000,000}$

33) a) $\dfrac{300}{761}$ b) $\dfrac{461}{761}$

35) $\dfrac{4}{5}$

Solutions to Workbook 3.3A

1)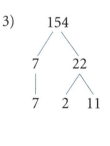

$340 = 2^2 \cdot 5 \cdot 17$

3)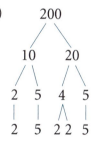

$154 = 7 \cdot 2 \cdot 11$

5)

$200 = 2^3 \cdot 5^2$

7)
```
        45
       /  \
      9    5
     / \   |
    3   3  5
```
11)
```
        99
       /  \
      9    11
     / \   |
    3   3  11
```
9)
```
      62
     /  \
    2    31
```

$$45 = 3^2 \cdot 5 \qquad 62 = 2 \cdot 31 \qquad 99 = 3^2 \cdot 11$$

13)
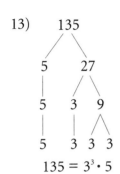
```
        135
       /   \
      5     27
      |    / \
      5   3   9
      |   |  / \
      5   3 3   3
```
$$135 = 3^3 \cdot 5$$

15)
```
      231
     /   \
    3     77
    |    / \
    3   7   11
```
$$231 = 3 \cdot 7 \cdot 11$$

17)
```
         16
        /  \
       4    4
      / \  / \
     2   2 2  2
```
$$16 = 2^4$$

19)
```
        80
       /  \
      8    10
     / \   / \
    4   2 2   5
   / \  |  |  |
  2  2  2  2  5
```
$$80 = 2^4 \cdot 5$$

21) prime   23) prime

Solutions to Workbook 3.3B

1) $16 = 2^4$ 3) $39 = 3(13)$ 5) $50 = 5^2 \cdot 2$ 7) $48 = 2^4 \cdot 3$

9) $100 = 2^2 \cdot 5^2$ 11) $153 = 3^2 \cdot 17$ 13) $36 = 3^2 \cdot 2^2$ 15) $64 = 2^6$

17) 29 is prime   19) $80 = 2^4 \cdot 5$ 21) $117 = 3^2 \cdot 13$ 23) $180 = 2^2 \cdot 3^2 \cdot 5$

25) $125 = 5^3$ 27) $147 = 7^2 \cdot 3$ 29) $300 = 2^2 \cdot 5^2 \cdot 3$ 31) $165 = 5(3)(11)$

33) prime   35) prime

Solutions to Workbook 3.4

1) $\dfrac{95}{100} = \dfrac{5(19)}{5(20)} = \dfrac{19}{20}$

3) $\dfrac{48}{144} = \dfrac{4(12)}{4(36)} = \dfrac{12(1)}{12(3)} = \dfrac{1}{3}$

5) $\dfrac{88}{99} = \dfrac{8(11)}{9(11)} = \dfrac{8}{9}$

7) $\dfrac{39}{52} = \dfrac{3(13)}{4(13)} = \dfrac{3}{4}$

9) $\dfrac{34}{102} = \dfrac{2(17)}{2(51)} = \dfrac{1(17)}{3(17)} = \dfrac{1}{3}$

11) $\dfrac{17}{49}$

13) $\dfrac{40}{50} = \dfrac{4}{5}$

15) $\dfrac{100}{240} = \dfrac{10}{24} = \dfrac{2(5)}{2(12)} = \dfrac{5}{12}$

17) $\dfrac{108}{235} = \dfrac{9(12)}{9(15)} = \dfrac{3(4)}{3(5)} = \dfrac{4}{5}$

19) $\dfrac{237}{555} = \dfrac{3(79)}{3(185)} = \dfrac{79}{185}$

21) $\dfrac{213}{300} = \dfrac{3(71)}{3(100)} = \dfrac{71}{100}$

23) $\dfrac{81}{405} = \dfrac{9(9)}{9(45)} = \dfrac{9(1)}{9(5)} = \dfrac{1}{5}$

25) $\dfrac{840}{910} = \dfrac{84}{91} = \dfrac{7(12)}{7(13)} = \dfrac{12}{13}$

27) $\dfrac{65}{286} = \dfrac{5(13)}{22(13)} = \dfrac{5}{22}$

29) $\dfrac{161}{483} = \dfrac{7(23)}{7(69)} = \dfrac{23(1)}{23(3)} = \dfrac{1}{3}$

Solutions to Workbook 3.5

1) $\dfrac{1}{4}\left(\dfrac{1}{3}\right) = \dfrac{1}{12}$

3) $\dfrac{1}{8}\left(\dfrac{3}{5}\right) = \dfrac{3}{40}$

5) $\dfrac{1}{11}\left(\dfrac{7}{9}\right) = \dfrac{7}{99}$

7) $\dfrac{9}{10}\left(\dfrac{3}{4}\right) = \dfrac{27}{40}$

9) $\dfrac{5}{6}\left(\dfrac{1}{9}\right) = \dfrac{5}{54}$

11) $\dfrac{3}{4}\left(\dfrac{1}{2}\right) = \dfrac{3}{8}$

13) $\dfrac{3}{4}\left(\dfrac{3}{10}\right) = \dfrac{9}{40}$

15) $\dfrac{1}{2}\left(\dfrac{3}{7}\right) = \dfrac{3}{14}$

17) $\dfrac{7}{8}\left(\dfrac{3}{8}\right) = \dfrac{21}{64}$

19) $\dfrac{1}{10}\left(\dfrac{3}{8}\right) = \dfrac{3}{80}$

21) $\dfrac{5}{8}\left(\dfrac{1}{2}\right) = \dfrac{5}{16}$

23) $\dfrac{1}{3}\left(\dfrac{1}{4}\right) = \dfrac{1}{12}$

Solutions to Workbook 3.6A

1) $\dfrac{5}{8}\left(\dfrac{12}{25}\right) = \dfrac{5(1)}{2(2)(2)} \times \dfrac{2(2)(3)}{5(5)} = \dfrac{3}{10}$

3) $\dfrac{4}{9}\left(\dfrac{15}{16}\right) = \dfrac{2(2)}{3(3)} \times \dfrac{3(5)}{2(2)(2)(2)} = \dfrac{5}{12}$

5) $\dfrac{14}{15}\left(\dfrac{5}{6}\right) = \dfrac{2(7)}{3(5)} \times \dfrac{5(1)}{2(3)} = \dfrac{7}{9}$

7) $\dfrac{11}{20}\left(\dfrac{4}{33}\right) = \dfrac{11(1)}{2(2)(5)} \times \dfrac{2(2)}{3(11)} = \dfrac{1}{15}$

9) $\dfrac{21}{50}\left(\dfrac{15}{28}\right) = \dfrac{3(7)}{(5)(5)(2)} \times \dfrac{3(5)}{2(2)(7)} = \dfrac{9}{40}$

11) $\dfrac{11}{12}\left(\dfrac{18}{33}\right) = \dfrac{11(1)}{2(2)(3)} \times \dfrac{3(3)(2)}{3(11)} = \dfrac{1}{2}$

13) $\dfrac{12}{21}\left(\dfrac{7}{8}\right)=\dfrac{\cancel{2}(\cancel{2})(\cancel{3})}{\cancel{3}(7)}\times\dfrac{\cancel{7}(1)}{\cancel{2}(\cancel{2})(2)}=\dfrac{1}{2}$

15) $\dfrac{27}{32}\left(\dfrac{40}{63}\right)=\dfrac{\cancel{3}(\cancel{3})(3)}{\cancel{2}(\cancel{2})(\cancel{2})(2)(2)}\times\dfrac{\cancel{2}(\cancel{2})(\cancel{2})(5)}{\cancel{3}(\cancel{3})(7)}=\dfrac{15}{28}$

17) $\dfrac{9}{16}\left(\dfrac{12}{21}\right)=\dfrac{3(3)}{\cancel{2}(\cancel{2})(2)(2)}\times\dfrac{\cancel{2}(\cancel{2})(\cancel{3})}{\cancel{3}(7)}=\dfrac{9}{28}$

19) $\dfrac{4}{15}\left(\dfrac{5}{16}\right)=\dfrac{\cancel{2}(\cancel{2})}{3(\cancel{5})}\times\dfrac{\cancel{5}(1)}{\cancel{2}(\cancel{2})(2)(2)}=\dfrac{1}{12}$

21) $\dfrac{4}{11}\left(\dfrac{3}{40}\right)=\dfrac{\cancel{2}(\cancel{2})}{11}\times\dfrac{3}{\cancel{2}(\cancel{2})(2)(5)}=\dfrac{3}{110}$

23) $\dfrac{7}{12}\left(\dfrac{18}{35}\right)=\dfrac{\cancel{7}}{2(\cancel{2})(\cancel{3})}\times\dfrac{3(\cancel{3})(\cancel{2})}{\cancel{7}(5)}=\dfrac{3}{10}$

Solutions to Workbook 3.6B

1) $\dfrac{5}{8}\left(\dfrac{4}{15}\right)=\dfrac{1}{6}$

3) $\dfrac{2}{7}\left(\dfrac{21}{26}\right)=\dfrac{3}{13}$

5) $\dfrac{9}{20}\left(\dfrac{5}{18}\right)=\dfrac{1}{8}$

7) $\dfrac{8}{15}\left(\dfrac{3}{4}\right)=\dfrac{2}{5}$

9) $\dfrac{3}{4}\left(\dfrac{8}{9}\right)=\dfrac{2}{3}$

11) $\dfrac{10}{13}\left(\dfrac{39}{50}\right)=\dfrac{3}{5}$

13) $\dfrac{1}{2}\left(\dfrac{4}{5}\right)=\dfrac{2}{5}$

15) $\dfrac{7}{8}\left(\dfrac{4}{21}\right)=\dfrac{1}{6}$

17) $\dfrac{8}{9}\left(\dfrac{27}{32}\right)=\dfrac{3}{4}$

19) $\dfrac{13}{22}\left(\dfrac{11}{52}\right)=\dfrac{1}{8}$

21) $\dfrac{5}{12}\left(\dfrac{18}{25}\right)=\dfrac{3}{10}$

23) $\dfrac{10}{21}\left(\dfrac{3}{5}\right)=\dfrac{2}{7}$

Solutions to Workbook 3.6C

1) $\dfrac{1}{2}\left(\dfrac{2}{3}\right)\dfrac{3}{4}=\dfrac{1}{4}$

3) $\dfrac{5}{9}\left(\dfrac{4}{5}\right)\left(\dfrac{3}{8}\right)=\dfrac{1}{6}$

5) $\dfrac{2}{5}\left(\dfrac{15}{24}\times\dfrac{2}{3}\right)=\dfrac{2}{5}\left(\dfrac{5}{12}\right)=\dfrac{1}{6}$

7) $\dfrac{9}{10}\left(\dfrac{5}{12}\times\dfrac{8}{9}\right)=\dfrac{9}{10}\left(\dfrac{10}{27}\right)=\dfrac{1}{3}$

9) $\dfrac{5}{6}\left(\dfrac{1}{3}\times\dfrac{9}{10}\right)=\dfrac{5}{6}\left(\dfrac{3}{10}\right)=\dfrac{1}{4}$

11) $\left(\dfrac{2}{7}\times\dfrac{3}{4}\right)\dfrac{14}{15}=\left(\dfrac{3}{14}\right)\left(\dfrac{14}{15}\right)=\dfrac{1}{5}$

13) $\dfrac{1}{4}\left(\dfrac{3}{8}\right)\left(\dfrac{\overset{3}{12}}{25}\right) = \dfrac{9}{200}$

15) $\dfrac{1}{2}\left(\dfrac{\overset{1}{3}}{5}\right)\left(\dfrac{5}{\underset{2}{6}}\right) = \dfrac{1}{4}$

17) $\dfrac{3}{4}\left(\dfrac{\overset{1}{5}}{\underset{1}{6}}\right)\left(\dfrac{\overset{1}{8}}{9}\right) = \dfrac{5}{9}$

19) $\dfrac{5}{8}\left(\dfrac{\overset{1}{3}}{\underset{2}{4}}\right)\left(\dfrac{2}{3}\right) = \dfrac{5}{8}\left(\dfrac{1}{2}\right) = \dfrac{5}{16}$

21) $\dfrac{9}{10}\left(\dfrac{5}{6}\times\dfrac{1}{4}\right) = \dfrac{\overset{3}{9}}{\underset{2}{10}}\left(\dfrac{\overset{1}{5}}{\underset{8}{24}}\right) = \dfrac{3}{16}$

23) $\left(\dfrac{5}{9}\times\dfrac{3}{4}\right)\dfrac{2}{5} = \left(\dfrac{\overset{1}{5}}{\underset{6}{12}}\right)\dfrac{\overset{1}{2}}{5} = \dfrac{1}{6}$

Solutions to Workbook 3.7A

1) $\left(\dfrac{2}{5}\right)^2 = \dfrac{4}{25}$

3) $\left(\dfrac{5}{8}\right)^2 = \dfrac{25}{64}$

5) $\dfrac{1}{4}\left(\dfrac{2}{7}\right)^2 = \dfrac{1}{\underset{2}{4}}\left(\dfrac{\overset{1}{2}}{7}\right)\left(\dfrac{\overset{1}{2}}{7}\right) = \dfrac{1}{49}$

7) $\dfrac{2}{7}\left(\dfrac{7}{10}\right)^2 = \dfrac{2}{7}\left(\dfrac{\overset{1}{7}}{\underset{5}{10}}\right)\left(\dfrac{7}{10}\right) = \dfrac{7}{50}$

9) $\left(\dfrac{5}{6}\right)^2\dfrac{1}{2} = \left(\dfrac{5}{6}\right)\left(\dfrac{5}{6}\right)\left(\dfrac{1}{2}\right) = \dfrac{25}{72}$

11) $\left(\dfrac{4}{7}\right)^2\dfrac{7}{10} = \left(\dfrac{4}{7}\right)\left(\dfrac{\overset{2}{4}}{7}\right)\left(\dfrac{\overset{1}{7}}{\underset{5}{10}}\right) = \dfrac{8}{35}$

13) $\left(\dfrac{1}{3}\right)^2 = \dfrac{1}{9}$

15) $\left(\dfrac{5}{9}\right)^2 = \dfrac{25}{81}$

17) $\dfrac{1}{2}\left(\dfrac{2}{3}\right)^2 = \dfrac{1}{\underset{1}{2}}\left(\dfrac{\overset{1}{2}}{3}\right)\left(\dfrac{2}{3}\right) = \dfrac{2}{9}$

19) $\dfrac{5}{8}\left(\dfrac{4}{9}\right)^2 = \dfrac{5}{\underset{2}{8}}\left(\dfrac{\overset{1}{4}}{9}\right)\left(\dfrac{4}{9}\right) = \dfrac{10}{81}$

21) $\dfrac{2}{9}\left(\dfrac{3}{4}\right)^2 = \dfrac{\overset{1}{2}}{\underset{3}{9}}\left(\dfrac{\overset{1}{3}}{\underset{2}{4}}\right)\left(\dfrac{3}{4}\right) = \dfrac{1}{8}$

23) $\left(\dfrac{5}{7}\right)^2\dfrac{14}{15} = \left(\dfrac{5}{7}\right)\left(\dfrac{\overset{1}{5}}{7}\right)\left(\dfrac{\overset{2}{14}}{\underset{3}{15}}\right) = \dfrac{10}{21}$

25) $\dfrac{3}{5}\left(\dfrac{4}{6}\right)^2 = \dfrac{3}{5}\left(\dfrac{2}{3}\right)^2 = \dfrac{\overset{1}{3}}{5}\left(\dfrac{4}{\underset{3}{9}}\right) = \dfrac{4}{15}$

27) $\left(\dfrac{2}{6}\right)^2\left(\dfrac{36}{40}\right) = \left(\dfrac{1}{3}\right)^2\left(\dfrac{9}{10}\right) = \left(\dfrac{1}{9}\right)\left(\dfrac{\overset{1}{9}}{10}\right) = \dfrac{1}{10}$

Solutions to Workbook 3.7B

1) $\left(\dfrac{1}{3}\right)^3 = \dfrac{1}{27}$

3) $\left(\dfrac{3}{10}\right)^3 = \dfrac{27}{1000}$

5) $\left(\dfrac{1}{6}\right)^3 = \dfrac{1}{216}$

7) $\dfrac{3}{5}\left(\dfrac{5}{6}\right)^3 = \dfrac{\overset{1}{3}}{\underset{1}{5}}\left(\dfrac{\overset{1}{5}}{\underset{2}{6}}\right)\left(\dfrac{5}{6}\right)\left(\dfrac{5}{6}\right) = \dfrac{25}{72}$

9) $\left(\dfrac{5}{6}\right)\left(\dfrac{3}{4}\right)^3 = \left(\dfrac{5}{6}\right)\left(\dfrac{\overset{1}{3}}{\underset{2}{4}}\right)\left(\dfrac{3}{4}\right)\left(\dfrac{3}{4}\right) = \dfrac{45}{128}$

11) $\left(\dfrac{5}{12}\right)^3 \dfrac{24}{25} = \left(\dfrac{\overset{1}{5}}{12}\right)\left(\dfrac{5}{12}\right)\left(\dfrac{5}{12}\right)\left(\dfrac{\overset{2\,1}{24}}{\underset{7}{35}}\right) = \dfrac{25}{504}$

13) $\left(\dfrac{1}{5}\right)^3 = \dfrac{1}{125}$

15) $\left(\dfrac{4}{5}\right)^3 = \dfrac{64}{125}$

17) $\left(\dfrac{5}{10}\right)^3 = \left(\dfrac{1}{2}\right)^3 = \dfrac{1}{8}$

19) $\dfrac{3}{4}\left(\dfrac{2}{3}\right)^3 = \dfrac{3}{4}\left(\dfrac{2}{3}\right)\left(\dfrac{2}{3}\right)\left(\dfrac{2}{3}\right) = \left(\dfrac{2}{9}\right)$

21) $\dfrac{15}{28}\left(\dfrac{2}{5}\right)^3 = \dfrac{\overset{3}{15}}{\underset{14}{28}}\left(\dfrac{\overset{1}{2}}{\underset{7}{5}}\right)\left(\dfrac{2}{5}\right)\left(\dfrac{2}{5}\right) = \dfrac{6}{175}$

23) $\left(\dfrac{3}{8}\right)^3 \dfrac{64}{81} = \left(\dfrac{3}{8}\right)\left(\dfrac{3}{8}\right)\left(\dfrac{3}{8}\right)\left(\dfrac{64}{81}\right) = \dfrac{1}{24}$

Solutions to Workbook 3.7C

1) $\dfrac{1}{2}$ 3) $\dfrac{1}{3}$ 5) $\dfrac{2}{5}$ 7) $\dfrac{5}{6}$ 9) $\dfrac{8}{11}$ 11) $\dfrac{1}{2}$

13) $\dfrac{3}{4}$ 15) $\dfrac{1}{2}$ 17) $\sqrt{\dfrac{12}{15}} = \sqrt{\dfrac{4}{25}} = \dfrac{2}{5}$ 19) $\sqrt{\dfrac{32}{98}} = \sqrt{\dfrac{16}{49}} = \dfrac{4}{7}$ 21) $\sqrt[3]{\dfrac{16}{54}} = \sqrt[3]{\dfrac{8}{27}} = \dfrac{2}{3}$

Solutions to Workbook 3.8

1) $\dfrac{2}{7} \div 3 = \dfrac{2}{7}\left(\dfrac{1}{3}\right) = \dfrac{2}{21}$

3) $\dfrac{9}{16} \div \dfrac{3}{4} = \dfrac{\overset{3}{9}}{\underset{4}{16}}\left(\dfrac{\overset{1}{4}}{\underset{1}{3}}\right) = \dfrac{3}{4}$

5) $\dfrac{5}{8} \div \dfrac{15}{16} = \dfrac{5}{8}\left(\dfrac{\overset{2}{16}}{\underset{3}{15}}\right) = \dfrac{2}{3}$

7) $\dfrac{5}{32} \div \dfrac{10}{42} = \dfrac{5}{32}\left(\dfrac{\overset{21}{42}}{\underset{2}{10}}\right) = \dfrac{21}{32}$

9) $0 \div \dfrac{11}{12} = 0$

11) $\dfrac{3}{8} \div \dfrac{9}{16} = \dfrac{\overset{1}{\cancel{3}}}{\underset{1}{\cancel{8}}}\left(\dfrac{\overset{2}{\cancel{16}}}{\underset{3}{\cancel{9}}}\right) = \dfrac{2}{3}$

13) $\dfrac{1}{4} \div 3 = \dfrac{1}{4}\left(\dfrac{1}{3}\right) = \dfrac{1}{12}$

15) $\dfrac{9}{16} \div \dfrac{2}{3} = \dfrac{9}{16}\left(\dfrac{3}{2}\right) = \dfrac{27}{32}$

17) $\dfrac{15}{20} \div 2 \div \dfrac{4}{5} = \dfrac{15}{\underset{4}{\cancel{20}}}\left(\dfrac{1}{2}\right)\left(\dfrac{\overset{1}{\cancel{5}}}{4}\right) = \dfrac{15}{32}$

19) $\dfrac{5}{15} \div \dfrac{1}{2} \div \dfrac{9}{10} = \dfrac{\overset{1}{\cancel{5}}}{\underset{3}{\cancel{15}}}\left(\dfrac{2}{1}\right)\left(\dfrac{10}{9}\right) = \dfrac{20}{27}$

21) $\dfrac{81}{100} \div \dfrac{18}{20} = \dfrac{\overset{9}{\cancel{81}}}{\underset{5}{\cancel{100}}}\left(\dfrac{\overset{1}{\cancel{20}}}{\underset{2}{\cancel{18}}}\right) = \dfrac{9}{10}$

23) $\dfrac{5}{14} \div \dfrac{25}{29} \div 2 = \dfrac{5}{14}\left(\dfrac{29}{\underset{5}{\cancel{25}}}\right)\left(\dfrac{1}{2}\right) = \dfrac{29}{140}$

Solutions to Workbook 3.9A

1) $\dfrac{3}{5}$

3) $\dfrac{6}{9} = \dfrac{2}{3}$

5) $\dfrac{5}{6}$

7) $\dfrac{25}{50} = \dfrac{1}{2}$

9) $\dfrac{6}{9} = \dfrac{2}{3}$

11) $\dfrac{1}{3}$

13) $\dfrac{10}{16} = \dfrac{5}{8}$

15) $\dfrac{8}{10} = \dfrac{4}{5}$

17) $\dfrac{10}{15} = \dfrac{2}{3}$

19) $\dfrac{7}{14} = \dfrac{1}{2}$

21) $\dfrac{15}{20} = \dfrac{3}{4}$

23) $\dfrac{8}{16} = \dfrac{1}{2}$

25) $\dfrac{9}{18} = \dfrac{1}{2}$

27) $\dfrac{7}{14} = \dfrac{1}{2}$

29) $\dfrac{4}{8} = \dfrac{1}{2}$

31) $\dfrac{20}{25} = \dfrac{4}{5}$

33) $\dfrac{25}{50} = \dfrac{1}{2}$

Solutions to Workbook 3.9B1

1) $\dfrac{2}{12} + \dfrac{1}{10} = \dfrac{1}{6} + \dfrac{1}{10} = \dfrac{5}{30} + \dfrac{3}{30} = \dfrac{8}{30} = \dfrac{4}{15}$

3) $\dfrac{6}{15} + \dfrac{7}{49} = \dfrac{2}{5} + \dfrac{1}{7} = \dfrac{14}{35} + \dfrac{5}{35} = \dfrac{19}{35}$

5) $\dfrac{2}{15} + \dfrac{1}{12} + \dfrac{4}{25} = \dfrac{40}{300} + \dfrac{25}{300} + \dfrac{48}{300} = \dfrac{113}{300}$

7) $\dfrac{2}{9} + \dfrac{3}{14} + \dfrac{1}{21} = \dfrac{28}{126} + \dfrac{27}{126} + \dfrac{6}{126} = \dfrac{61}{126}$

9) $\dfrac{5}{12} - \dfrac{1}{6} + \dfrac{1}{9} = \dfrac{60}{144} - \dfrac{9}{144} + \dfrac{16}{144} = \dfrac{67}{144}$

11) $\dfrac{19}{42} + \dfrac{13}{70} = \dfrac{19 \times 5}{42 \times 5} + \dfrac{13 \times 3}{70 \times 3} = \dfrac{95}{210} + \dfrac{39}{210} = \dfrac{134}{210} = \dfrac{67}{105}$

13) $\dfrac{3}{70} + \dfrac{5}{28} = \dfrac{3 \times 2}{70 \times 2} + \dfrac{5 \times 5}{28 \times 5} = \dfrac{6}{140} + \dfrac{25}{140} = \dfrac{31}{140}$

15) $\dfrac{13}{104} + \dfrac{13}{78} = \dfrac{13 \times 3}{104 \times 3} + \dfrac{13 \times 4}{78 \times 4} = \dfrac{39}{312} + \dfrac{52}{312} = \dfrac{91}{312} = \dfrac{7}{24}$

17) $\dfrac{1}{2} + \dfrac{1}{3} + \dfrac{1}{8} = \dfrac{1 \times 12}{2 \times 12} + \dfrac{1 \times 8}{3 \times 8} + \dfrac{1 \times 3}{8 \times 3} = \dfrac{12}{24} + \dfrac{8}{24} + \dfrac{3}{24} = \dfrac{23}{24}$

19) $\dfrac{2}{11} + \dfrac{3}{22} + \dfrac{4}{10} = \dfrac{2}{11} + \dfrac{3}{22} + \dfrac{2}{5} = \dfrac{2 \times 10}{11 \times 10} + \dfrac{3 \times 5}{22 \times 5} + \dfrac{2 \times 22}{5 \times 22} = \dfrac{20}{110} + \dfrac{15}{110} + \dfrac{44}{110} = \dfrac{79}{110}$

Solutions to Workbook 3.9B2

1) $\dfrac{1}{3} - \dfrac{1}{5} = \dfrac{5}{15} - \dfrac{3}{15} = \dfrac{2}{15}$

3) $\dfrac{4}{6} - \dfrac{1}{3} = \dfrac{2}{3} - \dfrac{1}{3} = \dfrac{1}{3}$

5) $\dfrac{6}{7} - \dfrac{3}{5} = \dfrac{30}{35} - \dfrac{21}{35} = \dfrac{9}{35}$

7) $\dfrac{4}{5} - \dfrac{3}{10} = \dfrac{8}{10} - \dfrac{3}{10} = \dfrac{5}{10} = \dfrac{1}{2}$

9) $\dfrac{5}{9} - \dfrac{1}{6} = \dfrac{10}{18} - \dfrac{3}{18} = \dfrac{7}{18}$

11) $\dfrac{4}{11} - \dfrac{2}{7} = \dfrac{28}{77} - \dfrac{22}{77} = \dfrac{6}{77}$

13) $\dfrac{9}{10} - \dfrac{2}{5} = \dfrac{9}{10} - \dfrac{4}{10} = \dfrac{5}{10} = \dfrac{1}{2}$

15) $\dfrac{1}{2} - \dfrac{5}{12} = \dfrac{6}{12} - \dfrac{5}{12} = \dfrac{1}{12}$

17) $\dfrac{2}{3} - \dfrac{2}{5} = \dfrac{10}{15} - \dfrac{6}{15} = \dfrac{4}{15}$

19) $\dfrac{9}{10} - \dfrac{3}{5} = \dfrac{9}{10} - \dfrac{6}{10} = \dfrac{3}{10}$

21) $\dfrac{5}{8} - \dfrac{1}{4} = \dfrac{5}{8} - \dfrac{2}{8} = \dfrac{3}{8}$

23) $\dfrac{12}{16} - \dfrac{1}{4} = \dfrac{3}{4} - \dfrac{1}{4} = \dfrac{2}{4} = \dfrac{1}{2}$

25) $\dfrac{9}{12} - \dfrac{1}{4} = \dfrac{3}{4} - \dfrac{1}{4} = \dfrac{2}{4} = \dfrac{1}{2}$

27) $\dfrac{3}{14} - \dfrac{1}{7} = \dfrac{3}{14} - \dfrac{2}{14} = \dfrac{1}{14}$

29) $\dfrac{7}{9} - \dfrac{1}{6} = \dfrac{14}{18} - \dfrac{3}{18} = \dfrac{11}{18}$

31) $\dfrac{4}{9} - \dfrac{3}{11} = \dfrac{44}{99} - \dfrac{27}{99} = \dfrac{17}{99}$

33) $\dfrac{9}{10} - \dfrac{1}{5} - \dfrac{1}{3} = \dfrac{9 \times 3}{10 \times 3} - \dfrac{1 \times 6}{5 \times 6} - \dfrac{1 \times 10}{3 \times 10} = \dfrac{27}{30} - \dfrac{6}{30} - \dfrac{10}{30} = \dfrac{11}{30}$

35) $\dfrac{69}{70} - \dfrac{3}{28} = \dfrac{69 \times 2}{70 \times 2} - \dfrac{3 \times 5}{28 \times 5} = \dfrac{138}{140} - \dfrac{15}{140} = \dfrac{123}{140}$

37) $\dfrac{55}{104} - \dfrac{11}{78} = \dfrac{55 \times 3}{312} - \dfrac{11 \times 4}{312} = \dfrac{165}{312} - \dfrac{44}{312} = \dfrac{121}{312}$

39) $\dfrac{40}{120} - \dfrac{3}{60} = \dfrac{20}{60} - \dfrac{3}{60} = \dfrac{17}{60}$

41) $\dfrac{27}{28} - \dfrac{7}{14} - \dfrac{2}{12} = \dfrac{27}{28} - \dfrac{1}{2} - \dfrac{1}{6} = \dfrac{27 \times 3}{28 \times 3} - \dfrac{1 \times 42}{2 \times 42} - \dfrac{1 \times 14}{6 \times 14} = \dfrac{81}{84} - \dfrac{42}{84} - \dfrac{14}{84} = \dfrac{25}{84}$

Solutions to Workbook 3.10

1) $\dfrac{3}{5} + \left(\dfrac{1}{4}\right) \div \dfrac{7}{8} = \dfrac{3}{5} + \dfrac{1}{4}\left(\dfrac{8}{7}\right) = \dfrac{3}{5} + \dfrac{2}{7} = \dfrac{21}{35} + \dfrac{10}{35} = \dfrac{31}{35}$

3) $\left(\dfrac{1}{3} + \dfrac{2}{5}\right)\left(\dfrac{7}{10} - \dfrac{2}{5}\right) = \left(\dfrac{5}{15} + \dfrac{6}{15}\right)\left(\dfrac{7}{10} - \dfrac{4}{10}\right) = \dfrac{11}{15}\left(\dfrac{3}{10}\right) = \dfrac{11}{50}$

5) $\dfrac{2}{3} - \left(\dfrac{1}{3}\right)^2 = \dfrac{2}{3} - \dfrac{1}{9} = \dfrac{6}{9} - \dfrac{1}{9} = \dfrac{5}{9}$

7) $\dfrac{5}{12} \div \left(\dfrac{3}{4}\right)^2 = \dfrac{5}{12}\left(\dfrac{16}{9}\right) = \dfrac{20}{27}$

9) $\dfrac{13}{15} - \left(\dfrac{1}{2}\right)^3 = \dfrac{13}{15} - \dfrac{1}{8} = \dfrac{104}{120} - \dfrac{15}{120} = \dfrac{89}{120}$

11) $\left(\dfrac{4}{5}\right)^2\left(\dfrac{1}{5}\right)^3 = \left(\dfrac{4}{5}\right)\left(\dfrac{4}{5}\right)\left(\dfrac{1}{5}\right)\left(\dfrac{1}{5}\right)\left(\dfrac{1}{5}\right) = \dfrac{16}{3125}$

13) $\dfrac{5}{8}\left(\dfrac{1}{4}\right) - \dfrac{1}{8} = \dfrac{5}{32} - \dfrac{1}{8} = \dfrac{5}{32} - \dfrac{4}{32} = \dfrac{1}{32}$

15) $\left(\dfrac{2}{3}\right)^2 \div \left(\dfrac{5}{6}\right)^2 = \dfrac{4}{9}\left(\dfrac{36}{25}\right) = \dfrac{16}{25}$

17) $\dfrac{1}{5} \div \left(\dfrac{1}{5} \div \dfrac{3}{5}\right) = \dfrac{1}{5} \div \left(\dfrac{1}{5}\cdot\dfrac{5}{3}\right) = \dfrac{1}{5} \div \dfrac{1}{3} = \dfrac{1}{5}\left(\dfrac{3}{1}\right) = \dfrac{3}{5}$

19) $\dfrac{\dfrac{5}{8} - \left(\dfrac{1}{4}\right)^2}{\dfrac{3}{4}} = \dfrac{\dfrac{5}{8} - \dfrac{1}{16}}{\dfrac{3}{4}} = \dfrac{\dfrac{10}{16} - \dfrac{1}{16}}{\dfrac{3}{4}} = \dfrac{\dfrac{9}{16}}{\dfrac{3}{4}} = \dfrac{9}{16}\left(\dfrac{4}{3}\right) = \dfrac{3}{4}$

21) $\dfrac{3}{10} \div \left(\dfrac{3}{5} \div \dfrac{4}{5}\right) = \dfrac{3}{10} \div \left(\dfrac{3}{5}\cdot\dfrac{5}{4}\right) = \dfrac{3}{10} \div \left(\dfrac{3}{4}\right) = \dfrac{3}{10}\left(\dfrac{4}{3}\right) = \dfrac{2}{5}$

Solutions to Workbook 3.11

1) $\dfrac{1}{4}+\dfrac{1}{5}=\dfrac{5}{20}+\dfrac{4}{20}=\dfrac{9}{20}$

3) $\dfrac{2}{3}\div 5=\dfrac{2}{3}\times\dfrac{1}{5}=\dfrac{2}{15}$

5) $\dfrac{40}{50}=\dfrac{4}{5}$

7) $\dfrac{3}{4}+\dfrac{1}{3}-\dfrac{3}{8}=\dfrac{18}{24}+\dfrac{8}{24}-\dfrac{9}{24}=\dfrac{17}{24}$ yards

9) $\dfrac{11}{12}-\dfrac{1}{3}=\dfrac{11}{12}-\dfrac{4}{12}=\dfrac{7}{12}$ acre

Solutions to Chapter 3 Review

1) $\left(\dfrac{1}{4}\right)^{3}=\dfrac{1}{4}\cdot\dfrac{1}{4}\cdot\dfrac{1}{4}=\dfrac{1}{64}$

2) $\dfrac{2}{7}+\dfrac{3}{7}=\dfrac{5}{7}$

3) $\dfrac{25}{30}=\dfrac{5}{6}$

4) $480=10(48)=(5)(2)(6)(8)=(5)(2)(2)(3)(2)(2)(2)=(2^{5})(3)(5)$

5) $\dfrac{1}{6}\div\dfrac{2}{3}=\dfrac{1}{\overset{}{\underset{2}{6}}}\times\overset{1}{\dfrac{3}{2}}=\dfrac{1}{4}$

6) $\overset{1}{\dfrac{3}{8}}\times\dfrac{5}{\underset{3}{9}}=\dfrac{5}{24}$

7) $\dfrac{1}{2}-\dfrac{1}{3}=\dfrac{1\times 3}{2\times 3}-\dfrac{1\times 2}{3\times 2}=\dfrac{3}{6}-\dfrac{2}{6}=\dfrac{1}{6}$

8) $\left(\dfrac{4}{7}\right)^{2}\dfrac{7}{12}=\dfrac{4}{7}\times\dfrac{\overset{1}{4}}{\underset{1}{7}}\times\dfrac{\overset{1}{7}}{\underset{3}{12}}=\dfrac{4}{21}$

9) (a) $\dfrac{14}{21}=\dfrac{2}{3}$ (b) $\dfrac{7}{21}=\dfrac{1}{3}$

10) (a) $\dfrac{2}{5}$ (b) 2 is the numerator; 5 is the denominator.

11) (a) $\dfrac{5}{8}$ (b) 5 is the numerator; 8 is the denominator.

12) $\dfrac{1}{4}$

13) $\dfrac{4}{24}=\dfrac{1}{6}$

14) $\dfrac{2}{15}+\dfrac{3}{5}=\dfrac{2}{15}+\dfrac{3\times 3}{5\times 3}=\dfrac{2}{15}+\dfrac{9}{15}=\dfrac{11}{15}$

15) $\dfrac{3}{8}\div\dfrac{3}{4}=\dfrac{\overset{1}{3}}{\underset{2}{8}}\times\dfrac{\overset{1}{4}}{\underset{1}{3}}=\dfrac{1}{2}$

16) $\dfrac{1}{2}+\dfrac{1}{4}+\dfrac{1}{8}=\dfrac{1\times 4}{2\times 4}+\dfrac{1\times 2}{4\times 2}+\dfrac{1}{8}=\dfrac{4}{8}+\dfrac{2}{8}+\dfrac{1}{8}=\dfrac{7}{8}$

17) $\dfrac{4}{5} - \dfrac{3}{4} = \dfrac{4\times 4}{5\times 4} - \dfrac{3\times 5}{4\times 5} = \dfrac{16}{20} - \dfrac{15}{20} = \dfrac{1}{20}$

18) $\left(\dfrac{1}{5}\right)^2 \div \left(\dfrac{3}{10}\right)^2 = \dfrac{1}{5}\times\dfrac{1}{5} \div \left(\dfrac{9}{100}\right) = \dfrac{1}{\overset{}{\underset{1}{25}}}\times\dfrac{\overset{4}{100}}{9} = \dfrac{4}{9}$

19) $\dfrac{2}{5}\times\dfrac{\overset{1}{2}}{3}\times\dfrac{1}{\underset{1}{2}} = \dfrac{2}{15}$

20) $\dfrac{1}{9}\times\dfrac{\overset{1}{3}}{\underset{1}{5}}\times\dfrac{\overset{1}{5}}{\underset{2}{6}} = \dfrac{1}{18}$

21) $\dfrac{7}{8} - \dfrac{3}{4} = \dfrac{7}{8} - \dfrac{3\times 2}{4\times 2} = \dfrac{7}{8} - \dfrac{6}{8} = \dfrac{1}{8}$

22) $\dfrac{1}{6} + \dfrac{1}{4} + \dfrac{3}{8} = \dfrac{1\times 4}{6\times 4} + \dfrac{1\times 6}{4\times 6} + \dfrac{3\times 3}{8\times 3} = \dfrac{4}{24} + \dfrac{6}{24} + \dfrac{9}{24} = \dfrac{19}{24}$

23) $\dfrac{2}{9} \div \dfrac{4}{11} = \dfrac{\overset{1}{2}}{9}\times\dfrac{11}{\underset{2}{4}} = \dfrac{11}{18}$

24) $\dfrac{9}{10} - \dfrac{3}{10} = \dfrac{6}{10} = \dfrac{3}{5}$

25) (a) $\dfrac{12}{28} = \dfrac{6}{14} = \dfrac{3}{7}$ (b) $\dfrac{16}{28} = \dfrac{8}{14} = \dfrac{4}{7}$

26) $\dfrac{1}{3}\left(\dfrac{\overset{1}{4}}{\underset{1}{5}}\right)\left(\dfrac{\overset{1}{5}}{\underset{2}{8}}\right) = \dfrac{1}{6}$ 27) $\dfrac{4}{\underset{3}{9}}\left(\dfrac{\overset{4}{12}}{13}\right) = \dfrac{16}{39}$

28)

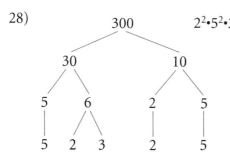

$2^2 \cdot 5^2 \cdot 3$

29) $\dfrac{1}{4}\left(\dfrac{1}{2}\right) = \dfrac{1}{8}$

30) $\dfrac{1}{16} + \dfrac{1}{8}$

$\dfrac{1}{16} + \dfrac{2}{16} = \dfrac{3}{16}$

31) $\dfrac{105}{945}$ $\qquad$ $\dfrac{5(7)(3)}{5(3)(3)(3)(7)} = \dfrac{1}{9}$

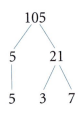

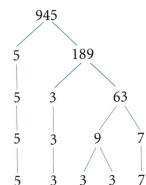

```
        105                          945
       /   \                        /   \
      5     21                     5     189
      |    / \                     |    /   \
      5   3   7                    5   3     63
                                   |   |    /  \
                                   5   3   9    7
                                       |  / \   |
                                       5 3   3  7
                                       3 3
```

32)
```
            1728
           /    \
          4      432        = 2^6 • 3^3
         / \    /   \
        2   2  4     108
        |   | / \   /   \
        2   2 2  2 2     54
        |   | |  | |    /  \
        2   2 2  2 2   9    6
                      / \  / \
                     3  3 2   3
```

$= 2^6 \cdot 3^3$

33) $\dfrac{5}{9}\left(\dfrac{2}{5}\right)^{3} = \dfrac{5}{9}\left(\dfrac{2}{5} \cdot \dfrac{2}{5} \cdot \dfrac{2}{5}\right) = \dfrac{8}{225}$

34) $\dfrac{4}{5} \times \dfrac{5}{12} = \dfrac{1}{3}$

35) $9 + 6 + 10 = 25$ coins

(a) $\dfrac{9}{25}$ $\qquad$ (b) $\dfrac{6}{25}$ $\qquad$ (c) $\dfrac{10}{25} = \dfrac{2}{5}$

36) $\dfrac{1}{4}\left(\dfrac{4}{5}\right)^{2} - \left(\dfrac{1}{5}\right)^{2} = \left(\dfrac{1}{4} \cdot \dfrac{16}{25}\right) - \dfrac{1}{25} = \dfrac{4}{25} - \dfrac{1}{25} = \dfrac{3}{25}$

37) $\left(\dfrac{1}{8}\right)^2 \left(\dfrac{2}{3}\right)^3 = \left(\dfrac{1}{8}\cdot\dfrac{1}{8}\right)\left(\dfrac{2}{3}\cdot\dfrac{2}{3}\cdot\dfrac{2}{3}\right) = \dfrac{1}{\underset{8}{64}}\cdot\dfrac{\overset{1}{8}}{27} = \dfrac{1}{216}$

38) $\dfrac{16}{25} \div \left(\dfrac{4}{5}\right)^2 = \dfrac{16}{25} \div \left(\dfrac{4}{5}\times\dfrac{4}{5}\right) = \dfrac{16}{25} \div \dfrac{16}{25} = 1$

39) $\dfrac{2}{3} - \dfrac{3}{5} + \dfrac{1}{25} = \dfrac{2\times25}{3\times25} - \dfrac{3\times15}{5\times15} + \dfrac{1\times3}{25\times3} = \dfrac{50}{75} - \dfrac{45}{75} + \dfrac{3}{75} = \dfrac{8}{75}$

40) $\dfrac{2}{3} \div \dfrac{3}{4} = \dfrac{2}{3} \times \dfrac{4}{3} = \dfrac{8}{9}$

41) $\dfrac{2}{\dfrac{1}{8} - \left(\dfrac{1}{2}\right)^3} = \dfrac{2}{\dfrac{1}{8} - \left(\dfrac{1}{2}\times\dfrac{1}{2}\times\dfrac{1}{2}\right)} = \dfrac{2}{\dfrac{1}{8} - \dfrac{1}{8}} = \dfrac{2}{0}$ is undefined

42) $\dfrac{3}{8} \div \dfrac{4}{5} \div \dfrac{9}{10} = \dfrac{3}{8}\left(\dfrac{5}{\underset{2}{4}}\right)\left(\dfrac{\overset{5}{10}}{\underset{3}{9}}\right) = \dfrac{25}{48}$

43) $\left(\dfrac{1}{3} - \dfrac{2}{9}\right) + \dfrac{2}{81} \div \left(\dfrac{3}{9}\right)^2 = \left(\dfrac{1\times3}{3\times3} - \dfrac{2}{9}\right) + \dfrac{2}{81} \div \left(\dfrac{1}{3}\right)^2 = \left(\dfrac{3}{9} - \dfrac{2}{9}\right) + \dfrac{2}{81} \div \left(\dfrac{1}{3}\times\dfrac{1}{3}\right) =$

$\dfrac{1}{9} + \dfrac{2}{81} \div \dfrac{1}{9} = \dfrac{1}{9} + \dfrac{2}{\underset{9}{81}} \times \dfrac{\overset{1}{9}}{1} = \dfrac{1}{9} + \dfrac{2}{9} \times \dfrac{1}{1} = \dfrac{1}{9} + \dfrac{2}{9} = \dfrac{3}{9} = \dfrac{1}{3}$

44) $\dfrac{1}{2}\left(\dfrac{2}{7} \div \dfrac{6}{7}\right)^2 + \dfrac{2}{3} = \dfrac{1}{2}\left(\dfrac{\overset{1}{2}}{\underset{1}{7}}\times\dfrac{\overset{1}{7}}{\underset{3}{6}}\right)^2 + \dfrac{2}{3} = \dfrac{1}{2}\left(\dfrac{1}{3}\right)^2 + \dfrac{2}{3} = \dfrac{1}{2}\left(\dfrac{1}{3}\times\dfrac{1}{3}\right) + \dfrac{2}{3} =$

$\dfrac{1}{2} \times \dfrac{1}{9} + \dfrac{2}{3} = \dfrac{1}{18} + \dfrac{2}{3} = \dfrac{1}{18} + \dfrac{2\times6}{3\times6} = \dfrac{1}{18} + \dfrac{12}{18} = \dfrac{13}{18}$

45) $\left(\dfrac{2}{3}\right)^3 \div \left(\dfrac{2}{3}\right)^2 \div \left(\dfrac{2}{3}\right) = \left(\dfrac{2}{3}\cdot\dfrac{2}{3}\cdot\dfrac{2}{3}\right) \div \left(\dfrac{2}{3}\cdot\dfrac{2}{3}\right) \div \dfrac{2}{3} = \left(\dfrac{8}{27}\right) \div \left(\dfrac{4}{9}\right) \div \dfrac{2}{3} = \dfrac{8}{27}\times\dfrac{9}{4}\times\dfrac{3}{2} = 1$

46) $\dfrac{4}{16} \div \left(\dfrac{3}{4}\right)^2 = \dfrac{1}{4} \div \dfrac{9}{16} = \dfrac{1}{\underset{1}{4}} \times \dfrac{\overset{4}{16}}{9} = \dfrac{4}{9}$

47) $\sqrt{\dfrac{100}{144}} = \dfrac{10}{12} = \dfrac{5}{6}$ 48) $\sqrt[3]{\dfrac{27}{1000}} = \dfrac{3}{10}$

49) $\sqrt{\dfrac{121}{196}} = \dfrac{11}{14}$ 50) $\sqrt[3]{\dfrac{16}{54}} = \sqrt[3]{\dfrac{8}{27}} = \dfrac{2}{3}$

Chapter 4

Solutions to Workbook 4.1

1) $\dfrac{12}{11} = 1\dfrac{1}{11}$

3) $\dfrac{17}{12} = 1\dfrac{5}{12}$

5) $\dfrac{27}{2} = 13\dfrac{1}{2}$

7) $\dfrac{33}{5} = 6\dfrac{3}{5}$

9) $\dfrac{121}{11} = 11$

11) $\dfrac{16}{14} = \dfrac{8}{7} = 1\dfrac{1}{7}$

13) $\dfrac{3}{2} = 1\dfrac{1}{2}$

15) $\dfrac{80}{8} = 10$

17) $\dfrac{71}{8} = 8\dfrac{7}{8}$

19) $2\dfrac{4}{7} = \dfrac{18}{7}$

21) $6\dfrac{1}{5} = \dfrac{31}{5}$

23) $18\dfrac{3}{12} = 18\dfrac{1}{4} = \dfrac{73}{4}$

25) $8\dfrac{11}{13} = \dfrac{115}{13}$

27) $4\dfrac{3}{5} = \dfrac{23}{5}$

29) $10\dfrac{2}{5} = \dfrac{52}{5}$

31) $3\dfrac{6}{8} = 3\dfrac{3}{4} = \dfrac{15}{4}$

33) $9\dfrac{2}{9} = \dfrac{83}{9}$

35) $21\dfrac{2}{11} = \dfrac{233}{11}$

37) $5\dfrac{3}{8} = \dfrac{43}{8}$

39) $20\dfrac{1}{2} = \dfrac{41}{2}$

Solutions to Workbook 4.2

1) $1\dfrac{1}{5}\left(2\dfrac{1}{12}\right) = \dfrac{\cancel{6}^{1}}{\cancel{5}^{1}}\left(\dfrac{\cancel{25}^{5}}{\cancel{12}^{2}}\right) = \dfrac{5}{2} = 2\dfrac{1}{2}$

3) $1\dfrac{2}{10} \div 1\dfrac{1}{2} = 1\dfrac{1}{5} \div 1\dfrac{1}{2} = \dfrac{\cancel{6}^{2}}{5}\left(\dfrac{2}{\cancel{3}^{1}}\right) = \dfrac{4}{5}$

5) $\left(1\dfrac{1}{2}\right)^{3} = \left(\dfrac{3}{2}\right)^{3} = \dfrac{27}{8} = 3\dfrac{3}{8}$

7) $1\dfrac{1}{7} \div 2\dfrac{2}{7} = \dfrac{8}{7} \div \dfrac{16}{7} = \dfrac{\cancel{8}^{1}}{\cancel{7}^{1}}\left(\dfrac{\cancel{7}^{1}}{\cancel{16}^{2}}\right) = \dfrac{1}{2}$

9) $3\dfrac{1}{2} \cdot \dfrac{5}{14} = \dfrac{\cancel{7}^{1}}{2}\left(\dfrac{5}{\cancel{14}^{2}}\right) = \dfrac{5}{4} = 1\dfrac{1}{4}$

11) $4 \div 13\dfrac{1}{3} = 4 \div \dfrac{40}{3} = \dfrac{\cancel{4}^{1}}{1}\left(\dfrac{3}{\cancel{40}^{10}}\right) = \dfrac{3}{10}$

13) $3\dfrac{3}{4} \div \left(2\dfrac{1}{2}\right)^{2} = \dfrac{15}{4} \div \left(\dfrac{5}{2}\right)^{2} = \dfrac{15}{4} \div \dfrac{25}{4} = \dfrac{\cancel{15}^{3}}{\cancel{4}^{1}}\left(\dfrac{\cancel{4}^{1}}{\cancel{25}^{5}}\right) = \dfrac{3}{5}$

15) $1\dfrac{1}{5} \div 1\dfrac{2}{7} = \dfrac{6}{5} \div \dfrac{9}{7} = \dfrac{\cancel{6}^{2}}{5}\left(\dfrac{7}{\cancel{9}^{3}}\right) = \dfrac{14}{15}$

17) $1\dfrac{1}{8}\left(3\dfrac{1}{5}\right) \div 4\dfrac{4}{5} = \dfrac{9}{8}\left(\dfrac{16}{5}\right) \div \dfrac{24}{5} = \dfrac{\cancel{9}^{3}}{\cancel{8}^{1}}\left(\dfrac{\cancel{16}^{2}}{\cancel{5}^{1}}\right)\left(\dfrac{\cancel{5}^{1}}{\cancel{24}^{3}}\right) = \dfrac{3}{4}$

19) $2\dfrac{1}{2}\left(\dfrac{1}{2}\div 2\dfrac{1}{2}\right)=\dfrac{5}{2}\left(\dfrac{1}{2}\div\dfrac{5}{2}\right)=\dfrac{5}{2}\left(\dfrac{1}{2}\cdot\dfrac{2}{5}\right)=\dfrac{1}{2}$

21) $\dfrac{3}{4}\div 1\dfrac{4}{11}=\dfrac{3}{4}\div\dfrac{15}{11}=\dfrac{3}{4}\left(\dfrac{11}{15}\right)=\dfrac{11}{20}$

23) $7\dfrac{1}{5}\cdot\dfrac{5}{9}=\dfrac{36}{5}\left(\dfrac{5}{9}\right)=\dfrac{4}{1}=4$

25) $\left[1\dfrac{11}{12}\left(1\dfrac{1}{23}\right)\right](4)=\dfrac{23}{12}\left(\dfrac{24}{23}\right)\left(\dfrac{4}{1}\right)=\dfrac{8}{1}=8$

27) $\left(2\dfrac{3}{4}\times 3\dfrac{1}{3}\right)\div\left(\dfrac{3}{4}\right)^{2}=\left(\dfrac{11}{4}\times\dfrac{10}{3}\right)\div\left(\dfrac{9}{16}\right)=\dfrac{55}{6}\left(\dfrac{16}{9}\right)=\dfrac{440}{27}=16\dfrac{8}{27}$

Solutions to Workbook 4.3

1) $3\dfrac{2}{9}+2\dfrac{1}{5}=3\dfrac{10}{45}+2\dfrac{9}{45}=5\dfrac{19}{45}$

3) $4\dfrac{4}{7}+3\dfrac{2}{3}=4\dfrac{12}{21}+3\dfrac{14}{21}=7\dfrac{26}{21}=8\dfrac{5}{21}$

5) $11\dfrac{9}{10}-4\dfrac{6}{7}=11\dfrac{63}{70}-4\dfrac{60}{70}=7\dfrac{3}{70}$

7) $10\dfrac{7}{8}+9\dfrac{1}{6}=10\dfrac{21}{24}+9\dfrac{4}{24}=19\dfrac{25}{24}=20\dfrac{1}{24}$

9) $20\dfrac{2}{15}-10\dfrac{7}{20}=20\dfrac{8}{60}-10\dfrac{21}{60}=19\dfrac{68}{60}-10\dfrac{21}{60}=9\dfrac{47}{60}$

11) $14\dfrac{3}{10}+25\dfrac{4}{25}=14\dfrac{15}{50}+25\dfrac{8}{50}=39\dfrac{23}{50}$

13) $12\dfrac{1}{2}+4\dfrac{1}{2}=16\dfrac{2}{2}=16+1=17$

15) $31\dfrac{1}{26}-13\dfrac{12}{13}=31\dfrac{1}{26}-13\dfrac{24}{26}=30\dfrac{27}{26}-13\dfrac{24}{26}=17\dfrac{3}{26}$

17) $6\dfrac{2}{5}+3\dfrac{4}{15}+2\dfrac{5}{6}=6\dfrac{12}{30}+3\dfrac{8}{30}+2\dfrac{25}{30}=11\dfrac{45}{30}=12\dfrac{15}{30}=12\dfrac{1}{2}$

19) $3\dfrac{5}{6}+4\dfrac{11}{14}+2\dfrac{11}{21}=3\dfrac{35}{42}+4\dfrac{33}{42}+2\dfrac{22}{42}=9\dfrac{90}{42}=9+2\dfrac{6}{42}=11\dfrac{1}{7}$

21) $7\dfrac{7}{10}+6\dfrac{11}{20}+3\dfrac{14}{15}=7\dfrac{42}{60}+6\dfrac{33}{60}+3\dfrac{56}{60}=16\dfrac{131}{60}=16+2\dfrac{11}{60}=18\dfrac{11}{60}$

23) $1\dfrac{1}{2}+1\dfrac{2}{3}+1\dfrac{4}{15}=1\dfrac{15}{30}+1\dfrac{20}{30}+1\dfrac{8}{30}=3\dfrac{43}{30}=3+1\dfrac{13}{30}=4\dfrac{13}{30}$

25) $12\dfrac{1}{2}+10\dfrac{1}{2}=22\dfrac{2}{2}=22+1=23$

27) $12-6\dfrac{7}{8}=11\dfrac{8}{8}-6\dfrac{7}{8}=5\dfrac{1}{8}$

29) $4\dfrac{2}{3}+1\dfrac{5}{6}-5\dfrac{1}{6}=4\dfrac{4}{6}+1\dfrac{5}{6}-5\dfrac{1}{6}=5\dfrac{9}{6}-5\dfrac{1}{6}=\dfrac{8}{6}=\dfrac{4}{3}=1\dfrac{1}{3}$

31) $15\dfrac{5}{8}-9\dfrac{1}{3}+4\dfrac{5}{6}=15\dfrac{15}{24}-9\dfrac{8}{24}+4\dfrac{20}{24}=10\dfrac{27}{24}=10+1\dfrac{3}{24}=11\dfrac{1}{8}$

Solutions to Workbook 4.4

1) $2\dfrac{2}{3}+1\dfrac{3}{4}+2\dfrac{1}{2}=2\dfrac{8}{12}+1\dfrac{9}{12}+2\dfrac{6}{12}=5\dfrac{23}{12}=5+1\dfrac{11}{12}=6\dfrac{11}{12}$ feet

3) $2\dfrac{1}{4}+3\dfrac{7}{8}+2\dfrac{1}{2}+3\dfrac{3}{4}=2\dfrac{2}{8}+3\dfrac{7}{8}+2\dfrac{4}{8}+3\dfrac{6}{8}=10\dfrac{19}{8}=10+2\dfrac{3}{8}=12\dfrac{3}{8}$ miles

5) $\dfrac{\overset{25}{\cancel{125}}}{2}\left(\dfrac{51}{\underset{1}{\cancel{5}}}\right)=\dfrac{1275}{2}=637\dfrac{1}{2}$ lbs.

7) $3\dfrac{1}{6}\times18=\dfrac{19}{\underset{1}{\cancel{6}}}\left(\dfrac{\overset{3}{\cancel{18}}}{1}\right)=57$ inches

9) $4\dfrac{1}{4}\div\dfrac{1}{4}=\dfrac{17}{\underset{1}{\cancel{4}}}\left(\dfrac{\overset{1}{\cancel{4}}}{1}\right)=17$ shirts

11) $106\dfrac{2}{3}\div5\dfrac{1}{3}=\dfrac{\overset{20}{\cancel{320}}}{\underset{1}{\cancel{3}}}\left(\dfrac{\overset{1}{\cancel{3}}}{\underset{1}{\cancel{16}}}\right)=20$ days

13) $\left(45\dfrac{1}{2}\right)\left(31\dfrac{4}{5}\right)\left(1\dfrac{1}{4}\right) = \dfrac{91}{2}\left(\dfrac{159}{\cancel{5}}\right)\left(\dfrac{\cancel{5}^{\,1}}{4}\right) = \dfrac{14,469}{8} = 1,808\dfrac{5}{8}$ cubic feet

Solutions to Workbook 4.5

1) $\dfrac{\dfrac{2}{30}}{\dfrac{4}{9}} = \dfrac{\cancel{2}^{1}}{\cancel{30}_{10}}\left(\dfrac{\cancel{9}^{3}}{\cancel{4}_{2}}\right) = \dfrac{3}{20}$

3) $\dfrac{\dfrac{3}{14}-\dfrac{1}{7}}{\dfrac{1}{5}-\dfrac{1}{10}} = \dfrac{\dfrac{3}{14}-\dfrac{2}{14}}{\dfrac{2}{10}-\dfrac{1}{10}} = \dfrac{\dfrac{1}{14}}{\dfrac{1}{10}} = \dfrac{1}{\cancel{14}_{7}}\left(\dfrac{\cancel{10}^{5}}{1}\right) = \dfrac{5}{7}$

5) $\dfrac{1\dfrac{1}{3}+\dfrac{3}{4}}{\dfrac{7}{12}+\dfrac{5}{6}} = \dfrac{\dfrac{\cancel{12}^{4}}{1}\left(\dfrac{4}{\cancel{3}_{1}}\right)+\left(\dfrac{3}{\cancel{4}_{1}}\right)\dfrac{\cancel{12}^{3}}{1}}{\dfrac{\cancel{12}^{1}}{1}\left(\dfrac{7}{\cancel{12}_{1}}\right)+\left(\dfrac{5}{\cancel{6}_{1}}\right)\dfrac{\cancel{12}^{2}}{1}} = \dfrac{4(4)+3(3)}{1(7)+5(2)} = \dfrac{16+9}{7+10} = \dfrac{25}{17} = 1\dfrac{8}{17}$

7) $\dfrac{2\dfrac{1}{5}-\dfrac{7}{15}}{\dfrac{13}{25}-\dfrac{3}{10}} = \dfrac{\dfrac{\cancel{150}^{30}}{1}\left(\dfrac{11}{\cancel{5}_{1}}\right)-\left(\dfrac{7}{\cancel{15}_{1}}\right)\dfrac{\cancel{150}^{10}}{1}}{\dfrac{\cancel{150}^{6}}{1}\left(\dfrac{13}{\cancel{25}_{1}}\right)-\left(\dfrac{3}{\cancel{10}_{1}}\right)\dfrac{\cancel{150}^{15}}{1}} = \dfrac{30(11)-7(10)}{6(13)-3(15)} = \dfrac{330-70}{78-45} = \dfrac{260}{33} = 7\dfrac{29}{33}$

9) $\dfrac{2\dfrac{4}{5}-\dfrac{2}{3}}{1\dfrac{1}{4}+1\dfrac{1}{6}} = \dfrac{\dfrac{14}{5}-\dfrac{2}{3}}{\dfrac{5}{4}+\dfrac{7}{6}} = \dfrac{\dfrac{42}{15}-\dfrac{10}{15}}{\dfrac{15}{12}+\dfrac{14}{12}} = \dfrac{\dfrac{32}{\cancel{15}_{5}}}{\dfrac{29}{12}} = \dfrac{32}{\cancel{15}_{5}}\left(\dfrac{\cancel{12}^{4}}{29}\right) = \dfrac{128}{145}$

11) $\dfrac{1\dfrac{2}{5}}{\dfrac{2}{5}} = \dfrac{7}{\cancel{5}_{1}}\left(\dfrac{\cancel{5}^{1}}{2}\right) = \dfrac{7}{2} = 3\dfrac{1}{2}$

13) $\dfrac{\dfrac{\cancel{24}^{6}}{1}\left(\dfrac{3}{\cancel{4}_{1}}\right)+\left(\dfrac{1}{\cancel{2}_{1}}\right)\dfrac{\cancel{24}^{12}}{1}}{\dfrac{\cancel{24}^{3}}{1}\left(\dfrac{7}{\cancel{8}_{1}}\right)-\left(\dfrac{5}{\cancel{12}_{1}}\right)\dfrac{\cancel{24}^{2}}{1}} = \dfrac{6(3)+1(12)}{3(7)-5(2)} = \dfrac{18+12}{21-10} = \dfrac{30}{11} = 2\dfrac{8}{11}$

15)

$$\dfrac{\dfrac{\overset{3}{\cancel{48}}}{1}\left(\dfrac{13}{16}\right)-\left(\dfrac{3}{\cancel{4}}\right)\dfrac{\overset{12}{\cancel{48}}}{1}}{\dfrac{\overset{6}{\cancel{48}}}{1}\left(\dfrac{9}{\cancel{8}}\right)+\left(\dfrac{1}{\cancel{3}}\right)\dfrac{\overset{16}{\cancel{48}}}{1}}=\dfrac{3(13)-3(12)}{6(9)+1(16)}=\dfrac{39-36}{54+16}=\dfrac{3}{70}$$

17)

$$\dfrac{\dfrac{\overset{12}{\cancel{36}}}{1}\left(\dfrac{2}{\cancel{3}}\right)+\left(\dfrac{3}{\cancel{4}}\right)\dfrac{\overset{9}{\cancel{36}}}{1}}{\dfrac{\overset{4}{\cancel{36}}}{1}\left(\dfrac{2}{\cancel{9}}\right)+\left(\dfrac{1}{\cancel{6}}\right)\dfrac{\overset{6}{\cancel{36}}}{1}}=\dfrac{12(2)+3(9)}{4(2)-1(6)}=\dfrac{24+27}{8-6}=\dfrac{51}{2}=25\dfrac{1}{2}$$

19)

$$\dfrac{1\dfrac{1}{5}+\dfrac{2}{3}}{1\dfrac{1}{9}+1\dfrac{1}{15}}=\dfrac{\dfrac{\overset{9}{\cancel{45}}}{1}\left(\dfrac{6}{\cancel{5}}\right)+\left(\dfrac{2}{\cancel{3}}\right)\dfrac{\overset{15}{\cancel{45}}}{1}}{\dfrac{\overset{5}{\cancel{45}}}{1}\left(\dfrac{10}{\cancel{9}}\right)+\left(\dfrac{16}{\cancel{15}}\right)\dfrac{\overset{3}{\cancel{45}}}{1}}=\dfrac{9(6)+2(15)}{5(10)+16(3)}=\dfrac{54+30}{50+48}=\dfrac{84}{98}=\dfrac{6}{7}$$

Solutions to Workbook 4.6

1) $4+\left(7\dfrac{2}{3}\right)\left(2\dfrac{1}{4}\right)=4+\dfrac{23}{\cancel{3}}\left(\dfrac{\overset{3}{\cancel{9}}}{4}\right)=4+\dfrac{69}{4}=4+17\dfrac{1}{4}=21\dfrac{1}{4}$

3) $\left(\dfrac{1}{8}+\dfrac{2}{5}\right)\left(7\dfrac{1}{4}+2\dfrac{1}{6}\right)=\left(\dfrac{5}{40}+\dfrac{16}{40}\right)\left(7\dfrac{3}{12}+2\dfrac{2}{12}\right)=\left(\dfrac{21}{40}\right)\left(9\dfrac{5}{12}\right)=\left(\dfrac{21}{40}\right)\left(\dfrac{\overset{7}{\cancel{113}}}{\underset{4}{\cancel{12}}}\right)=\dfrac{791}{160}=4\dfrac{151}{160}$

5) $\left(\dfrac{5}{12}+\dfrac{1}{4}\right)\left(1\dfrac{4}{5}+1\dfrac{5}{8}\right)=\left(\dfrac{5}{12}+\dfrac{3}{12}\right)\left(1\dfrac{32}{40}+1\dfrac{25}{40}\right)=\left(\dfrac{\overset{2}{\cancel{8}}}{\underset{3}{\cancel{12}}}\right)\left(2\dfrac{57}{40}\right)=\dfrac{2}{3}\left(\dfrac{\overset{1}{\cancel{137}}}{\underset{20}{\cancel{40}}}\right)=\dfrac{137}{60}=2\dfrac{17}{60}$

7) $\dfrac{8}{9}+\dfrac{5}{64}\left(9\dfrac{3}{7}-1\dfrac{3}{7}\right)^{2}=\dfrac{8}{9}+\dfrac{5}{64}(8)^{2}=\dfrac{8}{9}+\dfrac{5}{\cancel{64}}\left(\dfrac{\overset{1}{\cancel{64}}}{1}\right)=\dfrac{8}{9}+5=5\dfrac{8}{9}$

9) $39 - \left(4\frac{4}{5}\right)\left(3\frac{3}{4}\right) = 39 - \left(\frac{\overset{6}{\cancel{24}}}{\cancel{5}_{1}}\right)\left(\frac{\overset{3}{\cancel{15}}}{\cancel{4}_{1}}\right) = 39 - 18 = 21$

11) $\left(\frac{4}{5} - \frac{2}{3}\right)\left(9\frac{4}{9} - 7\frac{1}{3}\right) = \left(\frac{12}{15} - \frac{10}{15}\right)\left(9\frac{4}{9} - 7\frac{3}{9}\right) = \left(\frac{2}{15}\right)\left(2\frac{1}{9}\right) = \left(\frac{2}{15}\right)\left(\frac{19}{9}\right) = \frac{38}{135}$

13) $\frac{4}{5} + \frac{2}{3}\left(4\frac{3}{4} + 3\frac{1}{4}\right)^{2} = \frac{4}{5} + \frac{2}{3}\left(7\frac{4}{4}\right)^{2} = \frac{4}{5} + \frac{2}{3}(8)^{2} = \frac{4}{5} + \frac{2}{3}\left(\frac{64}{1}\right) = \frac{4}{5} + \frac{128}{3} = \frac{4}{5} + 42\frac{2}{3} = \frac{12}{15} + 42\frac{10}{15}$

$= 42\frac{22}{15} = 42 + 1\frac{7}{15} = 43\frac{7}{15}$

15) $\left(\frac{7}{15} + \frac{2}{15}\right) \div \left(1\frac{2}{3} - \frac{4}{5}\right) = \left(\frac{7}{15} + \frac{6}{15}\right) \div \left(1\frac{10}{15} - \frac{12}{15}\right) = \left(\frac{13}{15}\right) \div \left(\frac{13}{15}\right) = 1$

17) $\frac{1}{16} + \left\{\left[14 - \left(2\frac{1}{2} - \frac{1}{2}\right)^{2}\right] \div 20\right\}^{4} = \frac{1}{16} + \left\{\left[14 - (2)^{2}\right] \div 20\right\}^{4}$

$= \frac{1}{16} + \left\{[14 - 4] \div 20\right\}^{4} = \frac{1}{16} + \left\{[10] \div 20\right\}^{4} = \frac{1}{16} + \left\{\frac{10}{20}\right\}^{4} = \frac{1}{16} + \left\{\frac{1}{2}\right\}^{4} = \frac{1}{16} + \frac{1}{16} = \frac{2}{16} = \frac{1}{8}$

19) $\left(4\frac{1}{3} - 2\frac{3}{9}\right) \div \left[\left(\frac{1}{4}\right)^{2} - \left(\frac{1}{2}\right)^{4}\right] = \left(4\frac{1}{3} - 2\frac{1}{3}\right) \div \left[\frac{1}{16} - \frac{1}{16}\right] = 2 \div 0$ is undefined

Solutions to Chapter 4 Review

1) $2\frac{1}{2} \div 3\frac{2}{3} = \frac{5}{2} \div \frac{11}{3} = \frac{5}{2} \times \frac{3}{11} = \frac{15}{22}$

2) $4\frac{1}{5} \div 0$ is undefined.

3) $0 \div 4\frac{1}{5} = 0$

4) $\left(1\frac{1}{2}\right)^{3} = \left(\frac{3}{2}\right)^{3} = \frac{27}{8} = 3\frac{3}{8}$

5) $1\frac{2}{3} + 2\frac{1}{8} = 1\frac{16}{24} + 2\frac{3}{24} = 3\frac{19}{24}$

6) $\frac{7}{3} = 2\frac{1}{3}$

7) $4\frac{3}{5} = \frac{23}{5}$

8) $12 \times 4\frac{3}{4} = \frac{\overset{3}{\cancel{12}}}{1} \times \frac{19}{\cancel{4}_{1}} = 57$ feet

9) $\dfrac{\frac{1}{3} + \frac{3}{4}}{\frac{5}{6} + \frac{1}{2}} = \dfrac{\frac{4}{12} + \frac{9}{12}}{\frac{10}{12} + \frac{6}{12}} = \dfrac{\frac{13}{12}}{\frac{16}{12}} = \frac{13}{\cancel{12}} \times \frac{\cancel{12}^{1}}{16} = \frac{13}{16}$

10) $\dfrac{1\frac{7}{8} + \frac{7}{12}}{1\frac{5}{6} - \frac{3}{8}} = \dfrac{\frac{\overset{3}{24}}{1}\left(\frac{15}{\underset{1}{8}}\right) + \left(\frac{7}{\underset{1}{12}}\right)\frac{\overset{2}{24}}{1}}{\frac{\overset{4}{24}}{1}\left(\frac{11}{\underset{1}{6}}\right) - \left(\frac{3}{\underset{1}{8}}\right)\frac{\overset{3}{24}}{1}} = \dfrac{45 + 14}{44 - 9} = \dfrac{59}{35} = 1\frac{24}{35}$

11) $\begin{aligned} 5\tfrac{3}{5} &= 5\tfrac{12}{20} \\ -2\tfrac{1}{4} &= 2\tfrac{5}{20} \\ \hline & \ \ 3\tfrac{7}{20} \end{aligned}$

12) $\left(1\frac{2}{7}\right)^2 = \left(\frac{9}{7}\right)^2 = \frac{81}{49} = 1\frac{32}{49}$

13) $1\frac{1}{3} \times 2\frac{2}{5} = \frac{4}{\underset{1}{3}} \times \frac{\overset{4}{12}}{5} = \frac{16}{5} = 3\frac{1}{5}$

14) $4\frac{4}{5} = \frac{24}{5}$

15) $3\frac{4}{7}\left(9\frac{4}{5}\right) = \frac{\overset{5}{25}}{\underset{1}{7}} \bullet \frac{\overset{7}{49}}{\underset{1}{5}} = 35$

16) $\dfrac{1\frac{1}{8} + 2\frac{2}{3}}{2\frac{3}{4} + 3\frac{1}{6}} = \dfrac{\frac{9}{8} + \frac{8}{3}}{\frac{11}{4} + \frac{19}{6}} = \dfrac{\frac{\overset{3}{24}}{1}\left(\frac{9}{\underset{1}{8}}\right) + \left(\frac{8}{\underset{1}{3}}\right)\frac{\overset{8}{24}}{1}}{\frac{\overset{6}{24}}{1}\left(\frac{11}{\underset{1}{4}}\right) + \left(\frac{19}{\underset{1}{6}}\right)\frac{\overset{4}{24}}{1}} = \dfrac{27 + 64}{66 + 76} = \dfrac{91}{142}$

17) $\frac{3}{\underset{1}{4}} \times \frac{\overset{12}{48}}{1} = 36$ people.

18) $\begin{aligned} 3\tfrac{1}{2} &= 3\tfrac{3}{6} \\ 4\tfrac{1}{3} &= 4\tfrac{2}{6} \\ 2\tfrac{2}{3} &= 2\tfrac{4}{6} \\ \hline 9 \quad & 9\tfrac{9}{6} = 9 + 1\tfrac{3}{6} = 10\tfrac{1}{2} \end{aligned}$ miles.

19) $\frac{12}{5} = 2\frac{2}{5}$

20) $2\frac{3}{4} \div 3\frac{5}{8} = \frac{11}{4} \div \frac{29}{8} = \frac{11}{\underset{1}{4}} \times \frac{\overset{2}{8}}{29} = \frac{22}{29}$

21) $\left(1\frac{2}{3} - \frac{1}{6}\right)^2 \div \left(\frac{1}{3}\right)^2 = \left(1\frac{4}{6} - \frac{1}{6}\right)^2 \div \frac{1}{9} = \left(1\frac{1}{2}\right)^2 \div \frac{1}{9} = \left(\frac{3}{2}\right)^2 \div \frac{1}{9} = \frac{9}{4} \div \frac{1}{9}$
$= \frac{9}{4} \times \frac{9}{1} = \frac{81}{4} = 20\frac{1}{4}$

22) $3\frac{1}{2} + 2\frac{3}{4} + 1\frac{3}{5} = 6 + \frac{1}{2} + \frac{3}{4} + \frac{3}{5} = 6 + \frac{10}{20} + \frac{15}{20} + \frac{12}{20} = 6\frac{37}{20}$
$= 6 + 1\frac{17}{20} = 7\frac{17}{20}$

23) $\frac{12}{7} = 1\frac{5}{7}$

24) $4\frac{2}{5} \times 3\frac{3}{4} = \frac{\overset{11}{22}}{\underset{1}{5}} \times \frac{\overset{3}{15}}{\underset{2}{4}} = \frac{33}{2} = 16\frac{1}{2}$

25) $5\frac{7}{8} \div 2\frac{3}{10} = \frac{47}{8} \div \frac{23}{10} = \frac{47}{\underset{4}{8}} \times \frac{\overset{5}{10}}{23} = \frac{235}{92} = 2\frac{51}{92}$

26) $\left(2\frac{1}{4} - \frac{5}{6}\right) \times \left(\frac{3}{4} - \frac{1}{2}\right)^2 = \left(\frac{9}{4} - \frac{5}{6}\right) \times \left(\frac{3}{4} - \frac{2}{4}\right)^2 = \left(\frac{27}{12} - \frac{10}{12}\right) \times \left(\frac{1}{4}\right)^2$
$= \left(\frac{17}{12}\right) \times \left(\frac{1}{16}\right) = \frac{17}{192}$

27) $\dfrac{\frac{4}{5}+\frac{3}{10}}{6-\frac{8}{15}} = \dfrac{\frac{8}{10}+\frac{3}{10}}{5\frac{7}{15}} = \dfrac{\frac{11}{10}}{\frac{82}{15}} = \frac{11}{10} \times \frac{15}{82} = \frac{33}{164}$

28) $10\frac{5}{8} = \frac{85}{8}$

29)
$$12\frac{1}{4} = 12\frac{3}{12} = 11\frac{15}{12}$$
$$-5\frac{5}{6} = 5\frac{10}{12} = 5\frac{10}{12}$$
$$\overline{\phantom{-5\frac{5}{6} = 5\frac{10}{12} =}\;6\frac{5}{12}}$$

30) $\frac{2}{3} + \frac{1}{5}\left(2\frac{1}{3}+1\frac{2}{3}\right)^2 = \frac{2}{3}+\frac{1}{5}(4)^2 = \frac{2}{3}+\frac{1}{5}(16) = \frac{2}{3}+\frac{16}{5} = \frac{2}{3}+3\frac{1}{5}$
$= \frac{10}{15}+3\frac{3}{15} = 3\frac{13}{15}$

31) $2\frac{7}{12}+4\frac{1}{3} = $
$$2\frac{7}{12} = 2\frac{7}{12}$$
$$\underline{4\frac{1}{3} = 4\frac{4}{12}}$$
$$6\frac{11}{12}$$

32) $16 \div 2\frac{2}{3} = \frac{16}{1} \div \frac{8}{3} = \frac{16}{1} \times \frac{3}{8} = 6$

Chapter 5

Solutions to Workbook 5.1

1) hundredths' place; $0.4689 = \frac{4}{10} + \frac{6}{100} + \frac{8}{1,000} + \frac{9}{10,000}$.

Four thousand, six hundred eighty-nine ten thousandths

3) tenths' place; $4.689 = 4 + \frac{6}{100} + \frac{8}{100} + \frac{9}{1,000}$.

Four and six hundred eighty-nine thousandths

5) thousandths' place; $0.04689 = \frac{4}{100} + \frac{6}{1,000} + \frac{8}{10,000} + \frac{9}{100,000}$.

Four thousand, six hundred eighty-nine hundred thousandths

7) ones' place; $46.89 = 40 + 6 + \frac{8}{10} + \frac{9}{100}$.

Forty-six and eighty-nine hundredths

9) tenths' place; $0.6 = \frac{6}{10}$.

Six tenths

11) ten thousandths' place; $0.0006 = \frac{6}{10,000}$.

Six ten thousandths

13) hundreds' place; $652.002 = 600 + 50 + 2 + \frac{2}{1,000}$. 15) 0.03

six hundred fifty-two and two thousandths

17) 0.003 19) 0.908

six hundred fifty-two and two thousandths

21) 0.0333   23) 0.246, 2.0988, 2.47

25) 0.67, 0.087, 0.00921   27) 0.001, 0.01, 0.1, 1.1

Solutions to Workbook 5.2

1) 4.7   3) 0.00079   5) 10

7)

Number	Tenth	Hundredth	Thousandth
14.491	14.5	14.49	14.490
1.9746	2.0	1.97	1.975
0.48888	0.5	0.49	0.489
0.1209	0.1	0.12	0.121
0.09090909	0.1	0.09	0.091

9) a) 1.0   b) 0.9   c) 5.6   d) 0.8   e) 0.1

11) a) 9.910   b) 99.099   c) 990.990   d) 9909.900

13) a) 248.09926   b) 99.0098   c) 12,093.7896   d) 88.95999

15) a) $200   b) $209   c) $209.1   d) $209.09

Solutions to Workbook 5.3

1) 138.568   3) 713.471   5) 0.00882   7) 7.5969

9) 6.8931   11) 5.08   13) 87.12   15) 0.2061

17) 0.2119   19) 0.02079   21) 22.0034   23) 13.741

25) 2.6709   27) 45.111

Solutions to Workbook 5.4

1) 0.315   3) 4.5576   5) 3.171

7) 3.2175   9) 15.37   11) 909.6078   13) 20.361

15) $(1.1)(2.1)(3.01) = 2.31(3.01) = 6.9531$ 17) 28,993.5552

19) 8,827.6725

21) a) 4.56   b) 45.6   c) 456   d) 4,560

23) a) 345   b) 34.5   c) 3.45   d) 0.345   e) 0.0345   f) 0.00345

25) 0.09   27) 0.0009   29) 1.21   31) 4.1616

33) 0.01   35) 0.001   37) 0.0001   39) 0.0001

41) 0.000001   43) 1   45) 0.0000000001   47) 0.000000000001

Solutions to Workbook 5.5

1) 2.5   3) 21   5) 45   7) 6.2   9) 120

11) 0.6   13) 6   15) 7.108   17) 2.273   19) 38.462

21) 2.704   23) 124.183   25) 0.423   27) $8\overline{)1} = 0.125$ 29) $27\overline{)8} = 0.3$

31) a) 182.9034   b) 18.29034   c) 1.829034   d) 0.1829034

33) $(0.2 + 0.003) \div (0.33 - 0.001) = 0.62$

35) $145,678.91 - 35,484.88 + 16,580.47 - 212(500) = 20,774.53$

Solutions to Workbook 5.6

1) 6.0   3) 0.007   5) 0.09   7) 0.2   9) 0.8

11) 0.02   13) 0.5   15) .01   17) 0.7   19) 0.9

Solutions to Workbook 5.7

1) $4.56 + 3 = 7.56$

3) $0.04 \div 0.001 = 40$

5) $4.1 + 5.29 = 9.39$

7) $0.5[0.04 - 0.04] = 0.5[0] = 0$

9) $6.03 - 1.1 = 4.93$

11) $0.36 + 1.69 = 2.05$

13) $0.09 - 0.09 = 0$

15) $2.0625 - [0.16 \div 0.16] = 2.0625 - 1 = 1.0625$

17) $(0.299) + (.8)^2 = 0.299 + 0.64 = 0.939$

19) $2.1 + 0.3(0.6 - 0.06) = 2.1 + 0.3(0.54) = 2.1 + 0.162 = 2.262$

21) $(50)^2 + (2.1)^2 = 2,500 + 9.261 = 2,509.261$

23) $0.34[0.81 \div (.0081)]^2 = 0.34[100]^2 = 0.34[10,000] = 3,400$

Solutions to Workbook 5.8

1) 112.7

3) $3,116.98 + 4,908.22 - 3,792.11 = 4,233.09$

5) 115.736

7) 23.4

9) 23.71

11) $4.67(516) = 2,409.72$

13) $23.52(584) = 13,735.68$

15) $22.07(13.243) = 292.273$

17) $6.238(10.35) = 64.56$

19) $4.57(99.25) = \$471.44$

21) $\$313.63$

23) $70.194 - 64.632 = 5.562$

25) $125(9.29) = \$1161.25$

27) $1.4176(94.75) = 134.318$

29) $76 + 82 + 90 + 45 + 96 = 389$, $389 \div 5 = 77.8$ Tommy's average is 78.

   $45 + 66 + 88 + 90 + 100 = 389$, $389 \div 5 = 77.8$ Jeannie's average is the same as Tommy's average.

Solutions to Chapter 5 Review

1) $16,425.2$ 2) $129,878.99$ 3) one hundred nineteen and ninety-five hundredths

4) $14,567.9$

5) a) 5.3 b) 0.5 c) 0.1

6) a) 8 b) 0.8 c) 0.08

7) a) thousandths' place, two and three hundred forty-five thousandths.

$$2.345 = 2 + \frac{3}{10} + \frac{4}{100} + \frac{5}{1,000}$$

b) ones' place, five and nine hundred two thousandths, $5.902 = 5 + \frac{9}{10} + \frac{2}{1,000}$

c) tenths' place five hundred ninety-two thousandths, $0.592 = \frac{5}{10} + \frac{9}{100} + \frac{2}{1,000}$

d) tens' place, four hundred fifty and two hundredths, $450.02 = 400 + 50 + \frac{2}{100}$

8) a) 20   b) 460   c) 200

9) 340.889   10) 53,849.241   11) 171.9   12) 0.316

13)

Number	10	100	1,000	10,000
0.3	3	30	300	3,000
0.0003	0.003	0.03	0.3	3
3.33	33.3	333	3,330	33,300
0.303	3.03	30.3	303	3,030

14)

Number	10	100	1,000	10,000
0.3	0.03	0.003	0.0003	0.00003
300	30	3	0.3	0.03
3.33	0.333	0.0333	0.00333	0.000333
0.303	0.0303	0.00303	0.000303	0.0000303

15) 5.59   16) 12.4   17) $20.40   18) $1,521,003.06

19) $3,046.1005 = 3,000 + 40 + 6 + \dfrac{1}{10} + \dfrac{5}{10,000}$

20) a) ones' place   b) tenths' place
    c) thousandths' place   d) tens' place

21) 0.998 22) 67.4795 23) 113.7

24) $5\overline{)2} = 0.4$ 25) $2\overline{)5} = 2.5$ 26) 0.362

Chapter 6

Solutions to Workbook 6.1

1) 0.8 Terminating

3) 0.22 Terminating

5) 0 Terminating

7) 0.0077 Terminating

9) 5.9 Terminating

11) $0.\overline{3}$ Repeating

13) 6.25 Terminating

15) $1.\overline{2}$ Repeating

17) $0.5\overline{3}$ Repeating

19) 7 Terminating

21) $\dfrac{8}{10} = \dfrac{4}{5}$

23) $\dfrac{8}{1,000} = \dfrac{1}{125}$

25) $13\dfrac{125}{1,000} = 13\dfrac{1}{8}$

27) $3\dfrac{33}{100}$

29) $\dfrac{875}{10,000} = \dfrac{7}{80}$

31) $\dfrac{77}{100}$

33) $1\dfrac{67}{100}$

35) $66\dfrac{6}{10} = 66\dfrac{3}{5}$

37) $6\dfrac{2}{3}$

39) $0.8 + \dfrac{1}{5} = 0.8 + 0.2 = 1$ or $\dfrac{8}{10} + \dfrac{2}{10} = \dfrac{10}{10} = 1$

41) $40.5 + \dfrac{2}{3} = 40\dfrac{1}{2} + \dfrac{2}{3} = 40\dfrac{3}{6} + \dfrac{4}{6} = 40\dfrac{7}{6} = 40 + 1\dfrac{1}{6} = 41\dfrac{1}{6}$

43) $10.5 + 1\dfrac{1}{6} = 10\dfrac{1}{2} + 1\dfrac{1}{6} = 10\dfrac{3}{6} + 1\dfrac{1}{6} = 11\dfrac{4}{6} = 11\dfrac{2}{3}$

45) $(.3)^2 + \left(\dfrac{4}{5}\right)^2 = (.3)^2 + (.8)^2 = .09 + .64 = .73$

OR $\left(\dfrac{3}{10}\right)^2 + \left(\dfrac{4}{5}\right)^2 = \dfrac{9}{100} + \dfrac{16}{25} = \dfrac{9}{100} + \dfrac{64}{100} = \dfrac{73}{100}$ 47) $0.1 \div \dfrac{1}{9} = \dfrac{1}{10}\left(\dfrac{9}{1}\right) = \dfrac{9}{10}$

49) $\dfrac{2}{5} + (.1)^2 \div \left(\dfrac{1}{2}\right) = .4 + (.1)^2 \div .5 = .4 + .01 \div .5 = .4 + .02 = .42$ OR

$\left(\dfrac{2}{5}\right) + \left(\dfrac{1}{10}\right)^2 \div \dfrac{1}{2} = \dfrac{2}{5} + \left(\dfrac{1}{100}\right)\left(\dfrac{2}{1}\right) = \dfrac{2}{5} + \dfrac{1}{50} = \dfrac{20}{50} + \dfrac{1}{50} = \dfrac{21}{50}$

51) $12\dfrac{4}{5} \div 5\dfrac{3}{25} + 1.5 = 12.8 \div 5.12 + 1.5 = 2.5 + 1.5 = 4$

OR $\dfrac{64}{5}\left(\dfrac{25}{128}\right) + 1\dfrac{1}{2} = \dfrac{5}{2} + \dfrac{3}{2} = \dfrac{8}{2} = 4$

Solutions to Workbook 6.2

1) a) $\dfrac{50}{20} = \dfrac{5}{2}$, 5 wins to 2 losses

   b) $\dfrac{20}{50} = \dfrac{2}{5}$, 2 losses to 5 wins

   c) $\dfrac{50}{70} = \dfrac{5}{7}$, 5 wins to 7 games

3) a) $\dfrac{12}{48} = \dfrac{1}{4}$, 1 welder to 4 riveters

   b) $\dfrac{48}{12} = \dfrac{4}{1}$, 4 riveters to 1 welder

   c) $\dfrac{12}{60} = \dfrac{1}{5}$, 1 welder to 5 employees

5) $\dfrac{450}{9} = \dfrac{50}{1}$, 50 miles per hour

7) $\dfrac{120}{6} = \dfrac{20}{1}$, 20 scouts to 1 scout master

9) a) $\dfrac{80}{16} = \dfrac{5}{1}$, 5 days worked to 1 day sick

   b) $\dfrac{16}{80} = \dfrac{1}{5}$, 1 day sick to 5 days at work

11) a) $\dfrac{70}{30} = \dfrac{7}{3}$, 7 new to 3 used

   b) $\dfrac{30}{70} = \dfrac{3}{7}$, 3 used to 7 new

   c) $\dfrac{70}{100} = \dfrac{7}{10}$, 7 new to 10 cars

13) a) $\dfrac{10}{110} = \dfrac{1}{11}$, 1 officer to 11 enlisted

   b) $\dfrac{110}{10} = \dfrac{11}{1}$, 11 enlisted to 1 officer

   c) $\dfrac{10}{120} = \dfrac{1}{12}$, 1 officer to 12 soldiers

15) $\dfrac{2}{22} = \dfrac{1}{11}$, 1 hour to 11 lab reports or 11 lab reports per hour

17) $\dfrac{10}{320} = \dfrac{1}{32}$, 1 chapter every 32 days or $\dfrac{320}{10} = \dfrac{32}{1}$, 32 days per chapter

Solutions to Workbook 6.3

1) 5 feet 9 inches = 5(12) + 9 = 60 + 9 = 69 inches

3) $\dfrac{48}{12} = 4$ feet

5) $\dfrac{45}{36} = 1\dfrac{9}{36} = 1\dfrac{1}{4}$ yards

7) 5,280 − 10 = 5,270 feet

9) 2(5,280) + 100 = 10,560 + 100 = 10,660 feet

11) $\dfrac{25}{2,000} = .0125$ tons

13) $5(12) + 2 = 62$ inches, $\dfrac{62}{36} = \dfrac{31}{18} = 1\dfrac{13}{18}$ yards

15) $\dfrac{3}{4}\left(\dfrac{5,280}{1}\right) + 5 = 3,960 + 5 = 3,965$ feet

17) $12 + \dfrac{1}{2}\left(\dfrac{2000}{1}\right) = 12 + 1,000 = 1,012$ pounds

Solutions to Workbook 6.4

1) $1.99 \div 16 = \$.12$ per ounce

3) $29.50 \div 9.2 = \$3.21$ per yard

5) $16.39 \div 138 = \$.12$ per ounce

7) $31.25 \div 8.3m = \$3.77$ per yard

9) a) $1(16) + 4 = 20$, $1.69 \div 20 = \$.08$ per ounce

   b) 4 ounces $= \dfrac{4}{16} = \dfrac{1}{4} = .25$ pounds, $1.69 \div 1.25 = \$1.35$ per pound

11) $3(3) + 1 = 10$ feet. $10 \div 10 = \$1$ per foot

13) $8(16) + 7 = 128 + 7 = 135$ ounces, $596 \div 135 = \$4.41$ per ounce

15) a) $\dfrac{4}{16} = \dfrac{1}{4} = .25$ pounds, $5.99 \div 2.25 = \$2.66$ per pound

    b) $2(16) + 4 = 32 + 4 = 36$ ounces, $5.99 \div 36 = \$.17$ per ounce

17) $1.39 \div 12 = \$.12$ per ounce, $.99 \div 48 = \$.02$ per ounce. The 48 ounce can is the better buy.

19) $6.99 \div 24 = \$.29$ per exposure, $9.99 \div 36 = \$.28$ per exposure. The roll with 36 exposures is the better buy.

Solutions to Workbook 6.5

1) $a = \dfrac{28}{14} = 2$ 3) $b = \dfrac{36}{6} = 6$ 5) $x = \dfrac{70}{7} = 10$ 7) $y = \dfrac{66}{11} = 6$ 9) $c = \dfrac{7.7}{.1} = 77$

11) $y = \dfrac{7.04}{2.2} = 3.2$ 13) $a = \dfrac{102}{20.4} = 5$ 15) $x = \dfrac{.009}{.3} = .03$ 17) $x = \dfrac{20}{1}\left(\dfrac{5}{4}\right) = 25$

19) $t = \dfrac{12\frac{1}{4}}{3\frac{1}{2}} = \dfrac{49}{4}\left(\dfrac{2}{7}\right) = \dfrac{7}{2} = 3\dfrac{1}{2}$ 21) $s = \dfrac{1\frac{11}{25}}{1.2} = \dfrac{1.44}{1.2} = 1.2$ or $1\dfrac{1}{5}$

23) $x = \dfrac{3\frac{1}{5}}{2\frac{2}{10}} = \dfrac{\frac{16}{5}}{\frac{22}{10}} = \dfrac{16}{5}\left(\dfrac{5}{11}\right) = \dfrac{16}{11} = 1\dfrac{5}{11}$

25) $s = \dfrac{.3}{.9} = \dfrac{\frac{3}{10}}{\frac{9}{10}} = \dfrac{3}{10}\left(\dfrac{10}{9}\right) = \dfrac{1}{3}$

27) $y = \dfrac{6\frac{1}{4}}{25} = \dfrac{\frac{25}{4}}{\frac{25}{1}} = \dfrac{25}{4}\left(\dfrac{1}{25}\right) = \dfrac{1}{4}$ 29) $p = \dfrac{.001}{.1} = .01$

Solutions to Workbook 6.6

1) a) 3,6 means; 2,9 extremes   b) 1,16 means; 4,4 extremes

   c) 7,10 means; 5,14 extremes   d) x,y means; w,z extremes

3) $3(6) = 2m$ 5) $8(5) = 2r$ 7) $1(t) = 2(9)$ 9) $21a = 7(3)$
   $18 = 2m$ $40 = 2r$ $t = 18$ $21a = 21$
   $\dfrac{18}{2} = \dfrac{2m}{2}$ $\dfrac{40}{2} = \dfrac{2r}{2}$ $\dfrac{21a}{21} = \dfrac{21}{21}$
   $9 = m$ $20 = r$ $a = 1$

11) $\dfrac{\frac{5}{2}}{y} = \dfrac{5}{4}$ 13) $\dfrac{2}{.3} = \dfrac{y}{.2}$ 15) $\dfrac{2.5}{x} = \dfrac{75}{.3}$ 17) $\dfrac{1}{6}(.1) = \dfrac{1}{12}x$

$\dfrac{5}{2}\left(\dfrac{4}{1}\right) = 5y$ $\dfrac{2}{\frac{3}{10}} = \dfrac{y}{\frac{2}{10}}$ $2.5(.3) = 75x$ $\dfrac{1}{6}\left(\dfrac{1}{10}\right) = \dfrac{1}{12}x$

$10 = 5y$ $\dfrac{2}{1}\left(\dfrac{2}{10}\right) = \dfrac{3}{10}y$ $.75 = 75x$ $\dfrac{1}{60} = \dfrac{1}{12}x$

$2 = y$ $\dfrac{2}{5} = \dfrac{3}{10}y$ $\dfrac{.75}{75} = \dfrac{75x}{75}$ $\left(\dfrac{12}{1}\right)\left(\dfrac{1}{60}\right) = \left(\dfrac{12}{1}\right)\left(\dfrac{1}{12}\right)x$

$\dfrac{10}{3}\left(\dfrac{2}{5}\right) = \dfrac{10}{3}\left(\dfrac{3}{10}\right)y$ $.01 = x$ $\dfrac{1}{5} = x$

$\dfrac{4}{3} = 1\dfrac{1}{3} = y$

19) $\dfrac{\frac{11}{12}}{t} = \dfrac{\frac{41}{12}}{41}$ 21) $\dfrac{3}{14} = \dfrac{x}{40}$ 23) a) $\dfrac{\frac{1}{2}}{10} = \dfrac{2\frac{1}{2}}{x}$ b) $\dfrac{\frac{1}{2}}{10} = \dfrac{x}{15}$

$\dfrac{11}{12}\left(\dfrac{41}{1}\right) = \dfrac{41}{12}t$ $3(40) = 14x$ $\dfrac{1}{2}x = \dfrac{5}{2}\left(\dfrac{10}{1}\right)$ $\dfrac{1}{2}\left(\dfrac{15}{1}\right) = 10x$

$$\frac{12}{41}\left(\frac{11}{12}\right)\left(\frac{41}{1}\right)=\frac{12}{41}\left(\frac{41}{12}\right)t \qquad 120=14x \qquad \frac{1}{2}x=25 \qquad \frac{15}{2}=10x$$

$$11=t \qquad \frac{120}{14}=x \qquad \frac{2}{1}\left(\frac{1}{2}\right)x=\frac{2}{1}\left(\frac{25}{1}\right) \qquad \frac{1}{10}\left(\frac{15}{2}\right)=\left(\frac{1}{10}\right)\frac{10}{1}x$$

$$x=\frac{60}{7}=8\frac{4}{7} \qquad\qquad x=50 \qquad\qquad x=\frac{3}{4}$$

25) $\dfrac{4}{69}=\dfrac{6}{x}$ 27) $\dfrac{3}{4}=\dfrac{x}{480}$ 29) $\dfrac{5}{6\frac{1}{2}}=\dfrac{12}{x}$ 31) $\dfrac{15}{100}=\dfrac{165}{x}$

$4x=6(69)$ $\qquad$ $3(480)=4x$ $\qquad$ $5x=\dfrac{13}{2}\left(\dfrac{12}{1}\right)$ $\qquad$ $15x=165(100)$

$4x=414$ $\qquad$ $1,440=4x$ $\qquad$ $5x=78$ $\qquad$ $15x=16,500$

$x=\$103.50$ $\qquad$ $\dfrac{1440}{4}=360=x$ $\qquad$ $x=\dfrac{78}{5}=15\dfrac{3}{5}$ $\qquad$ $x=\dfrac{16,500}{15}=1,100$

33) $\dfrac{5}{2.5}=\dfrac{26}{x}$ $\qquad\qquad$ 35) $\dfrac{5}{2.75}=\dfrac{120}{x}$

$\quad 5x=26(2.5)$ $\qquad\qquad\qquad$ $5x=120(2.75)$

$\quad 5x=65$ $\qquad\qquad\qquad\qquad$ $5x=330$

$\quad x=13$ $\qquad\qquad\qquad\qquad\quad$ $x=66$

Solutions to Chapter 6 Review

1) $.2=\dfrac{2}{10}=\dfrac{1}{5}$ $\qquad\qquad$ 2) $.02=\dfrac{2}{100}=\dfrac{1}{50}$ $\qquad\qquad$ 3) $.002=\dfrac{2}{1,000}=\dfrac{1}{500}$

4) $.0002=\dfrac{1}{10,000}=\dfrac{1}{5,000}$ $\qquad$ 5) $2.2=2\dfrac{2}{10}=2\dfrac{1}{5}$ $\qquad$ 6) $.091=\dfrac{91}{1,000}$

7) $34.0001=34\dfrac{1}{10,000}$ $\qquad$ 8) $100.6=100\dfrac{6}{10}=100\dfrac{3}{5}$ $\qquad$ 9) $.1$ terminating

10) $.01$ terminating $\qquad$ 11) $.001$ terminating $\qquad$ 12) $.0001$ terminating

13) $.\overline{3}$ repeating $\qquad$ 14) $.8462$ continuous $\qquad$ 15) $.\overline{6}$ repeating

16) $4.\overline{1}$ repeating $\qquad$ 17) $.4$ terminating $\qquad$ 18) $.004$ terminating

19) 6.1905 continuous $\qquad$ 20) $3.0\overline{4}$ repeating

21) $2.5 + .1 = 2.6$

OR $2\dfrac{1}{2} + \dfrac{1}{10} = 2\dfrac{5}{10} + \dfrac{1}{10} = 2\dfrac{6}{10} = 2\dfrac{3}{5}$

22) $\left(\dfrac{1}{5}\right)^3 + \dfrac{1}{3} = \dfrac{1}{125} + \dfrac{1}{3} = \dfrac{3}{375} + \dfrac{125}{375} = \dfrac{128}{375}$

23) $2.1 \div .7 = 3$ OR $2\dfrac{1}{10} \div \dfrac{7}{10} = \dfrac{21}{10}\left(\dfrac{10}{7}\right) = 3$

24) $\dfrac{1}{3}\left(\dfrac{6}{5}\right) = \dfrac{2}{5}$

25) $\dfrac{16}{9}\left(\dfrac{3}{2}\right)^4 = \dfrac{16}{9}\left(\dfrac{81}{16}\right) = 9$

26) $.5625 \div .75^2 = .5625 \div .5625 = 1$ OR $\dfrac{9}{16} \div \dfrac{9}{16} = 1$

27) $200 \div 12 = 16\dfrac{8}{12} = 16\dfrac{2}{3}$ m.p.g.

28) a) $\dfrac{12}{10} = \dfrac{6}{5}$, 6 men to 5 women b) $\dfrac{10}{12} = \dfrac{5}{6}$, 5 women to 6 men

29) $\dfrac{450}{6} = \dfrac{150}{2} = \dfrac{75}{1}$, 75 m.p.h.

30) $65 \div 5 = 13$ lbs. per month 31) a) $\dfrac{20}{180} = \dfrac{1}{9}$, 1 defective to 9 non-defective b) $\dfrac{180}{20} = \dfrac{9}{1}$, 9 non-defective to 1 defective c) $\dfrac{20}{200} = \dfrac{1}{10}$, 1 defective in a shipment of 10. 32) $x = \dfrac{3.3}{3} = 1.1$

33) $x = \dfrac{.1}{10} = .01$ 34) $y = \dfrac{12}{\frac{3}{2}} = \dfrac{12}{1}\left(\dfrac{2}{3}\right) = 8$

35) $r = \dfrac{27}{\frac{3}{4}} = \dfrac{27}{1}\left(\dfrac{4}{3}\right) = 36$ 36) $z = \dfrac{75}{.25} = 300$ 37) $s = \dfrac{\frac{1}{3}}{.6} = \dfrac{\frac{1}{3}}{\frac{6}{10}} = \dfrac{1}{3}\left(\dfrac{5}{3}\right) = \dfrac{5}{9}$ 38) $\dfrac{144}{36} = 4$

39) $12(11.5) = 138$ 40) $5,280 + 700 = 5,980$ feet 41) $\dfrac{192}{16} - 12$ 42) $3.5(2,000) = 7,000$ lbs.

43) $\dfrac{78}{12} = 6\dfrac{1}{2}$ 44) $\dfrac{1.89}{64} = \$.03$ per ounce 45) $\dfrac{15}{3.75} = \$4$ per lb.

46) a) $\dfrac{20}{6\frac{2}{3}} = \dfrac{20}{1}\left(\dfrac{3}{20}\right) = 3$ dollars per yard. b) $\dfrac{20}{20} = \$1$ per foot

47) a) $\dfrac{15.89}{10\frac{11}{16}} = \$1.49$ per pound b) $\dfrac{15.89}{171} = \$.09$ per ounce.

48) a) $\dfrac{3}{16} = .1875$, $\dfrac{12,040}{2.1875} = \$5,504$ per pound

   b) $2(16) + 3 = 15$ ounces, $\dfrac{12,040}{35} = \$344$ per ounce

49) $3(8) = 24$ ounces, $\dfrac{1.55}{24} = \$.06$ per ounce, $3(10) = 30$ ounces, $\dfrac{1.95}{30} = \$.07$ per ounce,
   The 3 eight-ounce cans are the better buy.

50) $\dfrac{3.5}{12} = \$.29$ per ounce, $\dfrac{4.25}{16} = \$.27$ per ounce. The 16-ounce box is the better buy.

51) $x = 4(7) = 28$

52) $3(12) = 4x$
   $36 = 4x$
   $x = 9$

53) $8t = 5(24)$
   $8t = 120$
   $t = \dfrac{120}{8} = 15$

54) $\dfrac{4}{19} = \dfrac{x}{60}$
   $4(60) = 19x$
   $240 = 19x$
   $x = \dfrac{240}{19} = 12\dfrac{12}{19}$

55) $\dfrac{5}{19} = \dfrac{12}{x}$
   $5x = 12(19)$
   $5x = 228$
   $x = \dfrac{228}{5} = \$45.60$

56) a) $\dfrac{\frac{1}{2}}{1} = \dfrac{3\frac{1}{2}}{x}$
   $\dfrac{1}{2}x = 1\left(3\dfrac{1}{2}\right)$
   $x = \left(\dfrac{2}{1}\right)\left(\dfrac{7}{2}\right) = 7$

56) b) $\dfrac{\frac{1}{2}}{1} = \dfrac{x}{12}$
   $\dfrac{1}{2}\left(\dfrac{12}{1}\right) = x$
   $6 = x$

57) $\dfrac{35}{3.8} = \dfrac{x}{8}$
   $35(8) = 3.8x$
   $280 = 3.8x$
   $x = \dfrac{280}{3.8} = \$73.68$

58) $\dfrac{1.39}{2} = .70$ per can
   $\dfrac{5.25}{5} = 1.05$ per can
   The 5 cans is the better buy.

59) $\dfrac{16}{10} = \dfrac{28}{x}$
   $16x = 10(28)$
   $16x = 280$
   $x = \dfrac{280}{16} = 17.5$

Chapter 7

Solutions to Workbook 7.1

1) $\dfrac{10}{100} = \dfrac{1}{10}$

3) $\dfrac{45}{100} = \dfrac{9}{20}$

5) $\dfrac{2}{100} = \dfrac{1}{50}$

7) $\dfrac{200}{100} = 2$

9) $.735 = \dfrac{735}{1,000} = \dfrac{147}{200}$

11) $.0009 = \dfrac{9}{10,000}$

13) $\dfrac{\frac{1}{4}}{100} = \dfrac{1}{4}\left(\dfrac{1}{100}\right) = \dfrac{1}{400}$

15) $\dfrac{3\frac{1}{6}}{100} = \dfrac{\frac{19}{6}}{\frac{100}{1}} = \dfrac{19}{6}\left(\dfrac{1}{100}\right) = \dfrac{19}{600}$

17) $\dfrac{3\frac{1}{8}}{100} = \dfrac{\frac{25}{8}}{\frac{100}{1}} = \dfrac{25}{8}\left(\dfrac{1}{100}\right) = \dfrac{1}{32}$

19) $\dfrac{1\frac{6}{9}}{100} = \dfrac{1\frac{2}{3}}{\frac{100}{1}} = \dfrac{\frac{5}{3}}{\frac{100}{1}} = \dfrac{5}{3}\left(\dfrac{1}{100}\right) = \dfrac{1}{60}$

21) $\dfrac{1}{10} = \dfrac{10}{100} = 10\%$

23) $\dfrac{9}{20} = \dfrac{45}{100} = .45 = 45\%$

25) $\dfrac{1}{50} = \dfrac{2}{100} = 2\%$

27) $\dfrac{2}{1} = \dfrac{200}{100} = 200\%$

29) $\dfrac{147}{200} = .735 = 73.5\%$

31) $\dfrac{9}{10,000} = .0009 = .09\%$ or $\dfrac{9}{100}\%$ 33) $\dfrac{1}{400} = .0025 = .25\%$ or $\dfrac{1}{4}\%$

35) $\dfrac{19}{600} = .031\overline{6} = 3\dfrac{1}{6}\%$

37) $\dfrac{1}{32} = .03125 = 3.125\%$ or $3\dfrac{1}{8}\%$

39) $\dfrac{1}{60} = .01\overline{6} = 1\dfrac{2}{3}\%$

41) 0.2

43) 0.25

45) 0.02

47) 0.00002

49) 3.0059

51) 5.2% = 0.052

53) 0.38

55) 10.875% = 0.10875

57) 35%

59) 350%

61) 0.35%

63) 1%

65) 100%

67) 50%

69) 75%

71) 30%

73) $66.\overline{6} = 66\dfrac{2}{3}\%$

75) $2,222.\overline{2}\% = 2,222\dfrac{2}{9}\%$

Solutions to Workbook 7.2

1) $x = .06(400)$

$x = 24$

3) $400x = 24$

$x = \dfrac{24}{400} = .06 = 6\%$

5) $24 = .06x$

$x = \dfrac{24}{.06} = 400$

7) $400x = 240$

$x = \dfrac{240}{400} = .6 = .60 = 60\%$

9) $\dfrac{x}{100} = \dfrac{2,400}{400}$

$400x = 240,000$

$x = \dfrac{240,000}{400} = 600\%$

11) $\dfrac{.6}{100} = \dfrac{x}{400}$

$240 = 100x$

$x = \dfrac{240}{100} = 2.4$

13) $\dfrac{36}{100} = \dfrac{324}{x}$

$36x = 32,400$

$x = \dfrac{32400}{36} = 900$

15) $\dfrac{x}{100} = \dfrac{187.5}{1,250}$

$1,250x = 18,750$

$x = \dfrac{187.5}{1,250} = 0.15 = 15\%$

17) $\dfrac{1}{4}\% = .25\% = .0025$

$x = .0025(25) = 0.0625$

19) $0.0625 = 25x$

$x = \dfrac{.0625}{25} = .0025 = .25\%$

21) $\dfrac{1}{3}\% = \dfrac{\frac{1}{3}}{100} = \dfrac{1}{3}\left(\dfrac{1}{100}\right) = \dfrac{1}{300}$

$x = \dfrac{1}{300}\left(\dfrac{900}{1}\right) = 3$

23) $450x = 50$

$\dfrac{50}{450} = \dfrac{1}{9} = 0.\overline{1} = 11.1\%$

Solutions to Workbook 7.3

1) a) $\dfrac{120}{150} = \dfrac{4}{5} = .8 = 80\%$ decrease

b) $\dfrac{100}{150} = \dfrac{2}{3} = .\overline{6} = 67\%$ decrease

c) $\dfrac{0}{150} = 0\%$ change

d) $\dfrac{50}{150} = \dfrac{1}{3} = .\overline{3} = 33\%$ increase

e) $\dfrac{300}{150} = 2 = 2.00 = 200\%$ increase

3) $\dfrac{100}{400} = \dfrac{1}{4} = .25 = 25\%$ decrease

5) $\dfrac{.5}{4} = .125 = 13\%$ increase

7) $\dfrac{11,000}{23,000} = 48\%$ increase

9) $\dfrac{12}{24} = .5 = .50 = 50\%$ increase

11) $\dfrac{7}{6} = 1.1\overline{6} = 117\%$ increase

13) $\dfrac{210}{450} = \dfrac{7}{15} = 47\%$ increase

15) $\dfrac{10}{78} = 13\%$ decrease   17) Larry: $\dfrac{7800}{76500} = 10\%$ decrease   Samantha: $\dfrac{8500}{93100} = 9\%$

Solutions to Workbook 7.4

1) $349.95(.05) = \$17.50, \$367.45$ 3) $45(.06) = \$2.70, \$47.70$

5) $\dfrac{550}{16,500} = .0\overline{3} = 3.\overline{3} = 3.\overline{3} = 3\dfrac{1}{3}\%$ 7) $.065(38,150) = \$2,479.75$ 9) $\dfrac{360}{7200} = 5\%$

11) $.45(50) = \$22.5, \$50 - \$22.5 = \$27.50$

13) $I = 2,000 \times .09 \times 2 = \$360$ 15) $I = \dfrac{6,000}{1} \times \dfrac{180}{360} \times \dfrac{8}{100} = \$240,$ She had to repay $6,240

17) $\dfrac{1,600}{2(5,000)} = .16 = 16\%$ 19) $19(.7) = \$13.30$ sale price, $13.3(.8) = \$10.64$ new low price

Solutions to Chapter 7 Review

1) $\dfrac{4}{5} = \dfrac{80}{100} = 80\%$ 2) $\dfrac{6}{12} = \dfrac{1}{2} = .5 = .50 = 50\%$ 3) $\dfrac{12}{14} = 86\%$ 4) $\dfrac{24}{100} = \dfrac{6}{25}$

5) $\dfrac{124}{100} = 1\dfrac{24}{100} = 1\dfrac{6}{25}$ 6) $.024 = \dfrac{24}{1,000} = \dfrac{3}{125}$ 7) $.056 = \dfrac{56}{1,000} = \dfrac{28}{500} = \dfrac{7}{125}$

8) $\dfrac{2\frac{2}{3}}{100} = \dfrac{\frac{8}{3}}{\frac{100}{1}} = \dfrac{8}{3}\left(\dfrac{1}{100}\right) = \dfrac{2}{75}$ 9) $\dfrac{6}{100} = \dfrac{5}{6}\left(\dfrac{1}{100}\right) = \dfrac{1}{120}$ 10) 5% 11) 550%

12) $1 = 1.00 = 100\%$ 13) 0.1 14) 0.001

15) 1 16) $0.29(100) = 29$ 17) $0.29(562) = 162.98$

18) $0.29(901) = 261.29$ 19) $0.4(64) = 25.6$ ounces

20) a) $0.12(625) = 75$ b) $\$625 + 75 = \$700$

21) $\dfrac{75}{200} = .375 = 37.5\%$ 22) $45 \div 0.3 = 150$ 23) $\dfrac{454.36}{2,295.03} = 20\%$

24) $\dfrac{3,475 - 3,050}{3,050} = \dfrac{425}{3,050} \approx 14\%$

25) $.055(1,490) = \$81.95$ tax, $\$1,490 + \$81.95 = \$1,571.95$

26) Tax $= .07(25.99) = \$1.82,$ Total Price $= \$25.99 + \$1.82 = \$27.81$

27) $\dfrac{800}{16,000} = \dfrac{1}{20} = .05 = 5\%$

28) $12,000\,x = 600, \quad x = \dfrac{600}{12,000} = .05 = 5\%$

29) $I = 6,000 \times 0.03 \times 1 = \$180$

30) $I = 3,000\left(\dfrac{4}{100}\right)\left(\dfrac{390}{360}\right) = 130, \quad \$3,000 + \$130 = \$3,130$

31) $\dfrac{5,000}{3(10,000)} = \dfrac{1}{6} = 17\%$

Chapter 8

Solutions to Workbook 8.1

1) 12

3) –12

5) –9

7) –3

9) –15

11) 5

13) –48

15) $\dfrac{40}{20} = 2$

17) 81

19) 200

21) 9,600

23) $\dfrac{196}{-98} = -2$

25) $-125 \div -125 = 1$

27) 16

29) –16

31) –64

33) –64

35) 1

37) –1

Solutions to Workbook 8.2

1) –16

3) 1

5) –19

7) 0

9) –10

11) $-9 + 4 = -5$

13) $-5 + 15 = 10$

15) $17 + 17 = 34$

17) $-8 + 5 - 4 = -12 + 5 = -7$

19) $-12 + 7 = -5$

21) $-9(-1) = 9$ 23) $-2(12 - 15) = -2(-3) = 6$ 25) $-2 + -2 = -4$

27) $\dfrac{-6(-2)}{-12} = \dfrac{12}{-12} = -1$

29) $\dfrac{-211}{-3 - (-3)} = \dfrac{-211}{-3 + 3} = \dfrac{-211}{0}$ is undefined

31) $[-9 - (-2)(-7)] = [-9 - 14] = -23$

33) $[-33 - (-64 \div (-4)^3)] = [-33 - (-64 \div -64)] = [-33 - 1] = -34$

Solutions to Workbook 8.3

1) $2,011 + 396 + 1,135 - 214 - 481 - 459 = \$2,388$

3) $-49 + 17 = -32$ He owes \$32.

5) $0 - 5(100) = -\$500$

7) $-34 - 13 = -47$ You owe \$47.

9) $14 - (-17) = 14 + 17 = 31$ degrees

11) $-450 - 173 + 120 = -503$, 503 feet below the ocean's surface.

13) $485 - (-540) = 1,025$ feet apart.

15) $100,000 - 6(40,000) = -140,000$, The net worth is $- \$140,000$.

17) $250 - 85 - 270 - 900 + 335 = -\$650$

Solutions to Chapter 8 Review

1) -60 2) 7 3) -7

4) 60 5) 6 6) -6

7) $\dfrac{-2(-1)}{1} = 2$ 8) $(-6)(-6) = 36$ 9) $-(6)(6) = -36$

10) -21 11) $-75 + 64 = -11$ 12) $-4 - (-7) = -4 + 7 = 3$

13) -48 14) $2 - 6(-64) = 2 + 384 = 386$ 15) $200 - 500 + 150 = -150$

16) $-470 + 50 + 175 = -245$ 17) $6 + 3 = 9$ 18) $-6 + 3 = -3$

19) $2{,}000 - 450 - 1{,}234 - 1{,}345 = -\$1{,}029$ 20) $-3 - (-3) = -3 + 3 = 0$

21) $\dfrac{-9 - (-3) \div -1}{-(2)(2)} = \dfrac{-9 - 3}{-4} = \dfrac{-12}{-4} = 3$

22) $11 + [40 \div (-8) \div (1)] = 11 + [-5 \div 1] = 11 - 5 = 6$ 23) $16 \div 1 = 16$

24) 1 25) $(-2)^2 \div (0) = 4 \div 0$ is undefined

26) $-22 - (-13) = 13 - (-22) = 13 + 22 = 35$ degrees 27) $129 - 150 - 215 = -\$236$

Chapter 9

Solutions to Workbook 9.1

1) a) $P = 4(6) = 24$ feet, $A = 6^2 = 36$ square feet

   b) $P = 4(3) = 12$ inches, $A = 3^2 = 9$ square inches

   c) $P = 4(2.8) = 11.2$ cm., $A = (2.8)^2 = 7.84$ cm. squared

d) $P = 4\left(3\dfrac{2}{3}\right) = \dfrac{4}{1}\left(\dfrac{11}{3}\right) = \dfrac{44}{3} = 14\dfrac{2}{3}$ yards, $A = \left(3\dfrac{2}{3}\right)^2 = \left(\dfrac{11}{3}\right)^2 = \dfrac{121}{9} = 13\dfrac{4}{9}$ square yards

3) a) $A = \dfrac{1}{2}\left(\dfrac{3}{1}\right)\left(\dfrac{8}{1}\right) = 12$ square inches

   b) $A = 0.5(10.2)(8.5) = 43.35$ square inches

   c) $A = \dfrac{1}{2}\left(2\dfrac{2}{5}\right)\left(2\dfrac{1}{2}\right) = \dfrac{1}{2}\left(\dfrac{12}{5}\right)\left(\dfrac{5}{2}\right) = 3$ square miles

5) a) $P = 2 + 2\dfrac{1}{4} + 1\dfrac{1}{4} + 1\dfrac{1}{2}\left(2\dfrac{1}{4} - \dfrac{3}{4}\right) + \dfrac{3}{4} + \dfrac{3}{4} = 8\dfrac{1}{2}$ inches

   $A = \left(\dfrac{3}{4}\right)\left(\dfrac{3}{4}\right) + \left(1\dfrac{1}{4}\right)\left(2\dfrac{1}{4}\right) = \left(\dfrac{3}{4}\right)\left(\dfrac{3}{4}\right) + \left(\dfrac{5}{4}\right)\left(\dfrac{9}{4}\right) = \dfrac{9}{16} + \dfrac{45}{16} = \dfrac{54}{16} = \dfrac{27}{8} = 3\dfrac{3}{8}$ square inches

   b) $P = 4.75 + 1.5 + 4.5 + 3 + 5.4 + 4.5 = 23.65$ feet

   $A = .5(1.5)(4.5) + 4.5(4.5) + .5(3)(4.5) = 30.375$ square feet

Solutions to Workbook 9.2

1) $r = \dfrac{4}{2} = 2$ yards   3) $d = 2(10) = 20$ km.   5) $r = 1.75$, $C = 2(3.16)(1.75) = 10.99$ yards

7) $C = 2(3.14)(3) = 18.84$ inches   9) $r = 5$, $A = 3.14\,(5)^2 = 3.14(25) = 78.5$

11) $r = \dfrac{508.68}{2(3.14)} = 81$ miles   13) $A = 3.14\,(2)^2 + .5(6)(4) + .5(6)(4) = 36.56$

Solutions to Workbook 9.3

1) $V = \left(1\dfrac{1}{4}\right)^3 = \left(\dfrac{5}{4}\right)^3 = \dfrac{125}{64} = 1\dfrac{61}{64}$ cubic yards   3) $V = (3)^3 =$ cubic inches

5) $V = 5(4)(5) = 100$ cubic inches   7) $7.9(4.6)(6.2) = 225.308$ cubic yards

Solutions to Workbook 9.4

1) Total Area = 230(650) = 149,500 square yards, Area of each lot = $\dfrac{149,500}{4} = 37,375$ square yards

3) A = 3.14 (30)^2 = 2,826, A = 3.14 (20)^2 = 1,256, The area of garden is increased by: 2,826 − 1,256 = 1,570 square yards

5) A = .5(29)(17) = 246.5 square feet,
Cost = 3.75(246.5) = $924.38

7) A = 2(4) + 4(5) + 2(2) = 32 square yards,
Cost = 7.5(32) = $240

9) A = 38(12) = 456, A = 45(12) = 540, The area is increased by:
540 − 456 = 84 square inches

11) P = 2l + 2w, 22 = 2(5) + 2w, 22 = 10 + 2w, 2w = 12,
w = 6

13) P = 4s, 32 = 4s, s = 8 inches 15) V = 150(50)(6) = 45,000 cubic feet

17) 4 yards = 12 feet, V = 12^3 = 1,728 cubic feet, C = 5.75(1,728) = $9,936, 3 yards = 9 feet,
V = 9^3 = 729 cubic feet, C = 5.75(729) = $4,191.75, $9,936 − $4,191.75 = $5,744.25 more

Solutions to Chapter 9 Review

1) P = 4(12) = 48 A = (12)^2 = 144 2) P = 4(5) = 20 feet A = (5)^2 = 25 square feet

3) $P = 4\left(6\dfrac{3}{4}\right) = \dfrac{4}{1}\left(\dfrac{27}{4}\right) = 27$ inches $A = \left(6\dfrac{3}{4}\right)^2 = \left(\dfrac{27}{4}\right)^2 = \dfrac{729}{16} = 45\dfrac{9}{16}$ square inches

4) P = 4(15.3) = 61.2 yards A = (15.3)^2 = 234 square yards 5) P = 2(12) + 2(9) = 42

A = 12(9) = 108 6) $P = 2(14) + \left(8\dfrac{5}{6}\right) = 28 + \dfrac{2}{1}\left(\dfrac{53}{6}\right) = 28 + \dfrac{53}{3} = 28 + 17\dfrac{2}{3} = 45\dfrac{2}{3}$ inches

$A = 14\left(8\dfrac{5}{6}\right) = \dfrac{14}{1}\left(\dfrac{53}{6}\right) = \dfrac{371}{3} = 123\dfrac{2}{3}$ square inches

7) P = 2(13.4) + 2(10.7) = 48.2 yards, A = 13.4(10.7) = 143.38 square yards,

8) $P = 2\left(8\dfrac{1}{2}\right) + 2\left(4\dfrac{2}{3}\right) = \dfrac{2}{1}\left(\dfrac{17}{2}\right) + \dfrac{2}{1}\left(\dfrac{14}{3}\right) = 17 + \dfrac{28}{3} = 17 + 9\dfrac{1}{3} = 26\dfrac{1}{3}$ yards,

$A = lw = \left(8\dfrac{1}{2}\right)\left(4\dfrac{2}{3}\right) = \left(\dfrac{17}{2}\right)\left(\dfrac{14}{3}\right) = \dfrac{238}{6} = 39\dfrac{4}{6} = 39\dfrac{2}{3}$ square feet

9) $P = 9 + 14 + 16 = 39$, $A = \dfrac{1}{2}\left(\dfrac{9}{1}\right)\left(\dfrac{15}{1}\right) = \dfrac{135}{2} = 67\dfrac{1}{2}$

10) $P = 2\frac{1}{4} + 3\frac{1}{4} = 8\frac{2}{4} = 8\frac{1}{2}$ inches, $A = \frac{1}{2}\left(3\frac{1}{4}\right)\left(\frac{2}{1}\right) = \frac{1}{2}\left(\frac{13}{4}\right)\left(\frac{2}{1}\right) = \frac{13}{4} = 3\frac{1}{4}$ square inches

11) $P = 2 + 4 + 1 + 1 + 1 + 3 = 12$ inches, $A = 2(3) + 1(1) = 7$ square inches

12) $P = 4 + 8 + 4 + 2 + 4.5 + 4.5 + 2 = 29$, $A = 4(8) + .5(4)(4) = 40$

13) $P = 4(2) + 2(3.14)(.5) = 8 + 3.14 = 11.14$, $A = 2^2 + 3.14(.5)^2 = 4 + .785 = 4.785$

14) $r = 6.4 \div 2 = 3.2$, $C = 2(3.14)(3.2) = 20.096$ feet

15) $r = 3.5 \div 2 = 1.75$, $C = 2(3.14)(1.75) = 10.99$ miles   16) $C = 2(3.14)(8) = 50.24$ yards

17) $C = 2(3.14)(8.3) = 52.124$ 18) $C = 2(3.14)r$, $6.8 = 6.28r$, $r = \frac{6.8}{6.28} = 1.0828025$ inches

19) $C = (3.14)d$, $20.4 = (3.14)d$, $d = \frac{20.4}{3.14} \approx 6.5$ feet

20) $A = 3.14(15.2)^2 = 3.14(231.04) = 725.4656$ square yards

21) $r = \frac{88}{7} \times \frac{1}{2} = \frac{44}{7}$, $A = \frac{22}{7}\left(\frac{44}{7}\right)^2 = \frac{22}{7}\left(\frac{1936}{49}\right) = \frac{42,592}{343} = 124.17$ square feet

22) $\frac{15.4}{2} = 7.7$, $A = 3.14(7.7)^2 = 186.1706$ square inches   23) $A = 3.14r^2$, $314 = 3.14r^2$, $\frac{314}{3.14} = r^2$

$100 = r^2$, $r = 10$ yards

24) $C = \pi d$, $C = (3.14)d$, $15.7 = (3.14)d$, $15.7 = (3.14)d$, $d = \frac{15.7}{3.14 = 5} = 5$ inches

25) $A = 3.14(19)^2 = 1,133.54$, $A = 3.14(15)^2 = 706.5$, Area of path $= 1,133.54 - 706.5 = 427.04$ square feet

26) $18(12) = 216$ inches, $15(12) = 180$ inches, $A = 216(180) = 38,880$ square inches,

$8(8) = 64$ square inches per tile, Number of tiles $= \frac{38,880}{64} = 607.5$

27) $A = .5(39.5)(19.7) = 389.075$, $C = 4.35(389.075) = \$1,692.48$

28) $P = 100 + 150 + 100 + 200 + 200 + 350 = 1,100$ yards, $C = 15.85(1,100) = \$17,435$

29) $A = 100(150) + 200(200) = 15,000 + 40,000 = 55,000$ square yards,
$C = 8.5(55,000) = \$467,500$

30) Area of the field $= \left(178\frac{2}{3}\right)\left(114\frac{5}{8}\right) = \left(\frac{536}{3}\right)\left(\frac{917}{8}\right) = \frac{61,439}{3} = 20,479\frac{2}{3}$,

Area of each section $= \frac{61,439}{3} \div 3 = \frac{61,439}{3}\left(\frac{1}{3}\right) = \frac{61,439}{9} = 6,826\frac{5}{9}$ square yards

31) $A = 43(12) = 516$ square inches, $A = 54(12) = 660$ square inches, The increase in the area $= 660 - 516 = 144$ square inches.   32) $A = 3.14(15.375)^2 = 742.26$ square feet, $A = 3.14(12.25)^2 = 471.2$ square feet, $742.26 - 471.2 = 271.06$ square feet

33) $V = 8^3 = 512$ cubic feet, $C = 2(512) = \$1,024$

34) $V = 3(4)(2) = 24$ cubic inches, $C = 2.75(24) = \$66$

35) $V = 25(20)(15) = 7,500$ cubic yards, $V = (20)^3 = 8,000$ cubic yards. The cube has the greater volume, $(8,000 - 7,500 = 500$ cubic yards$)$.